Springer Proceedings in Mathematics

Volume 15

Springer Proceedings in Mathematics

Springer Proceedings in Mathematics features volumes of selected contributions from workshops and conferences in all areas of current research activity in mathematics. Besides an overall evaluation, at the hands of the publisher, of the interest, scientific quality, and timeliness of each proposal, every individual contribution is refereed to standards comparable to those of leading mathematics journals. This series thus proposes to the research community well-edited and authoritative reports on newest developments in the most interesting and promising areas of mathematical research today.

Series is continued with title "Springer Proceedings in Mathematics & Statistics". The new series starts with volume 19.

For further volumes:
http://www.springer.com/series/8806
and http://www.springer.com/series/10533

Vincenzo Capasso • Misha Gromov •
Annick Harel-Bellan • Nadya Morozova •
Linda Louise Pritchard
Editors

Pattern Formation in Morphogenesis

Problems and Mathematical Issues

Springer

Editors
Vincenzo Capasso
Dipartimento di Matematica, MIRIAM
Università Milano-Bicocca
Milano
Italy

Misha Gromov
Institut des Hautes Études Scientifiques
Bures-sur-Yvette
France

Annick Harel-Bellan
Nadya Morozova
Linda Louise Pritchard
Laboratoire Epigenetique et Cancer
CNRS FRE 3377
CEA Saclay
France

ISSN 2190-5614 ISSN 2190-5622 (electronic)
ISBN 978-3-642-20163-9 ISBN 978-3-642-20164-6 (eBook)
DOI 10.1007/978-3-642-20164-6
Springer Heidelberg New York Dordrecht London

Library of Congress Control Number: 2012948003

Printed on acid-free paper

Springer is part of Springer Science+Business Media (www.springer.com)

Preface

The present book can be considered as having been "motivated by an IHES Workshop" which was a unique event aimed at creating a venue for communication between biologists and mathematicians. The *Interdisciplinary Workshop on Pattern Formation in Morphogenesis* took place in 2010 at the IHES (Institut des Hautes Études Scientifiques) in Bures-sur-Yvette, France. Organized by Vincenzo Capasso, Misha Gromov, Annick Harel-Bellan, and Nadya Morozova, it was supported by the IHES programme for mathematical biology research.

Morphogenesis—the process of pattern formation during the development of living organisms—is a broad field, including areas ranging from physics to chemical and biological systems. At the same time, morphogenesis represents both a rich source of interesting mathematical challenges and perhaps the most intriguing process in living organisms remaining to be explored. The goal of the IHES Workshop was to generate an interdisciplinary interaction space for discussing the problems of Pattern Formation in Morphogenesis. To that end, leading biologists working on various aspects of Developmental Biology, i.e., embryology, genetics, molecular biology, and cell biology, were invited to discuss their own work as well as to describe the prospects of their specific fields along with the basic knowledge necessary to understand them. The specific idea of the Workshop was to create a frame of productive discussions, in the course of which invited mathematicians would apply the knowledge gained from the biologists for the generalization and mathematical formalization of morphogenetic processes.

This volume collects a set of mathematical contributions presented as a follow-up to discussions arising both during and after the workshop regarding key issues and perspectives on scientific themes related to pattern formation in morphogenesis. The biological presentations at the workshop, the necessary source of the biological input for this activity, were also reworked for inclusion in an independent part. Thus, the book contains three diverse parts: "Biological Background," which consists of adaptations of the biological presentations; "Mathematical Models," motivated by both biological presentations and discussions; and "Ideas, Hypotheses and Suggestions," for possible mathematical formalization (or, in other words: "Biology", "Mathematical Biology," and "Theoretical Biology" parts).

We hope that this volume will help to launch a new set of fruitful discussions in the field of mathematical formalization of Pattern Formation in Morphogenesis.

Vincenzo Capasso
Misha Gromov
Annick Harel-Bellan
Nadya Morozova
Linda Louise Pritchard

Contents

Why Would a Mathematician Care About Embryology?

Misha Gromov

Reason number one was given by Lewis Wolpert:

> It is not birth, marriage, or death, but gastrulation, which is truly the most important time in your life.

But mathematicians do not care about things and people just because they are "important." (If mathematicians study the "marriage problem," it is not for the reasons that motivate normal people to get married.)

Let us try again: Embryonic development is, structurally speaking, a damn interesting process, something that is very hard to understand.

This is better. We, mathematicians, are supposed to understand everything worth understanding and be able to solve any hard problem deserving a solution.

Tell me – a nut-cracker mathematician exclaims – just tell me in rigorous mathematical terms what the main problem in embryology is, and I will solve it.

OK, here is the problem: *What is the mathematical formulation of the main problem(s) in embryology?*

There are two opposite directions to take for a mathematician who enters a field in biology. The most productive one is to concentrate on a particular class of phenomena, learn as many specific details as possible from biologists, and work out a specific model in agreement with the data. You will find several fascinating talks along these lines in these pages.

But this does not lead, at least not directly, to the solution of the main problem. No matter how many successful models you have, there remains a lingering feeling of being a blind man touching different parts of an elephant. What is the elephant?

M. Gromov (✉)
Institut des Hautes Études Scientifiques, France
e-mail: gromov@ihes.fr

V. Capasso et al. (eds.), *Pattern Formation in Morphogenesis*, Springer Proceedings in Mathematics 15, DOI 10.1007/978-3-642-20164-6_1,

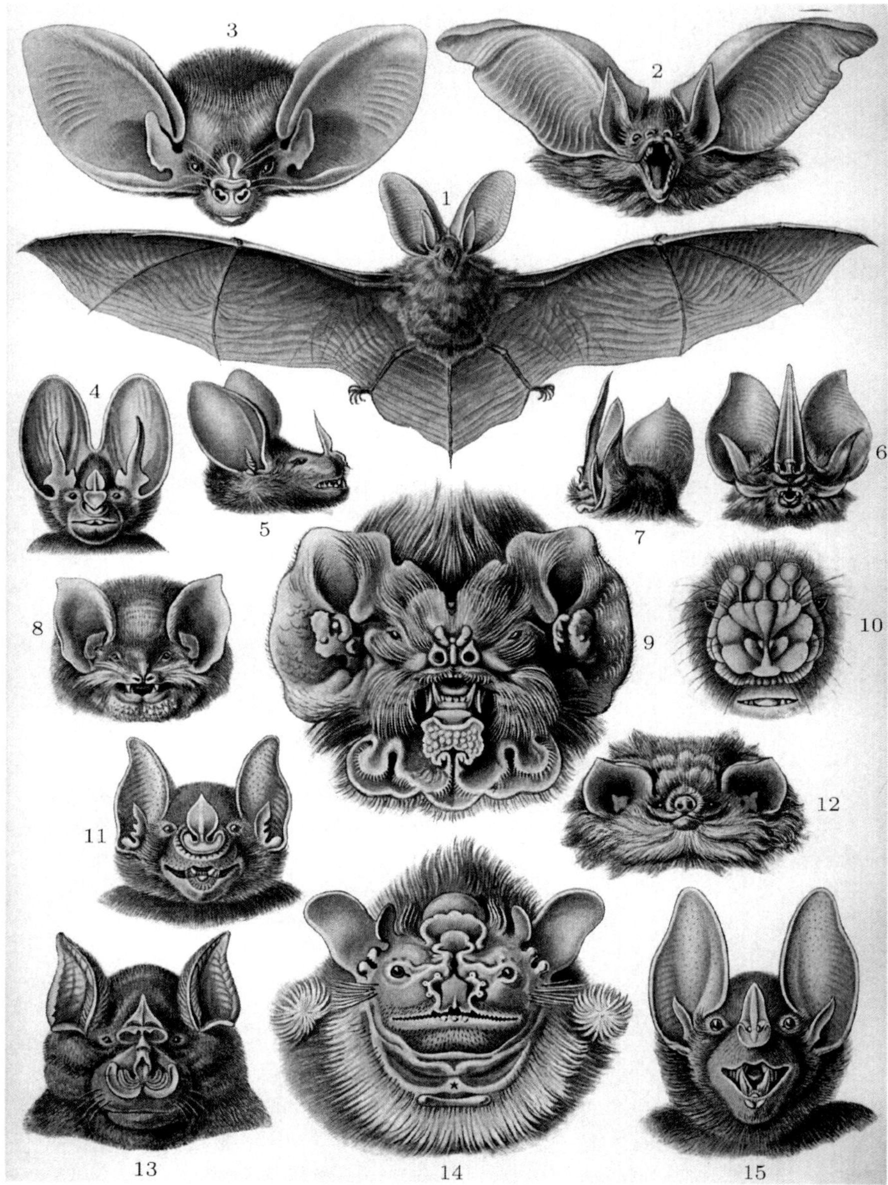

Fig. 1 Bats by Ernst Haeckel. Lithograph from "Kunstformen der Natur" (1900) -"Chiroptera", Plate 67-Vampyrus, 1-15: Plecotus auritus; Nyctophilus australis; Megaderma trifolium; Vampyrus auritus; Lonchorhina aurita; Natalus stramineus; Mormops blainvillei; Anthops ornatus; Phyllostoma hastatum; Furipterus coerulescens; Rhinolophus equinus; Centurio flavigularis; Vampyrus spectrum (permission to copy under the terms of the GNU Free Documentation License)

René Thom suggested an approach in the framework of the catastrophe theory, but, from a biologist's point of view, that was rather more about the shadow of an elephant on a wall, not about the real animal.

Earlier, John von Neumann proposed a model of "morphogenesis" of automata, but one still cannot adequately *formulate* the theorem that von Neumann proved.

It seems that the general ideas of von Neumann and of Thom, imaginative as they are, still fall short of addressing the main problem.

But where should a mathematician go, where can we start from?

I think all you can do is to proceed like a child learning a language, a child mathematician, though you be biology-blind as a bat (Fig. 1): you ask questions and listen to the echoes – the answers you get from biologists. You do not understand the "meaning" of the words, you are not supposed to "see" the objects they represent, but the familiar music of logic starts reverberating in your ears. If this "music" metamorphoses inside your head into a mathematically consistent arrangement of patterns, you might call it "life" or "the real world."

Your grandiose dreams of seeing this world in the light of day through the eyes of an eagle soaring high in the sky may never be fulfilled; yet, even if you are not quite a bat (the reader is a mammal, I presume), you will at least enjoy an erratic chiropteran flight in the moonless night.

Part I
Biological Background

Preface to the Biological Part

This Part includes biological reviews on selected problems in morphogenesis, based on the corresponding presentations at the IHES Workshop 2010. These biological presentations provided a framework of biological knowledge necessary for subsequent discussions with mathematicians about mathematical formalization of pattern formation. Presentations, devoted to different slices of investigations on morphogenesis (embryology, genetics, cell biology, molecular biology, etc.) were next reworked by the speakers, and arranged as chapters, each reviewing the corresponding biological problem or direction of research. The final version of each chapter was strongly influenced by the fruitful discussions with mathematicians, both in the framework of the Workshop, and after it in the course of further communications. Thus, some reviews mostly follow the structure of the corresponding talk at the Workshop, whereas others are significantly restructured or abridged. However, all of them present, together with the general background of the problem and the most important recent data in the field, an overview of the current work of the author that has implications for morphogenesis.

Pattern Formation in Regenerating Tissues

Andrea Hoffmann and Panagiotis A. Tsonis

1 Tissue Repair Versus Pattern Formation in Regenerating Tissues

To understand the concept of pattern formation and its role in regeneration, the basic differences between repair and regeneration need to be explained. Repair is defined as tissue restoration of a damaged tissue, without organized patterning. For example, vertebrates including humans are capable of reconstituting a functional liver following removal of up to 70 % of the original liver mass [1, 2]. Thus, liver repair includes reconstitution of the same volume but it lacks reconstruction of the same tissue pattern. Similarly, wound healing of the skin includes tissue repair by formation of a scar tissue that lacks some of the characteristic features of the original tissue. These are both examples for tissue repair that lack true regeneration or patterning. When looking at regeneration of whole body parts in salamanders, the whole limb is restored into its original form. In this case, the re-establishment of patterns is necessary because otherwise regeneration is meaningless.

2 Models of Pattern Formation in Regenerative Tissues

Traditionally regenerating tissues have been favorable models in the theoretical and experimental approaches to pattern formation. Issues of the exact restoration of patterns during regeneration have been explicitly discussed by researchers such as T. H. Morgan in the nineteenth and the beginning of the twentieth century [3]. However, it was not until the mid 1960s that formulation of ideas to explain mechanisms of pattern formation were presented. Lewis Wolpert established the

A. Hoffmann • P.A. Tsonis (✉)
Department of Biology and Center for Tissue Regeneration and Engineering, University of Dayton, Dayton, OH 45469-2320, USA
e-mail: ptsonis1@udayton.edu

V. Capasso et al. (eds.), *Pattern Formation in Morphogenesis*, Springer Proceedings in Mathematics 15, DOI 10.1007/978-3-642-20164-6_2,

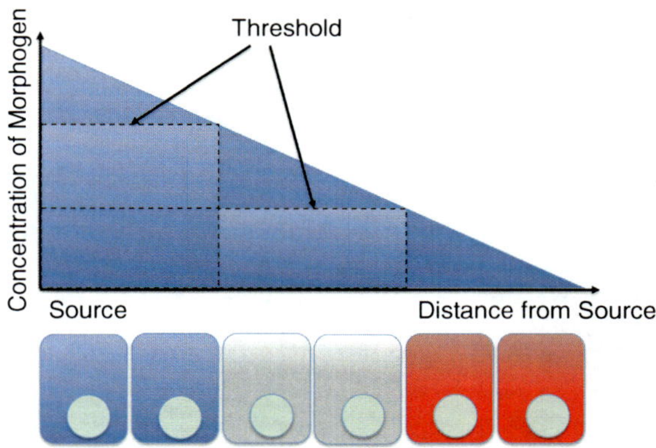

Fig. 1 The morphogen hypothesis (Adapted and modified from Jaeger et al. [43] according to Wolpert [4, 5]). Localized production of a morphogen by the blue source cells could be distributed in direction of the red cells by some form of active transport and other means according to a source and sink model. The concentration of the substance when produced at a fixed value would thus specify the cellular position when looking at the generated linear gradient. Cellular response is depending on a certain morphogen threshold that is defined by the distance of a cell towards the source

"Morphogen Hypothesis" by assuming production of different morphogens in a tissue system (Fig. 1) [4, 5]. He suggested that positional information along a certain axis can be provided by a gradient of a factor and that the cells will know their position because they will respond to a corresponding concentration of the morphogen. For instance, a suggested morphogen gradient that defines the proximal-distal axis in the regenerating newt limb could be one with higher concentration at the proximal regions and lower at the distal regions (Fig. 1).

A different set of ideas was presented by the "Polar Coordinate Model" for limb regeneration in the 1970s [6, 7]. The Polar Coordinate Model mainly emphasized that cell-to-cell communication is a necessary requirement for pattern formation and not morphogens. This model defines the site of the regenerating limb blastema using three-dimensional coordinates. By giving each position in the regenerating limb a coordinate Bryant et al. were able to successfully predict the outcome of regeneration as well as the polarity of limb following transplantation of blastemas from different positions of the limb.

In line with the Polar Coordinate Model and the Morphogen Hypothesis, Hans Meinhardt developed the "Boundary Model" in 1983 that applies the uni-dimensional Morphogen Hypothesis to all three-dimensions [8]. In detail, Meinhardt explains the establishment of coordinates/tissue identity within the embryonic growth axis by delineating distinct tissue boundaries between anterior primary competent cells (A) versus polarizing cells in the posterior (P), and cells of the dorsal (D) and ventral axis (V). Tissue identity is established by tissue area restricted morphogen production within well-defined borders that only allow unidirectional morphogen permeability. When two differently determined cell types meet at a certain tissue boundary, cooperative production of new morphogens in form of an area gradient can give information about the proximity of this cell and other cells of the tissue towards the tissue border.

Correspondingly, the established morphogen area gradient can provide directional information for tissue expansion in line with the embryonic growth axis. This model was also suggested to apply to limb regeneration when A/P and D/V area information is available to regenerating cells at the site of the limb blastema for production of a morphogen for induction of distalization.

These ideas eventually led to studies to identify such morphogens. The first candidate was retinoic acid [9–11]. During amphibian limb regeneration retinoic acid was found to reset the distal cell position to more proximal values resulting in proximalization of regeneration with additional humerus, radius and ulna. Similarly, during chick embryo development retinoic acid administration to the anterior limb bud results in duplication of all the distal limb structures along the antero-posterior axis [10, 12, 13]. Thus, the morphogenic effect of retinoic acid during development entails duplication along the antero-posterior axis compared to the proximalization effect of the proximal-distal axis during limb regeneration.

Subsequent investigations demonstrated the endogenous distribution of retinoic acid. While the group of Eichele et al. confirmed existence of a retinoic acid gradient that was higher in the posterior versus anterior chick limb bud [14, 15], Maden et al. also demonstrated existence of higher levels of the retinoic acid carrier protein CRABP in the anterior versus posterior chick limb bud suggesting a compensatory mechanism for gradient-established differences in retinoic acid levels [16, 17]. In addition, variable retinoic acid levels were found within regenerating amphibian limbs. For instance, in axolotls (*Ambystoma mexicanum*) higher retinoic acid levels were found in the posterior versus the anterior limb blastema, whereas the African clawed frog (*Xenopus laevis*) demonstrated no particular antero-posterior retinoic acid gradient [18, 19]. Also, analysis and cloning of retinoic acid receptor expression in newts did not demonstrate any kind of anterior-posterior gradient in receptor distribution throughout the limb blastema to explain the action of retinoic acid [20–25].

The actual existence of endogenous morphogen gradient molecules has recently gained again some momentum. In accordance with the "Source and Sink Model" that was originally created by Francis Crick and was recently reiterated by Schier et al. signalling proteins have been identified as the main morphogens during pattern formation in contrast to previously chemical molecules (Fig. 2) [26–28]. For example morphogens such as the signalling proteins FGF8 or DPP are either transported throughout the extracellular space by passive diffusion (Brownian motion) or active transport including transcytosis from cell to cell. It is now becoming more and more apparent that proteins diffuse as gradients in order to specify patterns.

3 Development Versus Regeneration

One of the major questions in regeneration research is how mechanisms of patterning compare with the ones during development. We will provide some insight into this question by considering the role and expression of key morphogenesis genes in limb and lens development and regeneration.

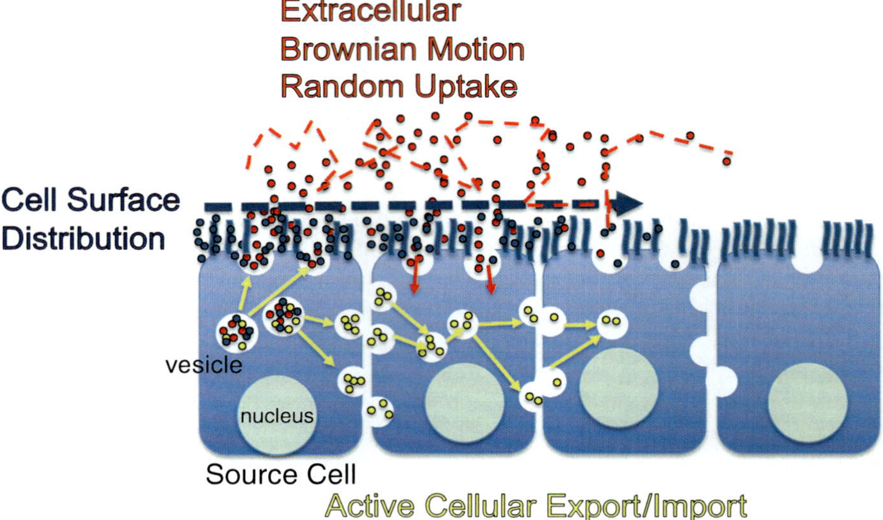

Fig. 2 Models of morphogen dispersal (Adapted and modified from Schier et al. [26]). Source cells harbor vesicles filled with morphogen molecules that fuse with the cell membrane and release their contents. Yu et al. [28] proposed that Brownian motion of molecules in the extracellular space leads to dispersal of the FGF8 morphogen as shown in *red*. Kicheva et al. suggested that repeated release and uptake by cells (transcytosis) leads to dispersal of the morphogen Dpp in the fly wing as delineated in *yellow* [44]. A few slowly diffusing FGF8 molecules are associated with carbohydrate cell surface distribution as delineated in *blue*. This cell-surface pool may contribute to long-range dispersal of FGF8

During limb regeneration in newts the regenerating blastema emerges from already patterned adult tissue. The adult limb of a newt, or a salamander has a defined skeletal pattern including humerus, ulna, radius, carpals and the fingers. Following surgical removal of the limb, the wound is covered by the wound epithelium within several hours. The differentiated tissue underneath the wound epithelium de-differentiates and forms the blastema by eliminating all previously established adult tissue characteristics. The de-differentiated blastema is then able to differentiate into the necessary cell types to reliable regenerate exactly the same structure [29]. With regards to gene expression pattern, research in axolotls clearly demonstrated existence of different patterns between amphibian limb regeneration and limb development. For instance the group of Gardiner et al. demonstrated that different members of the HoxA family, e.g., HoxA13 and HoxA9 that are expressed in cells of developing and regenerating limb buds do not follow the same spatial and temporal expression pattern during limb development and regeneration [30]. For instance, during limb development HoxA13 was suggested to specify the development of distal limb structures in comparison to HoxA9 that acts on proximal structure development. In contrast, HoxA13 and HoxA9 are both expressed at similar time points within the regenerating limb bud where they are important for the formation of distal limb structures.

Another interesting example is lens development and regeneration. To establish the very defined lens pattern with well arranged posterior-anterior polarity and fibers (Fig. 3A) lens development in vertebrates ensues by the contact of the surface ectoderm with the optic cup that leads to the formation of the lens placode (Fig. 3B). The lens placode then differentiates to form the distinct pattern of the lens. During vertebrate lens regeneration, the origin of the regenerating lens is different from the developing lens (Fig. 3C). For instance, in frogs the lens is regenerated from the inner layer of the cornea [31–33]. In contrast, lens regeneration in the newt is regenerated from dorsal iris pigmented epithelial cells that represent a completely different cell type [34, 35].

To explain these differences at the molecular level is challenging. The main regulatory pathway in the development of the eye is the *pax6/ey* subnetwork, which in vertebrates also includes the regulatory genes *six3* and *six6*. However, in some parts of the eye *six3* or *six6* depend on *pax6* and in the lens they do not [36, 37]. *Six3* regulation has been thought to be important for the induction of lens regeneration in newts [38]. Thus, it could be that differences in the regulatory role of the *pax6* subnetwork could account for the differences in the induction of lens regeneration.

In conclusion, both cell-to-cell communication and morphogens are likely to establish pattern formation during regeneration and development. The differences between differentiation and regeneration can be explained by the tissue origin, e.g., embryonic versus adult tissue patterning, and the corresponding differences in gene expression profiles between developing and regenerating tissues.

4 Role of Stem Cells in Regenerative Biology and Pattern Formation

Stem cells are defined as the cells that execute the repair of damaged tissue. To date, there remains the question if embryonic stem cells or induced reprogrammed stem cells are capable of building a structured organ or body part. Adult somatic fibroblasts cells of higher vertebrates, e.g., mouse, or humans were demonstrated to de-differentiate via a cocktail of four genes, *oct4*, *sox2*, *klf4*, and *c-myc* to become pluripotent stem cells [39]. When looking at the newt blastema, the newt somatic cells are capable of de-differentiation, similar to adult stem cells within higher vertebrates. However, blastema cells differ by being able to execute organ pattern formation that replaces only the part of limb that needs to be reproduced. When we examined expression of these four genes in newt limb blastema or lens regeneration only three of them, *sox2*, *klf4*, and *c-myc* were expressed but not *oct4* [40]. Oct4 has been heralded as *the* most important pluripotent stem cell maintaining factor. Correspondingly, the lack of *oct4* expression might be the difference between attaining the status of a stem cell and a de-differentiated cell in the newt blastema. In other words blastema cells (or any other regeneration-involved cell) cannot be pluripotent in order to ensure fidelity of regeneration. A recent study by Kragl et al. that used green fluorescence protein to trace axolotl

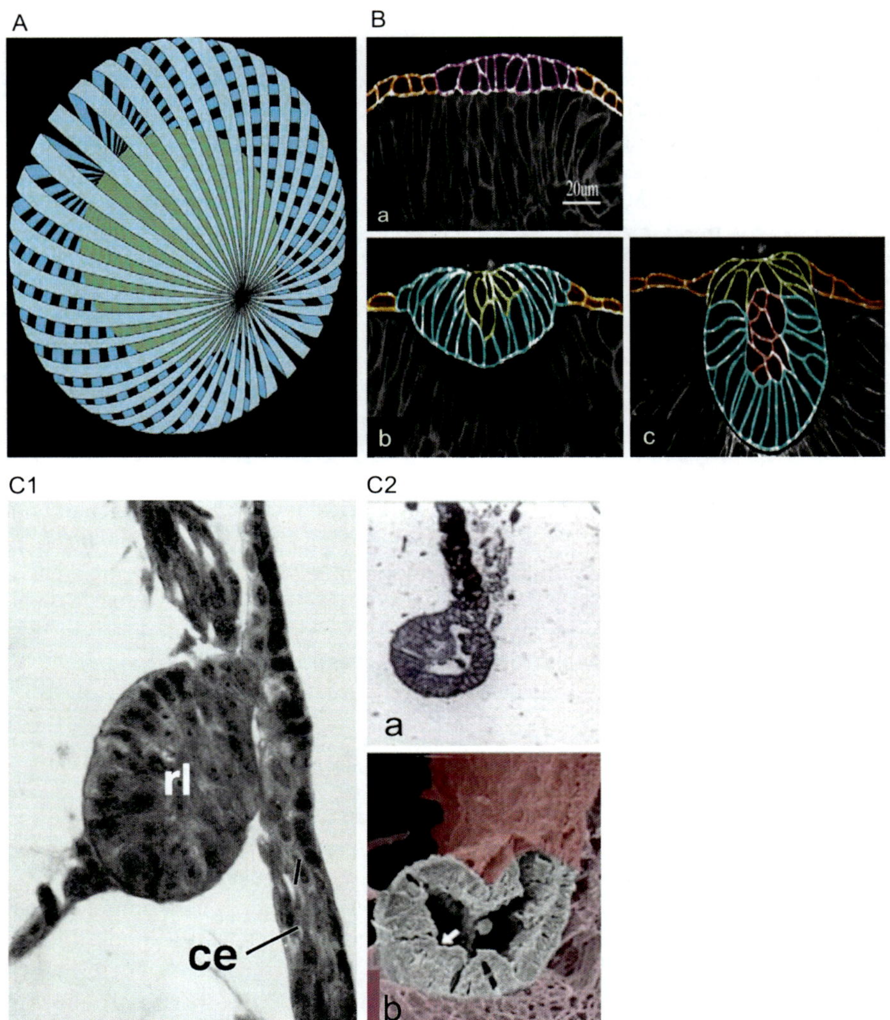

Fig. 3 Lens pattern formation during development and regeneration. (**A**) Complexity of Avian lens structure (umbilical suture type lens) (According to Kuszak et al. [45]). (**B**) Lens pattern formation during zebrafish development (According to Greiling and Clark [46]). Diagram representing the progressive stages of lens development in the zebrafish from the placode to spherical lens and the presumptive cellular differentiation at each stage. Cell membranes have been pseudocolored. *a* At approximately16 h post-fertilization (hpf), the cells in the surface ectoderm are *orange*, and the cells in the lens placode are *purple*. *b* At 18 hpf, elongating fiber-like cells are *blue*. *c* At 20 hpf, the cells in the organizing center (*red*) of the delaminating and elongated lens mass are surrounded by columnar primary fiber cells (*blue*). (**C**) Pattern formation during lens regeneration in amphibians (According to Henry and Tsonis [47]). **C1:** cornea-lens trans-differentiation in *X. laevis*: *ce* corneal epithelium*; rl* regenerating lens vesicle, **C2:** iris-lens trans-differentiation in newt: *a* phase-contrast image and *b* electron microscopy of lens vesicle formation (in *grey*) at day 14 lens vesicle elongating from iris stem followed by differentiation of lens fibers at the posterior part of the lens vesicle (*arrow*). Reproduced with permissions.

limb blastema cells revealed a distinct lineage-restriction (de-differentiation and re-differentiation potential) of defined cell types, e.g., muscle cells, epidermal cells, cartilage cells and Schwann cells [41]. In other an epidermal cell will de-differentiate according to the embryonic tissue origin and will give rise again to epidermis, a Schwann cell will dedifferentiate but will give rise to Schwann cells etc. In contrast, due to the lack of specific markers for connective tissue fibroblasts, there seems to be no lineage-restriction for fibroblast cells.

However, an intriguing study by Eiraku et al. comes to support the existence of an intrinsic positional memory within embryonic stem cells [42]. Mouse embryonic stem cell aggregates were able to generate the complex three-dimensional pattern of an optic cup in a self-directed fashion within an *in vitro* cell culture model that matched the *in vivo* temporal order of optic cup development. Following distinct cell culture conditions for initial induction of a neuroepithelium, a distally inward folded cup-like structure together with formation of a proximal pigment epithelium could be observed suggesting existence of an ES cell intrinsic self-organizing program for spatial pattern formation. In addition, the study also demonstrated that besides the stepwise domain–specific intrinsic information, soluble growth factors of local epithelia contribute to the tissue differentiation process. This study further supports the idea that certain cell types such as mesenchymal stem cells, blastema cells or fibroblast cells store intrinsic information for self-organized tissue pattern formation, that lacks explanation with current models on pattern formation.

In conclusion, this brief report touches upon tissue patterning during regeneration and the possibility that even dissociated cells can have intrinsic information to form patterns is elucidated. This provides new insights into the possible mechanisms of pattern formation, that most likely calls for a re-assessment of the theoretical models that have been used to explain the process. However, we do suggest that further studies on differences/similarities between stem cells and newt cells contributing to regeneration should be paramount for understanding the mechanisms of pattern formation.

References

1. Higgins GM, Anderson RM (1931) Experimental pathology of the liver. I. Restoration of the liver of the white rat following partial surgical removal. Arch Pathol 12:186–202
2. Michalopoulos GK, DeFrances MC (1997) Liver regeneration. Science 276:60–66
3. Morgan TH (1901) Regeneration. Macmillan, New York
4. Wolpert L (1969) Positional information and the spatial pattern of cellular differentiation. J Theor Biol 25:1–47
5. Wolpert L, Lewis JH (1975) Towards a theory of development. Fed Proc. Jan 34(1):14–20. Review
6. Bryant PJ, Bryant SV, French V (1977) Biological regeneration and pattern formation. Sci Am 237(66–76):81
7. French V, Bryant PJ, Bryant SV (1976) Pattern regulation in epimorphic fields. Science 193:969–981

8. Meinhardt H (1983) A boundary model for pattern formation in vertebrate limbs. J Embryol Exp Morphol 76:115–137
9. Maden M (1985) Retinoids and the control of pattern in regenerating limbs. Ciba Found Symp 113:132–155
10. Eichele G, Tickle C, Alberts BM (1985) Studies on the mechanism of retinoid-induced pattern duplications in the early chick limb bud: temporal and spatial aspects. J Cell Biol 101:1913–1920
11. Tickle C, Alberts B, Wolpert L, Lee J (1982) Local application of retinoic acid to the limb bond mimics the action of the polarizing region. Nature 296:564–566
12. Maden M, Summerbell D (1986) Retinoic acid-binding protein in the chick limb bud: identification at developmental stages and binding affinities of various retinoids. J Embryol Exp Morphol 97:239–250
13. Eichele G, Tickle C, Alberts BM (1984) Microcontrolled release of biologically active compounds in chick embryos: beads of 200-microns diameter for the local release of retinoids. Anal Biochem 142:542–555
14. Eichele G, Thaller C (1987) Characterization of concentration gradients of a morphogenetically active retinoid in the chick limb bud. J Cell Biol 105:1917–1923
15. Thaller C, Eichele G (1987) Identification and spatial distribution of retinoids in the developing chick limb bud. Nature 327:625–628
16. Maden M (1998) Retinoids as endogenous components of the regenerating limb and tail. Wound Repair Regen 6:358–365
17. Maden M, Ong DE, Summerbell D, Chytil F (1988) Spatial distribution of cellular protein binding to retinoic acid in the chick limb bud. Nature 335:733–735
18. Scadding SR, Maden M (1994) Retinoic acid gradients during limb regeneration. Dev Biol 162:608–617
19. Maden M (1997) Retinoic acid and its receptors in limb regeneration. Semin Cell Dev Biol 8:445–453
20. Giguere V, Ong ES, Evans RM, Tabin CJ (1989) Spatial and temporal expression of the retinoic acid receptor in the regenerating amphibian limb. Nature 337:566–569
21. Ragsdale CW Jr, Petkovich M, Gates PB, Chambon P, Brockes JP (1989) Identification of a novel retinoic acid receptor in regenerative tissues of the newt. Nature 341:654–657
22. Ragsdale CW Jr, Gates PB, Brockes JP (1992) Identification and expression pattern of a second isoform of the newt alpha retinoic acid receptor. Nucleic Acids Res 20:5851
23. Hill DS, Ragsdale CW Jr, Brockes JP (1993) Isoform-specific immunological detection of newt retinoic acid receptor delta 1 in normal and regenerating limbs. Development 117:937–945
24. Ragsdale CW Jr, Gates PB, Hill DS, Brockes JP (1993) Delta retinoic acid receptor isoform delta 1 is distinguished by its exceptional N-terminal sequence and abundance in the limb regeneration blastema. Mech Dev 42:113
25. Ragsdale CW Jr, Gates PB, Hill DS, Brockes JP (1993) Delta retinoic acid receptor isoform delta 1 is distinguished by its exceptional N-terminal sequence and abundance in the limb regeneration blastema. Mech Dev 40:99–112
26. Schier AF, Needleman D (2009) Developmental biology: Rise of the source-sink model. Nature 461:480–481
27. Crick F (1970) Diffusion in embryogenesis. Nature 225:420–422
28. Yu SR, Burkhardt M, Nowak M, Ries J, Petrasek Z, Scholpp S, Schwille P, Brand M (2009) Fgf8 morphogen gradient forms by a source-sink mechanism with freely diffusing molecules. Nature 461:533–536
29. Tsonis PA (1996) Limb regeneration. Cambridge University Press, New York
30. Gardiner DM, Blumberg B, Komine Y, Bryant SV (1995) Regulation of HoxA expression in developing and regenerating axolotl limbs. Development 121:1731–1741
31. Reeve JG, Wild AE (1981) Secondary lens formation from the cornea following implantation of larval tissues between the inner and outer corneas of *Xenopus laevis* tadpoles. J Embryol Exp Morphol 64:121–132

32. Bosco L, Venturini G, Willems D (1997) In vitro lens transdifferentiation of *Xenopus laevis* outer cornea induced by Fibroblast Growth Factor (FGF). Development 124:421–428
33. Henry JJ, Elkins MB (2001) Cornea-lens transdifferentiation in the anuran, *Xenopus tropicalis*. Dev Genes Evol 211:377–387
34. Del Rio-Tsonis K, Tsonis PA (2003) Eye regeneration at the molecular age. Dev Dyn 226 (2):211–224
35. Tsonis PA, Del Rio-Tsonis K (2004) Lens and retina regeneration: transdifferentiation, stem cells and clinical applications. Exp Eye Res 78(2):161–172
36. Donner AL, Maas RL (2004) Conservation and non-conservation of genetic pathways in eye specification. Int J Dev Biol 48(8–9):743–753
37. Wolf LV, Yang Y, Wang J, Xie Q, Braunger B, Tamm ER, Zavadil J, Cvekl A (2009) Identification of pax6-dependent gene regulatory networks in the mouse lens. PLoS One 4: e4159
38. Grogg MW, Call MK, Okamoto M, Vergara MN, Del Rio-Tsonis K, Tsonis PA (2005) BMP inhibition-driven regulation of six-3 underlies induction of newt lens regeneration. Nature 438 (7069):858–862
39. Takahashi K, Yamanaka S (2006) Induction of pluripotent stem cells from mouse embryonic and adult fibroblast cultures by defined factors. Cell 126:663–676
40. Maki N, Suetsugu-Maki R, Tarui H, Agata K, Del Rio-Tsonis K, Tsonis PA (2009) Expression of stem cell pluripotency factors during regeneration in newts. Dev Dyn 238:1613–1616
41. Kragl M, Knapp D, Nacu E, Khattak S, Maden M, Epperlein HH, Tanaka EM (2009) Cells keep a memory of their tissue origin during axolotl limb regeneration. Nature 460:60–65
42. Eiraku M, Takata N, Ishibashi H, Kawada M, Sakakura E, Okuda S, Sekiguchi K, Adachi T, Sasai Y (2011) Self-organizing optic-cup morphogenesis in three-dimensional culture. Nature 472:51–56
43. Jaeger J, Irons D, Monk N (2008) Regulative feedback in pattern formation: towards a general relativistic theory of positional information. Development 135:3175–3183
44. Kicheva A, Pantazis P, Bollenbach T, Kalaidzidis Y, Bittig T, Julicher F, Gonzalez-Gaitan M (2007) Kinetics of morphogen gradient formation. Science 315:521–525
45. Kuszak JR, Zoltoski RK, Sivertson C (2004) Fibre cell organization in crystalline lenses. Exp Eye Res 78:673–687
46. Greiling TM, Clark JI (2009) Early lens development in the zebrafish: a three-dimensional time-lapse analysis. Dev Dyn 238:2254–2265
47. Henry JJ, Tsonis PA (2010) Molecular and cellular aspects of amphibian lens regeneration. Prog Retin Eye Res 29:543–555

Gradients and Regulatory Networks of Wnt Signalling in *Hydra* Pattern Formation

Thomas W. Holstein

Abstract The Wnt/β-catenin pathway plays an important role in axis formation and axial patterning during metazoan development. Wnt genes are expressed around the blastopore of prebilaterian and bilaterian embryos and create a gradient of ligands governing the primary oral-aboral body axis. Although this polarised Wnt signalling along a primary body axis is a conserved property of metazoan development, the mechanisms governing localised Wnt expression and secretion are poorly understood. We study these questions in the freshwater polyp *Hydra*, a classical model system for the analysis of morphogenetic gradients and modelling of de novo pattern formation processes. New data emphasise the importance of gene regulatory networks for the formation of Wnt secreting signalling centres. It is proposed that a twist of transcriptional control and gradient formation was essential for the formation of Wnt signalling centres in metazoan axis formation.

1 Introduction

Wnt genes are specific for metazoans. No *Wnt* genes have been described so far from any unicellular eukaryotes, only certain modules of the Wnt signalling pathway have been identified in several protozoans including the cellular slime mould *Dictyostelium* [1]. This suggests that the emergence of *Wnt* genes was tightly coupled with the transition from unicellular to multicellular organisms and the evolutionary origin of metazoans [1, 2]. Orthologs of the main components of Wnt signalling pathway and a basic set of two to four Wnt ligands was identified in sponges and other prebilaterian animals [1]. In bilaterians, a major loss of *Wnt* gene

T.W. Holstein (✉)
Department for Molecular Evolution and Genomics, Centre for Organismal Studies (COS), Heidelberg University, Heidelberg, Germany
e-mail: holstein@uni-hd.de

V. Capasso et al. (eds.), *Pattern Formation in Morphogenesis*, Springer Proceedings in Mathematics 15, DOI 10.1007/978-3-642-20164-6_3,

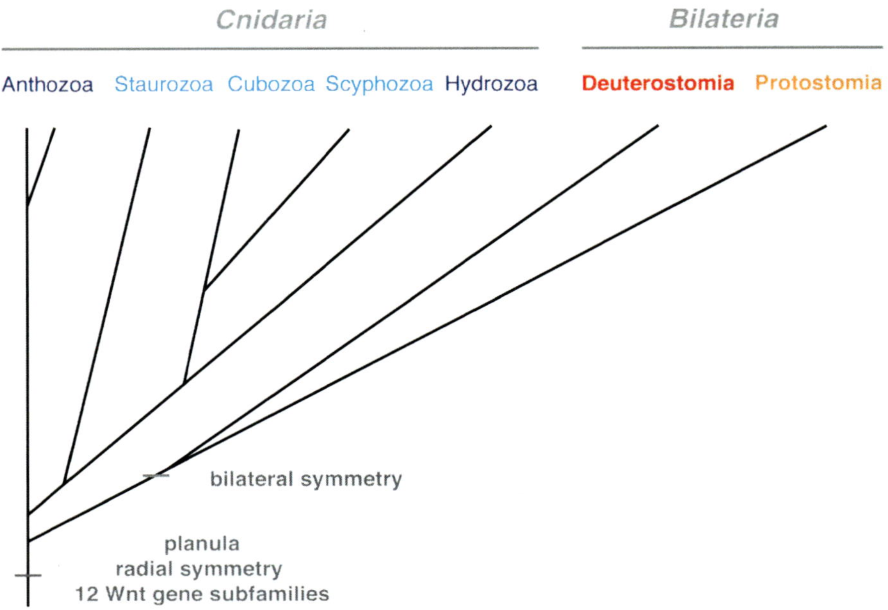

Fig. 1 Phylogenetic relationship of the phylum Cnidaria. The phylogenetic tree is based on the results of Collins et al. [5]. Sequenced genomes are available for *Nematostella* and *Hydra* [3, 4]

subfamilies occurred in various protostome lineages, while in deuterostomes this gene loss was minor [1, 2].

The first animals exhibiting the complete set of bilaterian *Wnt* genes are cnidarians. Cnidarians are more than 600 Myr old prebilaterians that exhibit a gene repertoire with 18,000–20,000 bona fide protein-coding genes [3, 4]. Cnidarians are the sister group to bilaterians (Fig. 1). The high genomic complexity of cnidarians is based on the presence of all major bilaterian gene families including those that encode for the major signalling pathways and transcription factor families. All bilaterian *Wnt* gene subfamilies are present in the genomes of two major cnidarian model systems (Fig. 2), i.e. the sea anemone *Nematostella vectensis* [6, 7] and the fresh water polyp *Hydra magnipapillata* [8].

Most cnidarian *Wnt* genes are expressed at the site of the blastopore, and they are required for gastrulation [1, 2, 9–12]. This localised Wnt/β-catenin signalling centre determines the orientation of the primary body axis in non-bilaterians and in bilaterian animals, where it forms the anterior-posterior axis [1, 2, 13–15]. Therefore, this function of Wnt/β-catenin signalling in gastrulation might be the first and primary function of this signalling pathway [1]. An important feature of embryonic *Wnt* gene expression in cnidarians is that most *Wnt* genes are expressed in overlapping domains along the oral-aboral axis of the embryo [6]. In contrast to the presence of a complete set of *Wnt* gene subfamilies in cnidarians, a true *Hox*-cluster is missing in these animals [16, 17]. Hence, we named the characteristic expression pattern of *Wnt* genes in

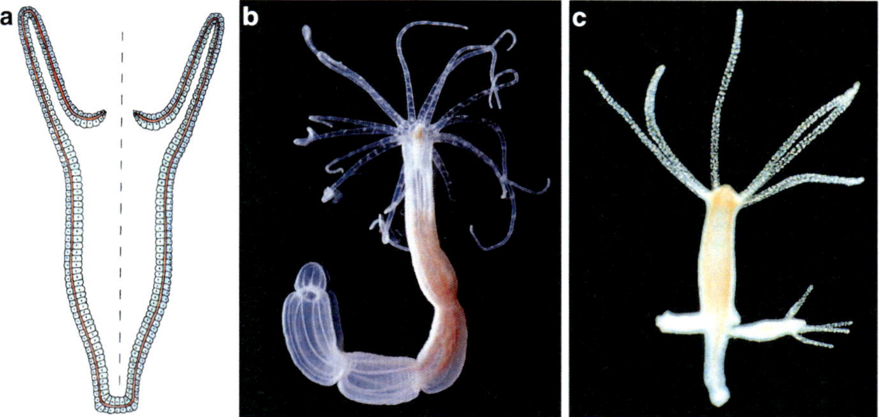

Fig. 2 Cnidarian bauplan. (**a**) Scheme of a diploblastic cnidarian polyp. (**b**) Micrograph of the starlet sea anemone *Nematostella vectensis*. (**c**) Micrograph of the freshwater polyp *Hydra vulgaris* with buds (b from Kusserow et al. [6] and c from Holstein et al. [2])

Nematostella the "Wnt-code" [18]. Interestingly two evolutionary clusters of *Wnt* genes have been identified (cluster 1 with *Wnt9* – *Wnt1* – *Wnt6* – *Wnt10* and cluster 2 with *Wnt5* – *Wnt7* – *Wnt4*) [1] suggesting that this ancient pattern of *Wnt* genes originating at the site of the blastopore has an instructive role in patterning the primary oral-aboral (i.e. anterior-posterior) body axis of metazoans.

While it is generally accepted that the Wnt signalling centre and a gradient of secreted Wnt ligands are shaped by Wnt antagonists, e.g. Dkk1/2/4 [18, 19], it is not known, which downstream target genes of Wnt signalling are responsible for the position-dependent differentiation patterns of *Hydra* and *Nematostella*. Transcriptome studies using *Wnt*/*β-catenin* over-expression can provide a comprehensive picture of Wnt target genes in *Hydra*.

2 The *Hydra* Head Organiser Is an Organiser Equivalent to the Blastoporal Organiser in Bilaterians

A source of Wnt ligands in many metazoan embryos is the blastoporal signalling centre. In cnidarians the blastoporal region of most embryos develops to the mouth region of the polyp. This mouth is located on the tip of the hypostome, a dome-like structure at the apical (oral) terminus of the body axis. A circle of five to seven tentacles forms at the base of the hypostome. Regarding morphogenetic signals, the hypostome has retained the signalling properties of the blastoporal signalling centre. In a classical experiment Ethel Browne demonstrated that tissue of the *Hydra* organiser induces a secondary body axis by recruiting host tissue when it is transplanted to an ectopic site [20].

The properties of this signalling centre are similar to those of the Spemann-Mangold organiser in amphibians; it was therefore designed as the "*Hydra* organiser" [1]. As pointed out previously [21] both organisers are not fully equivalent. A blastoporal ring serves as a source of Wnt ligands in both systems [1], but only in the amphibian Spemann-Mangold organiser a symmetry break creates a group of Chordin-secreting cells in the marginal zone. As a result of this commitment, mesodermal cells immigrate and generate a midline towards the anterior end and finally form the prechordal plate [21].

The morphogenetic properties of tissue exhibiting axis organiser properties in *Hydra* has been characterised on the cellular level [22]. Clusters consisting of only ten *Wnt* expressing cells can induce a secondary body axis. By comparison, single cells failed to form a signalling centre. The fact that a minimal size is required for the induction suggests that a community-effect between Wnt-signalling cells is required to establish a stabile signalling centre that successfully can induce new body axes [22].

3 An Autocatalytic Regulatory Network Maintains Wnt Signalling in the *Hydra* Head Organiser

Using regeneration experiments we discovered that Wnt3 is a putative master gene in *Hydra* axial patterning [8, 23]. A total of 11 Wnt genes was identified in *Hydra* [8]. During head regeneration Wnt3 was expressed within the first 30 min after the onset of head regeneration when a new organiser is formed [8, 24]. The other *Wnt* genes were also expressed during head regeneration and bud formation, *Hydra's* asexual form of reproduction, but significantly later. In the head regeneration-deficient mutant strain *reg-16* recombinant HyWnt3 protein increased the head formation capacity to that of wild-type strains.

The kinetics of *Wnt* gene expressions during head regeneration suggests that a cascade of consecutive Wnt activation accompanies regeneration (Fig. 3), and HyWnt3 begins this cascade [8]. Direct evidence exists for the expression of *Wnt5* and *Wnt8* that are expressed in regions, where bud and tentacle evaginations are initiated. These *Wnt* genes have functions similar to noncanonical *Wnt* genes in planar cell polarity movements of vertebrates. Interaction studies indicate that an activation of noncanonical Wnt signalling in *Hydra* requires Wnt/β-Catenin signalling, which acts in patterning the major oral-aboral body axis [25].

Ectopic *Wnt3* expression and axis formation can be induced by an elevated activation of Wnt/β-catenin signalling using alsterpaullone treatment, an inhibitor of GSK3 [19, 26], or by treatment with recombinant Wnt3 protein [8]. Since *Tcf* and *β-catenin* are synexpressed with *Wnt3* during organiser formation it was proposed that both genes are involved in the transcriptional auto-regulation of Wnt3 through a Wnt/β-catenin circuit [24]. An analysis of the cis-regulatory processes of *HyWnt3* transcription by using transgenic *Hydra* shows that HyWnt3 is regulated by two functionally distinct cis-regulatory elements located in the HyWnt3 promoter (Fig. 4). One of them, HyWnt3act, requires TCF-binding sites and thereby acts as an autoregulatory element.

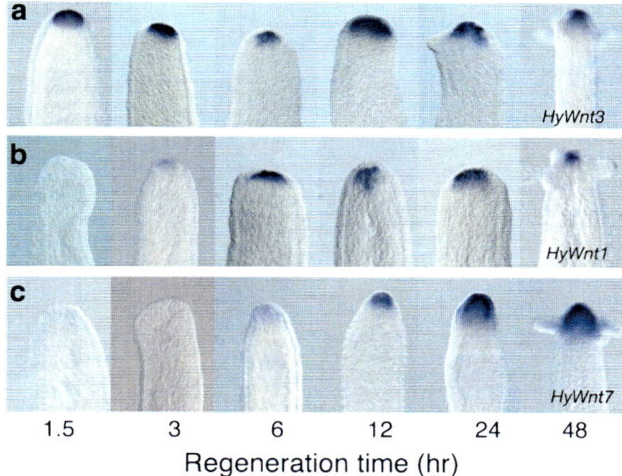

Fig. 3 Expression patterns of three *Wnt* genes (a–c) in *Hydra* during head regeneration. Animals were bisected below the head (80 % BL) at the times indicated (From Lengfeld et al. [8] with permission from the author)

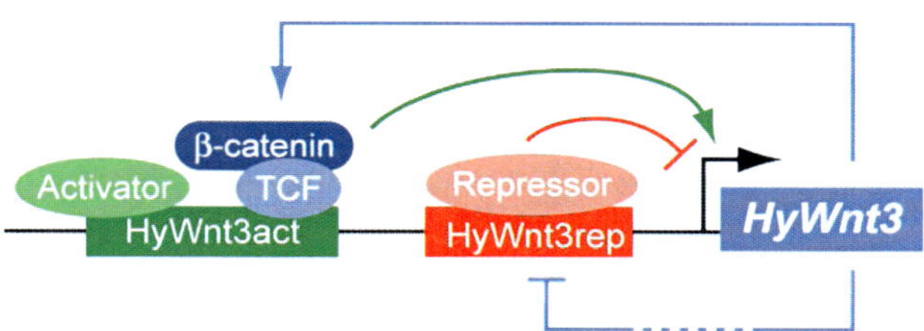

Fig. 4 Model for the transcriptional control of organiser specific *HyWnt3* expression. *HyWnt3* is controlled by two cis-regulatory elements (HyWnt3act, *green*, and HyWnt3rep, *red*), which are positively (*light blue arrow*) and negatively (*light blue bar*) regulated by Wnt/β -catenin signalling. The β-catenin/Tcf complex (*blue*) and putative activators (*light green*) bind to HyWnt3act, and their combinatorial inputs act in HyWnt3 transcription (*green arrow*). Potential repressors (*pink*) bind to HyWnt3rep and inhibit HyWnt3 expression (From Nakamura et al. [23] with permission from the author)

Thus, HyWnt3 can maintain its own expression by creating a positive feedback loop, which is reminiscent to the pattern formation model proposed by [27–29]. Further analyses of the HyWnt3 promoter identified an important repressor element. Deletion of the repressor element increased and expanded the eGFP expression outside the head organiser region dramatically [23].

These results demonstrate that a combination of autoregulation and repression has a crucial function for the establishment and maintenance of the localised *HyWnt3* expression in the head organiser. Transcriptional regulation of *Wnt3* expression is probably at the core of *Wnt*/β-catenin signalling in *Hydra*, and only complemented by Dkk1/2/4 [1]. This signalling centre may be independent of the function of extracellular Wnt antagonists, which could explain why Wnt antagonists are absent from several metazoan lineages [1, 23].

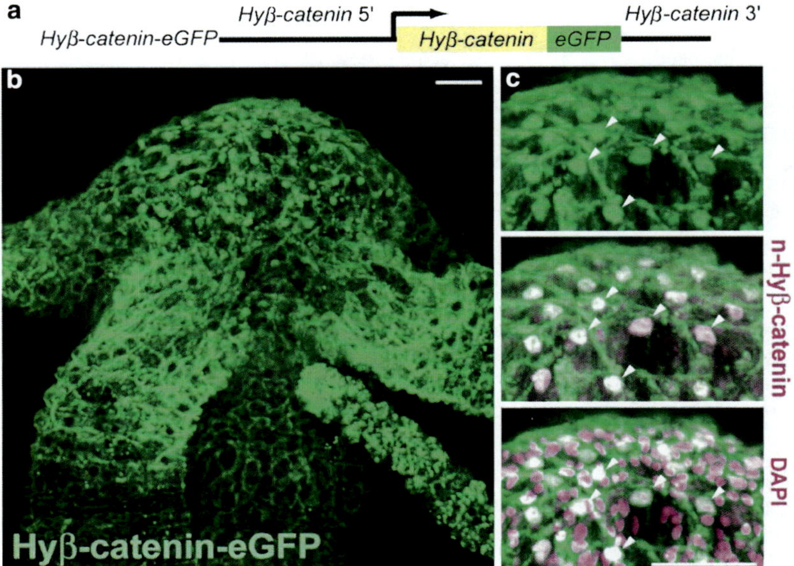

Fig. 5 ***Hydra* β-catenin**. (**a**) Schematic diagram of the Hyβ-catenin-EGFP construct. (**b**) Hyβ-catenin-EGFP is localized in the nuclei of cells at higher levels in the hypostome than in the body column. (**c**) Colocalization of Hyβ-catenin-EGFP with endogenous Hyβ-catenin detected with antinuclear Hyβ-catenin antibody (*arrowheads*). Note that Hyβ-catenin-EGFP is also associated with cell membranes (From Nakamura et al. [23] with permission from the author)

4 The Wnt Gradient Is Transformed into a Gradient of Stable β-Catenin Expression

Transplantation experiments in *Hydra* showed that the ability to form an organiser is graded over the entire body length [30, 31]. It was also proposed that a long-range gradient of inhibition is likely to establish a relatively more stable property graded across the tissue [21]. This graded and stable property was described as a gradient of source density [32].

Since the sufficiency of β-catenin signalling for providing the head organiser activity has been recently demonstrated by overexpression of a constitutively active form of Hyβ-catenin [23, 33], it is likely that this gradient is comprised by a gradient of nuclear β-catenin. High levels of nuclear β-catenin can be found in the organiser region of the hypostome (Fig. 5). β-catenin and Tcf transcripts are also broadly distributed along the oral–aboral axis, with higher levels in the hypostome than in the body column. It is also likely that the expanded eGFP expression after removal of the repressor element of the Wnt3 promoter (see above) corresponds to a long-range gradient of Wnt3 ligands activating nuclear β-catenin in a concentration-dependent manner.

5 Theses

Based on the recent progress in unravelling the function of Wnt signalling in the *Hydra* organiser the following main theses are put forward: (1) A Wnt/β-catenin signalling centre is responsible for the formation of the oral-aboral body axis in early multi-cellular animals (i.e. the first metazoans including *Hydra*), and a gradient of secreted Wnt ligands – shaped by Wnt antagonists (Dkk and others) – induces downstream target genes in a position-dependent manner. (2) The *Hydra* head organiser is an ancestral organiser equivalent to the blastoporal organiser in bilaterians. (3) An autocatalytic regulatory network maintains the *Hydra* head organiser, which includes a feedback loop of Wnt signalling and TCF binding sites in the Wnt promoter. Repressor elements in the Wnt promoter are required to suppress ectopic Wnt expression outside of the head organizer. (4) The Wnt gradient is transformed into a gradient of stable β-catenin expression, a parameter, which corresponds to an oral-aboral gradient value in *Hydra's* body called 'source density' in the classical Gierer-Meinhardt model.

References

1. Holstein TW (2012) The evolution of the Wnt pathway. Cold Spring Harb Perspect Biol (in press, doi: 10.1101/cshperspect.a007922)
2. Holstein TW, Watanabe H, Ozbek S (2011) Signaling pathways and axis formation in the lower metazoa. Curr Top Dev Biol 97:137–177
3. Chapman JA, Kirkness EF, Simakov O, Hampson SE, Mitros T, Weinmaier T, Rattei T, Balasubramanian PG, Borman J, Busam D, Disbennett K, Pfannkoch C, Sumin N, Sutton GG, Viswanathan LD, Walenz B, Goodstein DM, Hellsten U, Kawashima T, Prochnik SE, Putnam NH, Shu S, Blumberg B, Dana CE, Gee L, Kibler DF, Law L, Lindgens D, Martinez DE, Peng J, Wigge PA, Bertulat B, Guder C, Nakamura Y, Ozbek S, Watanabe H, Khalturin K, Hemmrich G, Franke A, Augustin R, Fraune S, Hayakawa E, Hayakawa S, Hirose M, Hwang JS, Ikeo K, Nishimiya-Fujisawa C, Ogura A, Takahashi T, Steinmetz PR, Zhang X, Aufschnaiter R, Eder MK, Gorny AK, Salvenmoser W, Heimberg AM, Wheeler BM, Peterson KJ, Bottger A, Tischler P, Wolf A, Gojobori T, Remington KA, Strausberg RL, Venter JC, Technau U, Hobmayer B, Bosch TC, Holstein TW, Fujisawa T, Bode HR, David CN, Rokhsar DS, Steele RE (2010) The dynamic genome of Hydra. Nature 464:592–596
4. Putnam NH, Srivastava M, Hellsten U, Dirks B, Chapman J, Salamov A, Terry A, Shapiro H, Lindquist E, Kapitonov VV, Jurka J, Genikhovich G, Grigoriev IV, Lucas SM, Steele RE, Finnerty JR, Technau U, Martindale MQ, Rokhsar DS (2007) Sea anemone genome reveals ancestral eumetazoan gene repertoire and genomic organization. Science 317:86–94
5. Collins AG, Schuchert P, Marques AC, Jankowski T, Medina M, Schierwater B (2006) Medusozoan phylogeny and character evolution clarified by new large and small subunit rDNA data and an assessment of the utility of phylogenetic mixture models. Syst Biol 55:97–115
6. Kusserow A, Pang K, Sturm C, Hrouda M, Lentfer J, Schmidt HA, Technau U, von Haeseler A, Hobmayer B, Martindale MQ, Holstein TW (2005) Unexpected complexity of the Wnt gene family in a sea anemone. Nature 433:156–160
7. Lee PN, Pang K, Matus DQ, Martindale MQ (2006) A WNT of things to come: evolution of Wnt signaling and polarity in cnidarians. Semin Cell Dev Biol

8. Lengfeld T, Watanabe H, Simakov O, Lindgens D, Gee L, Law L, Schmidt HA, Ozbek S, Bode H, Holstein TW (2009) Multiple Wnts are involved in Hydra organizer formation and regeneration. Dev Biol 330:186–199
9. Lee PN, Kumburegama S, Marlow HQ, Martindale MQ, Wikramanayake AH (2007) Asymmetric developmental potential along the animal-vegetal axis in the anthozoan cnidarian, Nematostella vectensis, is mediated by Dishevelled. Dev Biol
10. Momose T, Derelle R, Houliston E (2008) A maternally localised Wnt ligand required for axial patterning in the cnidarian *Clytia hemisphaerica*. Development 135:2105–2113
11. Momose T, Houliston E (2007) Two oppositely localised frizzled RNAs as axis determinants in a cnidarian embryo. PLoS Biol 5:e70
12. Wikramanayake AH, Hong M, Lee PN, Pang K, Byrum CA, Bince JM, Xu R, Martindale MQ (2003) An ancient role for nuclear beta-catenin in the evolution of axial polarity and germ layer segregation. Nature 426:446–450
13. De Robertis EM (2008) Evo-devo: variations on ancestral themes. Cell 132:185–195
14. Niehrs C (2010) On growth and form: a Cartesian coordinate system of Wnt and BMP signaling specifies bilaterian body axes. Development 137:845–857
15. Rentzsch F, Guder C, Vocke D, Hobmayer B, Holstein TW (2007) An ancient chordin-like gene in organizer formation of Hydra. Proc Natl Acad Sci U S A 104:3249–3254
16. Chourrout D, Delsuc F, Chourrout P, Edvardsen RB, Rentzsch F, Renfer E, Jensen MF, Zhu B, de Jong P, Steele RE, Technau U (2006) Minimal ProtoHox cluster inferred from bilaterian and cnidarian Hox complements. Nature 442:684–687
17. Technau U, Steele RE (2011) Evolutionary crossroads in developmental biology: cnidaria. Development
18. Guder C, Philipp I, Lengfeld T, Watanabe H, Hobmayer B, Holstein TW (2006a) The Wnt code: cnidarians signal the way. Oncogene 25:7450–7460
19. Guder C, Pinho S, Nacak TG, Schmidt HA, Hobmayer B, Niehrs C, Holstein TW (2006b) An ancient Wnt-Dickkopf antagonism in Hydra. Development 133:901–911
20. Browne EN (1909) The production of new hydranths in hydra by the insertion of small grafts. J Exp Zool 7:1–37
21. Meinhardt H (2012) Modeling pattern formation in hydra: a route to understanding essential steps in development. Int J Dev Biol
22. Technau U, Cramer von Laue C, Rentzsch F, Luft S, Hobmayer B, Bode HR, Holstein TW (2000) Parameters of self-organization in Hydra aggregates. Proc Natl Acad Sci U S A 97:12127–12131
23. Nakamura Y, Tsiairis CD, Ozbek S, Holstein TW (2011) Autoregulatory and repressive inputs localize Hydra Wnt3 to the head organizer. Proc Natl Acad Sci U S A 108:9137–9142
24. Hobmayer B, Rentzsch F, Kuhn K, Happel CM, von Laue CC, Snyder P, Rothbacher U, Holstein TW (2000) WNT signalling molecules act in axis formation in the diploblastic metazoan Hydra. Nature 407:186–189
25. Philipp I, Aufschnaiter R, Ozbek S, Pontasch S, Jenewein M, Watanabe H, Rentzsch F, Holstein TW, Hobmayer B (2009) Wnt/beta-catenin and noncanonical Wnt signaling interact in tissue evagination in the simple eumetazoan Hydra. Proc Natl Acad Sci U S A 106:4290–4295
26. Broun M, Gee L, Reinhardt B, Bode HR (2005) Formation of the head organizer in hydra involves the canonical Wnt pathway. Development 132:2907–2916
27. Gierer A, Meinhardt H (1972) A theory of biological pattern formation. Kybernetik 12:30–39
28. Meinhardt H (2004a) Different strategies for midline formation in bilaterians. Nat Rev Neurosci 5:502–510
29. Meinhardt H (2004b) Models for the generation of the embryonic body axes: ontogenetic and evolutionary aspects. Curr Opin Genet Dev 14:446–454
30. MacWilliams HK (1983a) Hydra transplantation phenomena and the mechanism of hydra head regeneration. I. Properties of the head inhibition. Dev Biol 96:217–238

31. MacWilliams HK (1983b) Hydra transplantation phenomena and the mechanism of Hydra head regeneration. II. Properties of the head activation. Dev Biol 96:239–257
32. Meinhardt H (1993) A model for pattern formation of hypostome, tentacles, and foot in hydra: how to form structures close to each other, how to form them at a distance. Dev Biol 157:321–333
33. Gee L, Hartig J, Law L, Wittlieb J, Khalturin K, Bosch TC, Bode HR (2010) Beta-catenin plays a central role in setting up the head organizer in hydra. Dev Biol 340:116–124

Mathematical Modeling of Planar Cell Polarity Signaling

Jeffrey D. Axelrod

The Planar Cell Polarity (PCP) system polarizes cells in diverse epithelial sheets along an axis orthogonal to their apical-basal axis, and is necessary for numerous physiological functions [1–3]. In *Drosophila*, the most thoroughly studied planar polarized tissue is the fly wing, in which each cell produces a trichome ("hair"), that in wild type, emerges from the distal side of the cell and points distally (Fig. 1) [4]. PCP mutants cause disruption of this pattern. In vertebrates, the PCP signaling modules identified in flies are conserved, and function together with additional regulators not present in flies [1, 3, 5, 6]. Defects in PCP result in a range of developmental anomalies and diseases including open neural tube defects, polycystic kidneys (reviewed in in [3, 6–8]), conotruncal heart defects [9–12], and disruption of sensory hair cell polarity causing deafness [13–23]. PCP is also believed to underlie the directed migration of malignant cells during invasion and metastasis [24–28]. Insufficient understanding of the PCP signaling modules and their interactions limits the potentially substantial opportunities for therapeutic intervention.

Studies in the fruitfly, *Drosophila*, have led to the concept of a modular system controlling PCP, in which functional modules each comprise a genetically and biochemically related unit [29]. Two principal PCP modules ("global" and "core") provide polarizing information, and target effector modules carry out morphological polarization [30]. Target effector modules for PCP signaling may be discrete systems that act within single cells to build polarized structures such as trichomes (hairs), or may be multicellular systems in which signals within and between cells contribute to patterning. For example, in the fly leg, a signaling module that regulates the development of multicellular mechanosensory bristles [31] is polarized by the PCP system.

J.D. Axelrod (✉)
Department of Pathology, Stanford University School of Medicine, Stanford, CA, USA
e-mail: jaxelrod@stanford.edu

V. Capasso et al. (eds.), *Pattern Formation in Morphogenesis*, Springer Proceedings in Mathematics 15, DOI 10.1007/978-3-642-20164-6_4,

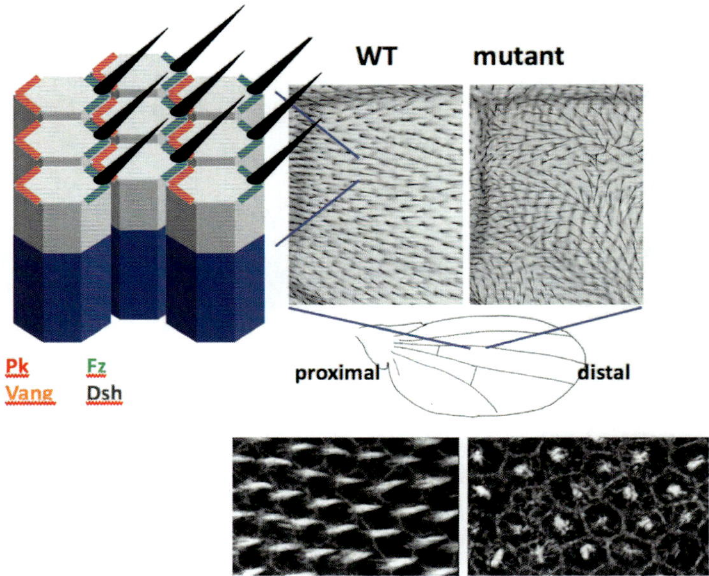

Fig. 1 Wild-type and mutant fly wings showing PCP. The cartoon shows hairs growing from the distal side, and asymmetric localization of PCP proteins. Below, Actin-stained pupal wings showing wildtype hairs (*left*) and mutant hairs growing from the center of the cell (*right*)

The core module acts both to amplify asymmetry, and to coordinate polarization between neighboring cells, producing a local alignment of polarity. Proteins in the core signaling module, including the serpentine receptor Frizzled (Fz) [32, 33], the multi-domain protein Dishevelled (Dsh) [34, 35], the Ankryin repeat protein Diego (Dgo) [36], the four-pass transmembrane protein Van Gogh (Vang; a.k.a. Strabismus) [37, 38], the Lim domain protein Prickle (Pk) [39] and the seven-transmembrane atypical cadherin Flamingo (Fmi; a.k.a. Starry night) [40, 41] (reviewed in [1]), adopt asymmetric subcellular localizations that predict the hair polarity pattern (Fig. 1) [36, 41–45]. These proteins communicate at cell boundaries, recruiting one group to the distal side of cells, and the other to the proximal side, through the function of an incompletely understood feedback mechanism, thereby aligning the polarity of adjacent cells [45, 46].

Considerable insight into the core PCP mechanism comes from studies of clones of cells either not expressing or overexpressing core PCP components. These clones display characteristic perturbations (or lack thereof) of cells in nearby wing tissue (referred to as domineering non-autonomy) [32, 34, 35, 37, 44, 46–48].

Prior to our work, the predominant model to explain the distribution of polarity information, and the domineering non-autonomy around mutant clones, relied on gradients of diffusible factors, referred to as "diffusible factor X" or "factor Z" models [2, 48–56], in which a diffusible factor produced at a localized source produces a gradient that in turn stimulates cells to produce a second, unidentified,

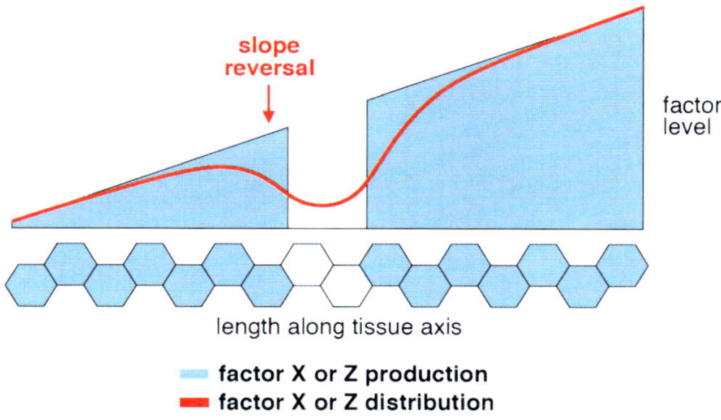

Fig. 2 Schematic of the factor X or Z models (Reprinted from [46])

factor X or Z in an amount proportional to the concentration of the original gradient. Factor X or Z would then also produce a gradient whose direction is sensed by individual cells, providing polarizing information. Mutant clones of core PCP components that produce domineering non-autonomy were proposed to either impair or augment the ability of cells to produce factor X or Z, thereby perturbing the gradient and causing local reversals of polarity (Fig. 2).

Based on molecular genetic analyses and the recognition that core PCP proteins segregate specifically to proximal or distal boundaries of cells during polarization, we proposed a contrasting model based on a feedback mechanism that communicates information between adjacent cell surfaces [45] (Fig. 3). The feedback model appeared to explain a range of phenotypes. A compilation of observations about the autonomy or non-autonomy of mutant clones resulted in a dataset that could be used to validate the biological model [46]. However, it became evident that intuition was insufficient to link molecular genetic interventions with the dataset of emergent tissue level patterns. Therefore, we, and others, have used mathematical modeling to better understand the properties of proposed biological signaling mechanisms (reviewed in 57).

Here, I describe how mathematical modeling has been used to (1) prove the feasibility of our local feedback model in robustly producing the range of clonal phenotypes observed, (2) predict additional features of the model that were then experimentally verified, and (3) validate the hypothesis that the apparently stochastic appearance of swirls in clones of cells mutant for the global component *fat* arise as a consequence of altered cell geometries affecting the ability of the core mechanism to propagate polarity between cells.

We developed a mathematical model of the core module, consisting of a set of linked, partial differential equations (PDEs) representing a reaction-diffusion system in which the states are the concentrations of core PCP proteins and proposed complexes [46]. The equations were used to represent protein concentrations on a discretized mesh as an abstraction of an epithelial cell sheet that could simulate a fly

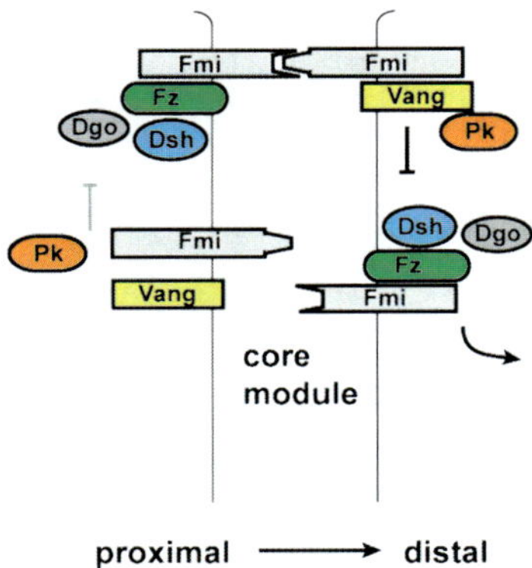

Fig. 3 Schematic of the intercellular feedback loop model. Protein interactions at a cell–cell interface are shown. Proteins assemble into an asymmetric complex, with distal proteins from one cell interacting with proximal proteins from a neighboring cell. Oppositely oriented complexes are mutually exclusive, such that with an input bias, the mechanism behaves as a bistable switch, accumulating strong asymmetry that is coordinated with neighboring cells

wing. The model was constructed in a manner that gave careful consideration to necessary assumptions resulting from gaps in understanding, and allowed us to test the sensitivity to those assumptions.

The first goal was to test the feasibility of the model. To do so, we solved the equations given a set of feature constraints representing the observed phenotypes. We found that the model was able to reproduce all characteristic PCP phenotypes (Fig. 4), and that the phenotypes were robust to specific parameter values, suggesting that the network architecture was the key feature allowing the system to function. This result proved the feasibility of the proposed model.

We also used the model to make predictions about the functions of specific alleles of *fz* that could be experimentally verified. We predicted that the cell autonomous alleles fz^{F31} and fz^{J22}, should retain their ability to interact with Vang in the neighboring cell, but lose their ability to recruit Dsh in the same cell. These predictions were tested and verified experimentally (Fig. 5).

Taken together, these results support the hypothesis that the feedback loop architecture, although a simplification of the true underlying biology, is a feasible one for explaining the core PCP mechanism [46, 58, 59]. A major part of this effort was in parameter identification: the computationally intensive Nelder-Mead simplex method was used in [46], while numerical methods based on optimal control were used in [59]. We have since updated our mathematical model of the core PCP module [46] to include active motion of molecules along microtubules, using rates of polarized transport of Fz and Dsh determined in [60] and [Germain and Axelrod, unpublished]. We have also begun development of a PDE model of the global module, in which we are investigating whether the known local interactions between *fj*, *ds*, and *ft* respond to the A/P gradients of *fj* and *ds* in such a way to produce cellular level asymmetry in these components.

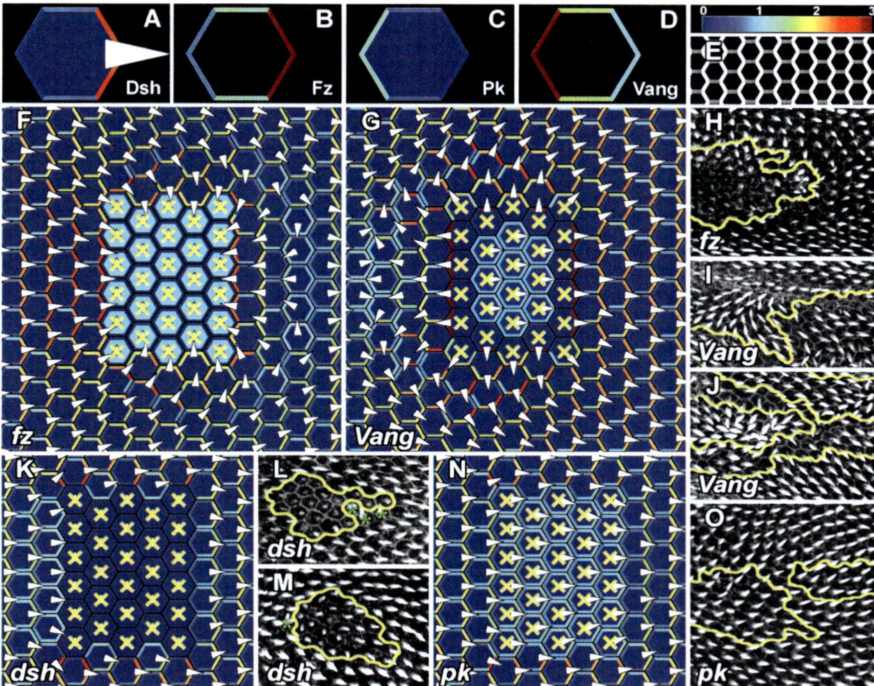

Fig. 4 Model simulation outputs, showing Dsh protein concentrations as a colorplot. Simulations of *fz*, *dsh*, *vang* and *pk* mutant clones, and corresponding examples of phalloidin stained clones are shown for comparison. Note that the appropriate domineering non-autonomy, or lack thereof, is captured by the model (Figure taken from [46])

In another use of the model, we wished to test the hypothesis that the apparently stochastic appearance of swirls in clones of cells mutant for the global component *fat* arise as a consequence of altered cell geometries affecting the ability of the core mechanism to propagate polarity between cells [58]. Biological experiments were consistent with this explanation, but we wished to use modeling to test whether this explanation was sufficient to account for the observations. To do so, the model was adapted to allow for simulations on cell grids extracted from experimental images. This required a number of modifications to the model implementation. Using the modified modeling framework, we found that in the *Drosophila* wing, a combination of cell geometry and non-autonomous signaling at clone boundaries determines the correct or incorrect polarity propagation in clones that lack Fat mediated global directional information. We suggest that this is important to allow for the development of pattern elements, such as veins, and for the sporadic occurrences of irregular geometry are obstacles to polarity propagation. Thus, in wild type tissue, broad distribution of the global directional cue combines with a local feedback mechanism to overcome irregularities in cell packing geometry during PCP signaling.

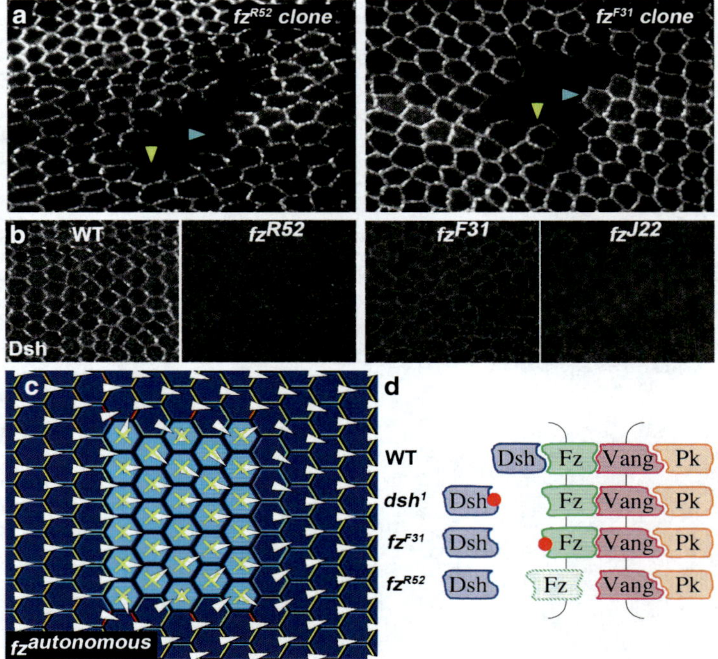

Fig. 5 Predictions and experimental validation of fz^{F31} and fz^{J22} behaviors. (**a**) fz^{R52} null clones do not recruit Vang from neighboring cells to the clone boundary (*left*) while fz^{F31} clones retain the ability to recruit Vang (*right*). (**b**) Both classes of *fz* allele lose the ability to recruit Dsh to the cell membrane. (**c**) Simulation of the *fz* autonomous clone phenotype. (**d**) Cartoon of lost and retained interactions for the two types of allele (Figure reproduced from [46])

We have shown through these examples the utility of a class of mathematical modeling for helping understand emergent biological phenotypes. Other mathematical modeling approaches have been applied to understanding PCP, and they are reviewed in [57].

References

1. Zallen JA (2007) Planar polarity and tissue morphogenesis. Cell 129:1051–1063
2. Adler PN (2002) Planar signaling and morphogenesis in *Drosophila*. Dev Cell 2:525–535
3. Simons M, Mlodzik M (2008) Planar cell polarity signaling: from fly development to human disease. Annu Rev Genet 42:517–540
4. Wong LL, Adler PN (1993) Tissue polarity genes of *Drosophila* regulate the subcellular location for prehair initiation in pupal wing hairs. J Cell Biol 123:209–221
5. Jones C, Chen P (2007) Planar cell polarity signaling in vertebrates. Bioessays 29:120–132
6. Wang Y, Nathans J (2007) Tissue/planar cell polarity in vertebrates: new insights and new questions. Development 134:647–658
7. Simons M, Walz G (2006) Polycystic kidney disease: cell division without a c(l)ue? Kidney Int 70:854–864

8. Wallingford JB (2006) Planar cell polarity, ciliogenesis and neural tube defects. Hum Mol Genet 15(Suppl 2):R227–R234
9. Phillips HM, Murdoch JN, Chaudhry B, Copp AJ, Henderson DJ (2005) Vangl2 acts via RhoA signaling to regulate polarized cell movements during development of the proximal outflow tract. Circ Res 96:292–299
10. Henderson DJ, Phillips HM, Chaudhry B (2006) Vang-like 2 and noncanonical Wnt signaling in outflow tract development. Trends Cardiovasc Med 16:38–45
11. Garriock RJ, D'Agostino SL, Pilcher KC, Krieg PA (2005) Wnt11-R, a protein closely related to mammalian Wnt11, is required for heart morphogenesis in Xenopus. Dev Biol 279:179–192
12. Phillips HM et al (2007) Disruption of planar cell polarity signaling results in congenital heart defects and cardiomyopathy attributable to early cardiomyocyte disorganization. Circ Res 101:137–145
13. Jones C et al (2008) Ciliary proteins link basal body polarization to planar cell polarity regulation. Nat Genet 40:69–77
14. Qian D et al (2007) Wnt5a functions in planar cell polarity regulation in mice. Dev Biol 306:121–133
15. Deans MR et al (2007) Asymmetric distribution of prickle-like 2 reveals an early underlying polarization of vestibular sensory epithelia in the inner ear. J Neurosci 27:3139–3147
16. Montcouquiol M et al (2006) Asymmetric localization of Vangl2 and Fz3 indicate novel mechanisms for planar cell polarity in mammals. J Neurosci 26:5265–5275
17. Davies A, Formstone C, Mason I, Lewis J (2005) Planar polarity of hair cells in the chick inner ear is correlated with polarized distribution of c-flamingo-1 protein. Dev Dyn 233:998–1005
18. Lu X et al (2004) PTK7/CCK-4 is a novel regulator of planar cell polarity in vertebrates. Nature 430:93–98
19. Montcouquiol M et al (2003) Identification of Vangl2 and Scrb1 as planar polarity genes in mammals. Nature 423:173–177
20. Wang Y, Guo N, Nathans J (2006) The role of Frizzled3 and Frizzled6 in neural tube closure and in the planar polarity of inner-ear sensory hair cells. J Neurosci 26:2147–2156
21. Wang J et al (2006) Dishevelled genes mediate a conserved mammalian PCP pathway to regulate convergent extension during neurulation. Development 133:1767–1778
22. Wang J et al (2005) Regulation of polarized extension and planar cell polarity in the cochlea by the vertebrate PCP pathway. Nat Genet 37:980–985
23. Curtin JA et al (2003) Mutation of Celsr1 disrupts planar polarity of inner ear hair cells and causes severe neural tube defects in the mouse. Curr Biol 13:1129–1133
24. Kuriyama S, Mayor R (2008) Molecular analysis of neural crest migration. Philos Trans R Soc Lond B Biol Sci 363:1349–1362
25. Katoh M (2005) WNT/PCP signaling pathway and human cancer (review). Oncol Rep 14:1583–1588
26. Coyle RC, Latimer A, Jessen JR (2008) Membrane-type 1 matrix metalloproteinase regulates cell migration during zebrafish gastrulation: evidence for an interaction with non-canonical Wnt signaling. Exp Cell Res 314:2150–2162
27. Weeraratna AT et al (2002) Wnt5a signaling directly affects cell motility and invasion of metastatic melanoma. Cancer Cell 1:279–288
28. Lee JH et al (2004) KAI1 COOH-terminal interacting tetraspanin (KITENIN), a member of the tetraspanin family, interacts with KAI1, a tumor metastasis suppressor, and enhances metastasis of cancer. Cancer Res 64:4235–4243
29. Tree DRP, Ma D, Axelrod JD (2002) A three-tiered mechanism for regulation of planar cell polarity. Semin Cell Dev Biol 13:217–224
30. Axelrod JD (2009) Progress and challenges in understanding planar cell polarity signaling. Semin Cell Dev Biol 20:964–971
31. Held LI Jr, Duarte CM, Derakhshanian K (1986) Extra tarsal joints and abnormal cuticular polarities in various mutants of *Drosophila melanogaster*. Roux's Arch Dev Biol 195:145–157

32. Vinson CR, Adler PN (1987) Directional non-cell autonomy and the transmission of polarity information by the *frizzled* gene of *Drosophila*. Nature 329:549–551
33. Vinson CR, Conover S, Adler PN (1989) A *Drosophila* tissue polarity locus encodes a protein containing seven potential transmembrane domains. Nature 338:263–264
34. Klingensmith J, Nusse R, Perrimon N (1994) The *Drosophila* segment polarity gene dishevelled encodes a novel protein required for response to the wingless signal. Genes Dev 8:118–130
35. Thiesen H et al (1994) *Dishevelled* is required during *wingless* signaling to establish both cell polarity and cell identity. Development 120:347–360
36. Feiguin F, Hannus M, Mlodzik M, Eaton S (2001) The ankyrin repeat protein Diego mediates Frizzled-dependent planar polarization. Dev Cell 1:93–101
37. Taylor J, Abramova N, Charlton J, Adler PN (1998) Van Gogh: a new *Drosophila* tissue polarity gene. Genetics 150:199–210
38. Wolff T, Rubin GM (1998) Strabismus, a novel gene that regulates tissue polarity and cell fate decisions in *Drosophila*. Development 125:1149–1159
39. Gubb D et al (1999) The balance between isoforms of the prickle LIM domain protein is critical for planar polarity in *Drosophila* imaginal discs. Genes Dev 13:2315–2327
40. Chae J et al (1999) The *Drosophila* tissue polarity gene starry night encodes a member of the protocadherin family. Development 126:5421–5429
41. Usui T et al (1999) Flamingo, a seven-pass transmembrane cadherin, regulates planar cell polarity under the control of Frizzled. Cell 98:585–595
42. Axelrod JD (2001) Unipolar membrane association of Dishevelled mediates Frizzled planar cell polarity signaling. Genes Dev 15:1182–1187
43. Bastock R, Strutt H, Strutt D (2003) Strabismus is asymmetrically localised and binds to Prickle and Dishevelled during *Drosophila* planar polarity patterning. Development 130:3007–3014
44. Strutt DI (2001) Asymmetric localization of Frizzled and the establishment of cell polarity in the *Drosophila* wing. Mol Cell 7:367–375
45. Tree DRP et al (2002) Prickle mediates feedback amplification to generate asymmetric planar cell polarity signaling. Cell 109:371–381
46. Amonlirdviman K et al (2005) Mathematical modeling of planar cell polarity to understand domineering nonautonomy. Science 307:423–426
47. Gubb D, Garcia-Bellido A (1982) A genetic analysis of the determination of cuticular polarity during development in *Drosophila melanogaster*. J Embryol Exp Morphol 68:37–57
48. Adler PN, Taylor J, Charlton J (2000) The domineering non-autonomy of frizzled and van Gogh clones in the *Drosophila* wing is a consequence of a disruption in local signaling. Mech Dev 96:197–207
49. Struhl G, Barbash DA, Lawrence PA (1997) Hedgehog acts by distinct gradient and signal relay mechanisms to organise cell type and cell polarity in the *Drosophila* abdomen. Development 124:2155–2165
50. Wehrli M, Tomlinson A (1998) Independent regulation of anterior/posterior and equatorial/polar polarity in the *Drosophila* eye; evidence for the involvement of Wnt signaling in the equatorial/polar axis. Development 125:1421–1432
51. Lawrence PA, Casal J, Struhl G (1999) The hedgehog morphogen and gradients of cell affinity in the abdomen of *Drosophila*. Development 126:2441–2449
52. Strutt D, Johnson R, Cooper K, Bray S (2002) Asymmetric localization of frizzled and the determination of notch-dependent cell fate in the *Drosophila* eye. Curr Biol 12:813–824
53. Lawrence PA, Casal J, Struhl G (2002) Towards a model of the organisation of planar polarity and pattern in the *Drosophila* abdomen. Development 129:2749–2760
54. Fanto M et al (2003) The tumor-suppressor and cell adhesion molecule Fat controls planar polarity via physical interactions with Atrophin, a transcriptional co-repressor. Development 130:763–774

55. Zeidler MP, Perrimon N, Strutt DI (1999) Polarity determination in the *Drosophila* eye: a novel role for unpaired and JAK/STAT signaling. Genes Dev 13:1342–1353
56. Lawrence PA, Casal J, Struhl G (2004) Cell interactions and planar polarity in the abdominal epidermis of *Drosophila*. Development 131:4651–4664
57. Axelrod JD, Tomlin CJ (2011) Modeling the control of planar cell polarity. Wiley Interdiscip Rev Syst Biol Med 3:588–605
58. Ma D et al (2008) Cell packing influences planar cell polarity signaling. Proc Natl Acad Sci U S A 105:18800–18805
59. Raffard R, Amonlirdviman K, Axelrod JD, Tomlin CJ (2008) An adjoint-based parameter identification algorithm applied to planar cell polarity signaling. IEEE Trans Automat Contr & IEEE Transactions on Circuits and Systems I: Regular Papers 53:109–121
60. Shimada Y, Yonemura S, Ohkura H, Strutt D, Uemura T (2006) Polarized transport of frizzled along the planar microtubule arrays in *Drosophila* wing epithelium. Dev Cell 10:209–222

Integrated Molecular Circuits for Stem Cell Activity in Arabidopsis Roots

B. Scheres

Pattern formation in biological systems has long been thought to emanate from hierarchical control systems. More recently, another view on the development of patterns has gained ground, where multilevel interactions (spanning molecular, cellular and organismal processes and allowing feedback between all levels) create different attractor states for the biological system. In moving towards attractor states, complicated patterns may ensue as an emergent property from simple interactions. We study these processes in plant development, because plant cells do not move relative to one another which facilitates analysis.

Plant stem cells reside in niches and are maintained by short-range signals emanating from organizing centres. The Arabidopsis *PLETHORA* genes encode transcription factors required for root stem cell specification [1, 2]. *PLT* expression is induced by the indolic hormone auxin, depends on auxin response factors and follows auxin accumulation patterns. The *PLT* gene clade extensively regulates expression of the PIN facilitators of polar auxin transport in the root and this contributes to a specific auxin transport route that maintains stem cells at the appropriate position [3]. Hence the system is wired with feedbacks at all levels. We are addressing the properties of the auxin-PLT-PIN feedback loop by gene and protein network analysis and computational modelling. The emerging picture is one in which flexible feedback circuits translate auxin accumulation into region- and cell type specification patterns.

Stem cells and their daughters in the root display specific asymmetric divisions at fixed locations. We also investigate how such divisions are spatially regulated with such precision. The SHORTROOT-SCARECROW transcription factor pathway plays a role in patterning the quiescent center and cortex/endodermis stem cells and provides mitotic potential to the stem cell daughters that form the proximal meristem. This activity involves the conserved RETINOBLASTOMA-

B. Scheres (✉)
Utrecht University, Utrecht, The Netherlands
e-mail: b.scheres@uu.nl

V. Capasso et al. (eds.), *Pattern Formation in Morphogenesis*, Springer Proceedings in Mathematics 15, DOI 10.1007/978-3-642-20164-6_5,

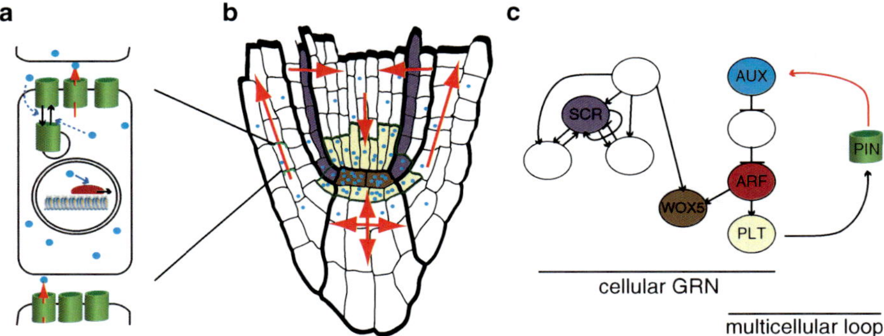

Fig. 1 The root stem cell niche area at different levels of organization. (**a**) Intracellular representation of polar transport (*red arrows*) of auxin (*blue dots*) by polarly localized PIN transmembrane proteins (*green pipes*). PIN proteins cycle between transport vesicles and plasma membrane and models postulating feedback from auxin levels to PIN recycling (*dotted blue arrows*) can generate polar PIN distributions at the cellular level and developmental patterns at the tissue level, see text. (**b**) Stereotyped polarization of PIN proteins leads to auxin transport (*green arrows*), accumulation of auxin (*blue dots*) and restriction of PLT expression (*yellow*) to stem cell niche. SCR (*purple*) is expressed in the layer surrounding the vasculature due to movement of its activator SHR from the stele and nuclear localization. WOX5 expression (*brown*) is dependent on SCR and auxin accumulation and restricted to the stem cell organizer. (**c**) Proposed cellular gene regulatory network with a PLT-PIN feedback loop at the tissue level, see text. *Black arrows*, proven interactions, *red arrow*, polar auxin transport (Copyrighted by Current Opinion in Genetics and Development)

RELATED (RBR) pocket protein [4], and we have established molecular links between the RBR pathway and SCARECROW action that form a feedback control system. In addition, RBR activity is modulated by auxin abundance, itself regulated through an intercellular distribution system, and by cell cycle progression. Formal analysis of this feedback circuit indicates that it acts as a bistable switch that ensures the occurrence of an asymmetric division at fixed positions. Our work illustrates how formative divisions that shape plant tissues can be robustly positioned by dynamic regulatory circuits that combine intracellular and extracellular loops (Fig. 1).

References

1. Aida M, Beis D, Heidstra R, Willemsen V, Blilou I, Galinha C, Nussaume L, Noh Y-S, Amasino R, Scheres B (2004) The *PLETHORA* genes mediate patterning of the *Arabidopsis* root stem cell niche. Cell 119:109–120
2. Galinha C, Hofhuis H, Luijten M, Willemsen V, Blilou I, Heidstra R, Scheres B (2007) PLETHORA proteins as dose-dependent master regulators of Arabidopsis root development. Nature 449:1053–1057

3. Grieneisen V, Xu J, Marée AFM, Hogeweg P, Scheres B (2007) Auxin transport is sufficient to generate maximum and gradient guiding root growth. Nature 449:1008–1013
4. Wildwater M, Campilho A, Perez-Perez JM, Heidstra R, Blilou I, Korthout H, Chatterjee J, Mariconti L, Gruissem W, Scheres B (2005) The *RETINOBLASTOMA-RELATED* gene regulates stem cell maintenance in Arabidopsis roots. Cell 123:1337–1349

The Mechanics of Tissue Morphogenesis

Thomas Lecuit

Abstract Cells and tissues display remarkable robustness, i.e., the ability to keep a polarized or vectorial organization, important for their physiological role. Meanwhile epithelia show tremendous plasticity, during embryonic development and organ regeneration, i.e., the capacity to adapt to intrinsic or extrinsic signals or perturbations. We are interested in deciphering basic principles of cell and tissue organization and dynamics with a special focus on the problem of how robustness and plasticity are jointly regulated. The main problem I will discuss is an example of tissue plasticity manifested in embryonic tissues. I will present our current research characterizing how the spatial distribution of tension in cells controls locally stereotyped cell shape changes and how these in turn are coordinated at the tissue level to produce tissue morphogenesis. The presentation will delineate (1) spatio-temporal patterns in cell dynamics driving tissue morphogenesis; (2) the subcellular force-generating systems driving these kinematic patterns; (3) how tension transmission affects this process. We will discuss how fluctuations in contractile activity are spatially organized to yield robust and reproducible symmetry-breaking in cell dynamics. The underlying theme will be to understand how tissue-level dynamics emerges from subcellular mechanics.

1 Introduction

As everybody knows, in living organisms, shapes are wonderful, very diverse, amazing, and it might take more than 20 pages just to try to illustrate the concepts of diversity; and the point is that underlying, or behind, this diversity of complexity,

---

Adapted from the talk given at the IHES Workshop on 11 January 2010.

T. Lecuit (✉)
Institut de Biologie du Développement de Marseille Luminy (IBDML), UMR 6216, CNRS-Université de la Méditerranée, Campus de Luminy, case 907, Marseille cedex 09 13288, France
e-mail: thomas.lecuit@ibdml.univmed.fr

V. Capasso et al. (eds.), *Pattern Formation in Morphogenesis*, Springer Proceedings in Mathematics 15, DOI 10.1007/978-3-642-20164-6_6,

we as biologists and scientists would like to understand what the underlying order is.

Henri Poincaré (1854–1912) said: "The researcher must organize; one does science with facts as one builds a house with stones; but an accumulation of facts is not a science, just as an accumulation of stones is not a house." And I think it suitable to quote a famous mathematician, who actually had a very strong statement on this, years ago – I don't know if he was talking about science in general, or mathematics in particular, but it certainly applies to biology, because we have such a diversity of shapes, structures, dynamics, and everything; thus I would like first to propose three major directions (topics) to try to describe the underlying rules of construction in the structures of embryos.

The first one is obvious, and I think it beautifully illustrates the concepts of pattern formation – and, in a way, of dominant inheritance. It has been very empowering to understand, or at least to establish the concept, that a small set of genes could work in concert in an order that is hierarchical to establish structures as beautiful as the stripes of a fly maggot. This is also very general – not about the specific mechanisms, but about the idea that idea that *a few genes work in concert in hierarchies that can be defined experimentally to organize patterns* in any organisms, from highly developed to underdeveloped organisms, or the other way around. And this understanding of course comes from the contributions of people like Ed Lewis and many others who have been working in vertebrates on how *Hox* genes organise patterns along the anterior-posterior axis. So this is almost trivial to us, but I think it was worthwhile starting with this. But first I would like to go into the history of the contributions of several people, trying to establish at least some other possible ways of describing structures that can be important to bear in mind when we want to understand rules of construction.

The second is the *contribution of mathematical description* to the definition of structures and patterns. Based on the works of Hans Meinhardt, we can realize that simple mathematical descriptions of relationships between activators and inhibitors might really potentially explain such different behaviours as patterns in shells, fishes, and beyond. Talking about the mathematical rules for description, it is interesting to quote the contribution of D'Arcy Thompson, the Scottish scientist – it's difficult to put him in a small box, as being either mathematician or biologist; he was just a scientist interested in living forms – who proposed, without any mechanism, that simple rules of construction could be derived based on growth patterns in the diversity of shapes in fishes, and skulls, etc., etc. We know nothing about how growth and pattern interplay to produce the diversity of shapes seen in the natural world, but I think it was interesting that it was proposed, long before patterning was known and established as a discipline, that the interplay of growth and pattern could be used to describe the resultant pattern of different structures.

The third topic is the *physical principles or constraints that underlie pattern formation*; and here I could quote many, but I would like to emphasize the contribution of, first, Joseph Antoine Ferdinand Plateau, a Belgian physicist, who reported that bubbles and fluids organise into three-dimensional structures that obey specific geometrical rules, where in two dimensions bubbles would organise with specific

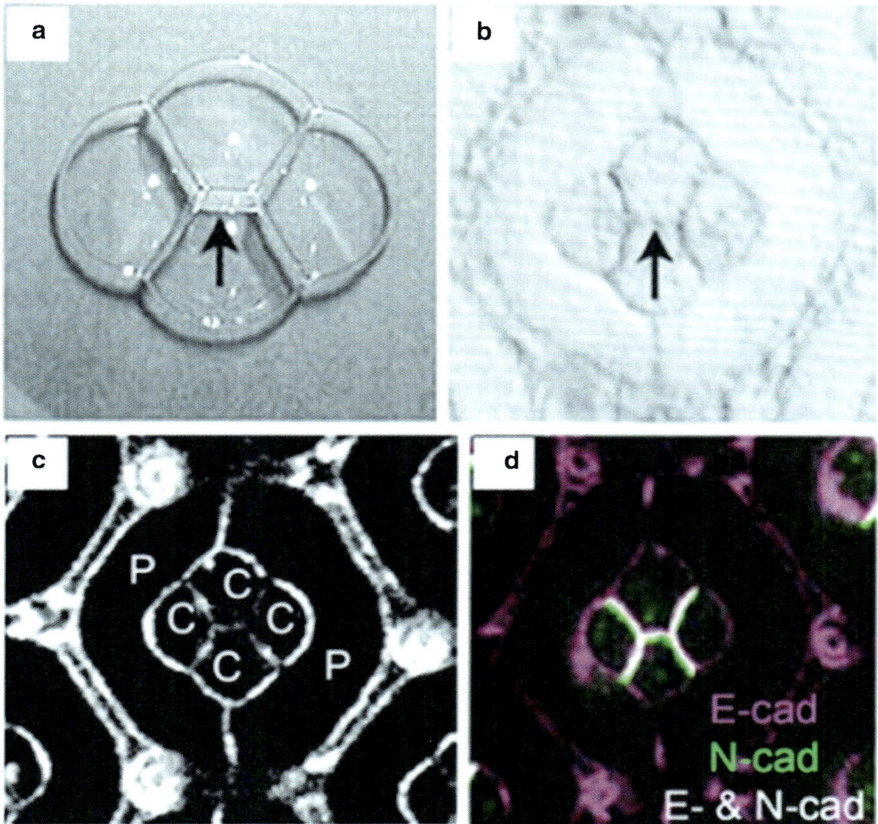

Fig. 1 Cell surface mechanics and the organization of static cell patterns. (**a**) Four soap bubbles intersect at 120° angles at three-way vertices. The *arrow* points to an interface at the center of the cluster. This configuration is the result of a minimization of contacts between bubbles and with the air. (**b**) Four cone cells in the *Drosophila ommatidium*, a cluster of cells in the retina. Cone cells develop specific contacts (*arrow*). The packing geometry is very similar to bubbles. (**c**) Confocal staining of an *ommatidium* showing the four cone cells (*C*) and the three surrounding primary pigment cells (*P*). (**d**) Staining showing the distribution of the cell adhesion molecules E-cadherin (E-cad, *purple*) and N-cadherin (N-cad, *green*) in an *ommatidium*. All cells express E-cad, but only cone cells express also N-cad. Because N-cad requires homophilic (i.e., N-cad/N-cad) interactions at contacting surfaces, N-cad is only present at the contacts between cone cells (shown in *white* because both E-cad and N-cad accumulate) (From Lecuit [5], with permission)

angles, and this could be explained following thermodynamic principles to organise fluids in two or three dimensions; and I would like to mention here, more than a century later, the beautiful work of Richard Carthew's lab on the retina in the fly eye.

There are four epithelial cells surrounded by the iris, or these primary pigment cells (Fig. 1b), and these four cells, not just in flies but also in all arthropods, have this very unique organisation, which is the so-called equatorial association, and two cells nearby. These four cells form specific patterns that can be mimicked if you use bubbles, in two dimensions; and Hayashi and Carthew showed that the specific geometrical organisation of epithelial cells could be explained by relatively simple thermodynamic constraints, controlled, in biology, essentially by adhesion molecules. And it has been appreciated by many people, e.g., D'Arcy Thompson, that cells in embryos in all organisms have specific rules of association that follow these principles.

I would like to end this introduction by putting together all these different levels of description. So if we want to understand how embryos and organs are organised in vertebrate and invertebrate organisms – and these are very complex dynamical structures, we need to understand how the whole matrix of genes, the signals and the patterning activity, with feedbacks, positive and negative, organise and set territories with specific behaviours. That's all very important, but we also need to take into account the physical substratum upon which these networks operate. I will call it, very broadly, cell and tissue mechanics. Of course a cell cannot migrate at just any rate; there are some fundamental limits to how fast a cell can migrate – that depends on what it's made of, and, also, the shape of cells cannot be randomly organised; there are some limits to the shape the cells can adopt, because there is an energetic cost to whatever shape you could adopt, and also because you probably need some particular constraints that cannot be biologically controlled. There are some fundamental limits, and it's really important for us to understand what are the very fundamental physical properties of cells that are going to canalise, or constrain, how gene networks control cell shape and tissue behaviour.

Another limit is cell division and tissue growth – just considering cell motility and cell deformation, a tissue cannot grow at an infinite rate, there are some limits to how fast the energy put into the system can translate into growth, which puts the system out of equilibrium, and it's really important to appreciate that as well.

Having these two important constraints, from the growth and also from the organisational dynamics of these systems, then we can begin to ask how signals control growth, on the one hand – this we know from many people working in different organisms – and how signals control tissue mechanics; how cells and genes control the basic properties of different tissues, but also how they can tune these properties to make them go into a particular direction.

You can also ask another set of questions: how cell and tissue mechanics controls, or limits, the rate at which cell division and growth affect the tissue – you can imagine, for example, that if you have an overgrowth, like a cancer, in a particular territory, it's going to constrain the tissue around it just because the cell has some boundaries; and if you have an overgrowth, it's just like having a hundred people in this small room: there would be some constraint among us. So clearly cell division, resulting in tissue growth, is going to have some impact on tissue mechanics; and, vice versa, you have many, many strong constraints in the tissue – maybe it's going to be more difficult for the cells to divide, or maybe you're going to skew or orient the division along a particular direction, which itself is a by-product of tissue mechanics.

So there are some very important questions to understand here, and this is clearly not within the realm of genetic control, but it's something that genetic control really has to take into account, if you want to understand ultimately how shapes are organised at the level of cells and tissues. This is something we have begun to work on in the past 2 or 3 years; and what I would like to do is first, sort of as a programme for the research, set this as an underlying philosophy of our study of morphogenesis. We're not interested just in tissue mechanics; many people have been interested in that for many years, with cell cultures and organisms. We want ultimately to integrate constraints and controls into the study of morphogenesis, and I think this is just the beginning of a programme rather than a completion.

2 Research Results

The system that we've been working on is the epithelial cells in the fly embryo, and what I would like to talk to you about is a particular dynamic process taking place in the early drosophila embryo, to try to illustrate what we do know about tissue mechanics, and how genes control and bias particular behaviours.

First, I would like to emphasise the fact that it has been appreciated more and more in the past few years that the ability of cells to differentiate in vitro, or also in vivo, is highly dependent upon the mechanical environments that these cells explore. And it is known that if you take fibroblasts and put them on a hard substratum, you can make osteoblasts; if you put them on slightly more compliant substrates, then they will make muscle; and if you put them in a very soft environment, then they will differentiate into something very soft. So clearly cells explore, sample their physical environment, measure it, and can take information from this substratum to "decide" to differentiate into a particular lineage. Of course that doesn't rule out a contribution of very important transcription factors to control that, but it does show that in normal settings and also in tumours, the physical environment greatly influences the differentiation potential of different cells.

In the early fly embryo (Fig. 2b), the tissue wants to elongate, just as any organism wants to do, and it cannot do it to the right because there is a shell, so instead the tissue's going to curl dorsally and then to elongate its back towards the neck, i.e., towards the left in the Figure. So there's about a twofold elongation; this process is known as germline extension, and in dark grey you have the region that elongates, about twofold in about 45 min.

This process is not driven by cell motility or by cell division or anything; it's driven essentially by a process called cell intercalation, which is shown when you label cells and photograph them at 30 min intervals in a time-lapse film. You can see here a blue, a pink, and a green row of cells; and after 30 min, cells have exchanged neighbors (Fig. 2c, right panel). The cells that at first touched each other no longer do so, and the blue and pink cells now touch each other, whereas they were not in contact initially.

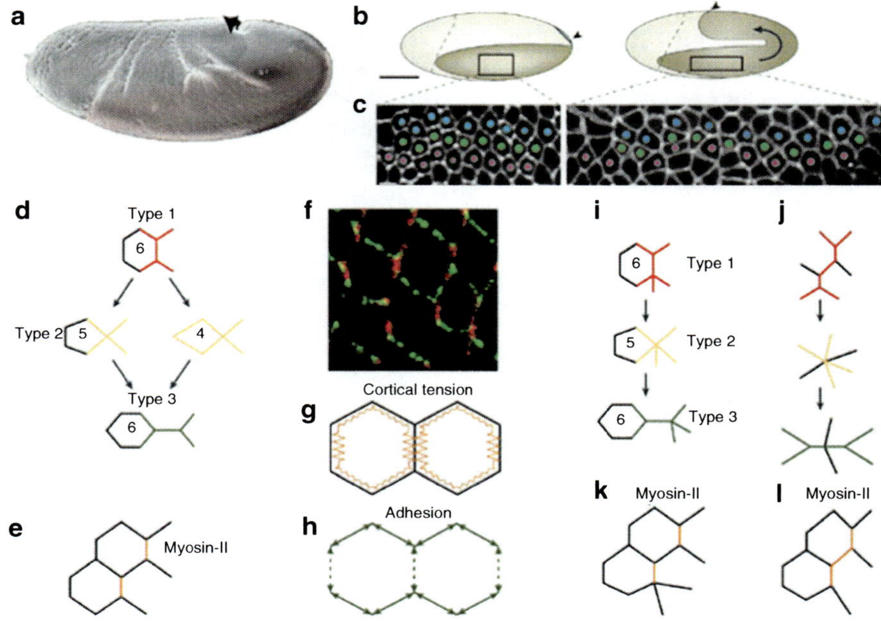

Fig. 2 Antero-posterior elongation of the *Drosophila melanogaster* germband epithelium (shown on a scanning electron micrograph (**a**) and schematic diagram (**b**)) is driven by intercalation. The posterior end (*arrowheads* in **a** and **b**) moves towards the anterior (*left*) on the dorsal side (*top*). During intercalation, epithelial cells (which have been marked with E-cadherin – green fluorescent protein and then labelled in different colours) exchange neighbours within 30 min (**c**). Cell intercalation depends on the irreversible change in the geometry of cell contacts, with shrinkage of type-1 junctions (*red*) to the type-2 configuration (*yellow*) and regrowth of type-3 junctions (*green*) (**d**). This planar polarized remodelling of cell contacts is driven by the polarized enrichment of myosin-II (*orange*) in type-1 junctions (**e**, **f**) and the downregulation of PAR-3 (shown in *green* in part f). Polarized remodelling of cell surface tension by an upregulation of cortical tension (*orange*) (**g**) and a downregulation of adhesion (*green*) (**h**) in shrinking junctions drives cell intercalation. Irregularities in the packing of cells (**i**, **j**) and in the distribution of myosin-II (**k**, **l**) during polarized junction remodelling causes the clustering of more than four cells in the type-2 configuration and the formation of transient fivefold (**i**) or sixfold (**k**) rosettes (From Lecuit and Lenne [7], with permission)

And this exchange of neighbors is not randomly organised, but consistently occurs along the anteroposterior axis where the cells have a motility which is biased along the dorsoventral axis, thus leading to this movement. So we want to understand how this occurs. To do that, we zoom on individual cells, and use each cell as a referential, and observe what happens around them during a time-lapse experiment. When we focus on the geometry of cell contacts, we see that the vertical junctions (Type 1 in Fig. 2) first shrink and produce a new configuration of tetrads of four cells, where cells have equal contacts. And then you have new contacts that form that are perpendicular to the initial orientation of the junction.

To summarize this, you can describe two parallel pathways that have common properties. The first is what I just described, a planar polarized irreversible change in cell contacts, composed of two steps, shrinking of one contact and regrowth at a perpendicular; there's a variation on the theme when more than four cells meet together, producing these five-way vertices, or rosettes, and this also resolves, producing contacts that are perpendicular to the initial orientation (Fig. 2i, j). The key point here is – and I emphasise it – that this reaction, so to say, is irreversible.

There is another term to describe this, and that is the concept of planar polarity. Any cell in the population, out of the five or six neighbors that it has, on average, has an ability to recognize the contacts toward the head and toward the tail of the embryo, and decide which of these contacts are going to disappear. And that's exactly what happens here, certain contacts disappear, and the other ones are maintained normally during the process. So any cell has an ability to recognize polarities or coordinates in the embryo, has a compass to measure "where's the head?", "where's the tail?", and say, "OK, these contacts need to shrink," and they are destroyed. And it is precisely because cells have this ability to read the global polarity of the tissue that they can perform this reaction, and can intercalate in a way that is spatially oriented, right? So it's *irreversible*, and *planar polarized*.

Now you might say, well, thinking about mechanisms, is it driven by outside forces, or is intercalation, as we say, an autonomous, intrinsic property of these cells that is not influenced at all by the surrounding tissue?

Well, we do know of an active process here (the details don't matter). So, can we genetically ablate it? Yes, we can. And when we do, intercalation proceeds normally. And not just does it occur normally, but the cumulative effect of all the local forces contributed by cells which locally remodel their contacts amounts to a net total force that is so big that if you close the gate at the end, you're going to create buckling in the tissue. And that's precisely what you see when you do the experiment. So after we had done these experiments, we were forced to conclude that there must be some autonomous property here, and we hypothesized that there is a local generator of forces at work. The molecules that control mechanics are ATPases, motors. At this point we concluded that there must be an intrinsic property; and we proposed that, for instance, the simplest scenario would be that each cell, or group of four cells, to take an even simpler case, would have the ability to remodel its contacts, and to provide the force that would be needed to do that. And if all of them did it at the same time, then you would have a macroscopic effect at the tissue level: it would elongate, and the force of elongation could be measured one way or another.

Going into the chemistry, and making a long story short, we were able to show that the non-muscle Myosin II, which is a major force-generating system acting on actin filaments to change shape, has a very unique and interesting localisation: it is (1) enriched at the junctions, precisely where we see these configurational changes; (2) it has planar polarity (Fig. 2e, f). You can see, in red in Fig. 2f, that the

Myosin II complex is not uniformly distributed: it has a bias along the vertical junctions, and you can measure that. In fact, it is enriched in the vertical junctions. And we can show, using genetics and pharmacological tricks to inactivate or reduce the amount of Myosin II, that it is essential for the processus of intercalation. So I think that here you have both the chemistry and the mechanics, because clearly this is a motor that can contribute the force to act locally to control the process.

Clearly these cells have an ability to read coordinates. The phenomenonological manifestation of that is the irreversible pattern of junction remodelling, and the chemical–mechanical presentation of that is Myosin II planar polarity.

During cell intercalation, cell contacts are maintained by E-cadherin molecules engaged in homophilic complexes and dynamically organized by F-actin. The actomyosin network and the adhesion molecules are mostly restricted to the plane of adherens junctions between basal and apical surfaces. The basal region most likely responds passively to cell deformation occurring in the plane of adherens junctions and has a minor mechanical role. Thus cell networks can be considered as two-dimensional and cell contacts as lines joining vertices. The extent of cell contacts is governed by adhesion molecules in dynamic interaction with the actin network. E-cadherin, responsible for intercellular adhesion, tends to increase cell contact (Fig. 2h), whereas a cortical tension controlled by the actomyosin network tends to reduces these contacts (Fig. 2g).

From a physical point of view, the length and orientation of cell interfaces, and therefore the cell network configuration, are defined by the tensions acting at cell interfaces (Fig. 3). We consider that biologically controlled changes of tensions bring cells into a series of configurations close to mechanical equilibrium, driving cell-network evolution and tissue dynamics. Teaming up with the physicists, we tested the hypothesis framed here, that the planar polarized distribution of Myosin II controls the anisotropic tension of the cortex.

If you have planar polarity of Myosin II, you're going to have this irreversible pattern of junction remodelling; and, in the context of rosettes, all you need to do is simply have more than two junctions together having more Myosin II. Then you bring several vertices together, and you will have the same type of remodelling, i.e., in both cases you have planar polarity of Myosin II. So can we test whether an anisotropy, or planar polarity, of Myosin II cortical tension is sufficient to produce what we see in vivo? Yes, we can do it, and I'm going to show you how we did in fact do it in collaboration with Pierre-François Lenne, physicist.

The first approach was a very simplistic, but quite efficient model of cell intercalation based on a thermodynamic model of tissue properties. Figure 3 describes it. We assume that the tensions at cell interfaces derive from a potential energy that determines different organizational states of the network. There are two types of tensions that could be described, talking about a thermodynamic description of tension. One is line tension, which is akin to what you see in soap bubbles; this is not dependent on the deformations of cells, it is only dependent on local properties of cells, whether adhesion or tension molecules, we don't know. And we always say that it depends on the lengths of the junctions, the modular factor, which is an anisotropic factor. The other one is an elastic tension, which is dependent upon

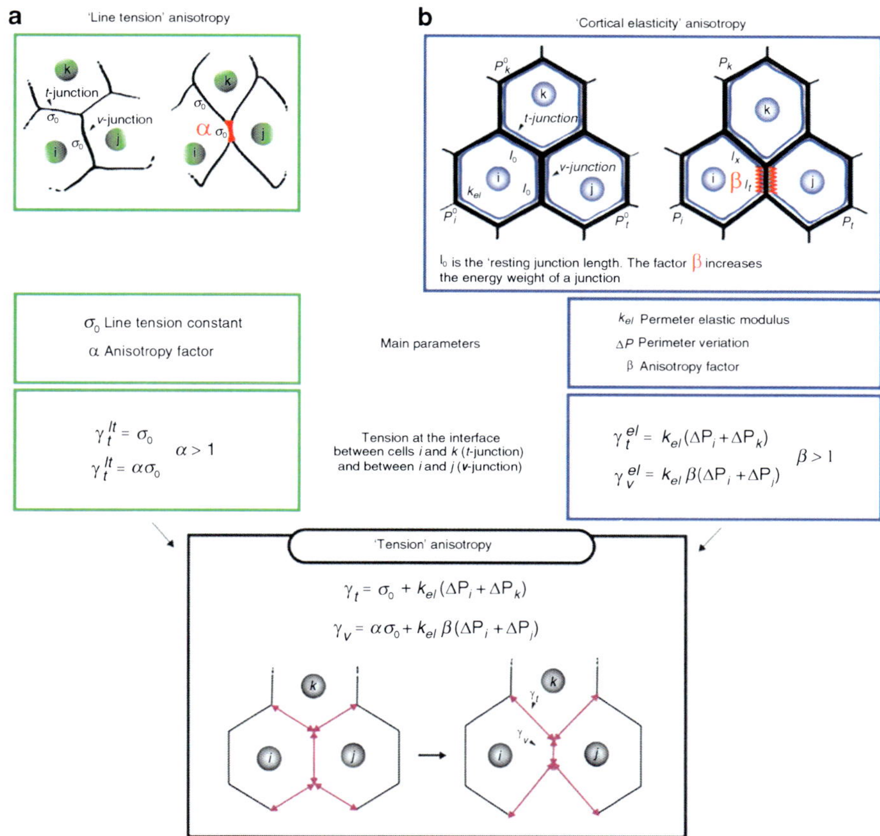

Fig. 3 Model of tension anisotropy and cell intercalation. During intercalation, the junctional cortex along *v*-junctions is enriched in myosin II (*red*). We assumed that this results in an increase in actomyosin contractility along *v*-junctions, compared with transverse junctions (*t*-junctions). The tension is said to be anisotropic. In the line tension model (**a**), line tension along *v*-junctions γ_v^{lt} is α times larger than along transverse junctions (t-junctions, γ_t^{lt}). Therefore a *v*-junction of a given length contributes more to mechanical energy than a *t*-junction of the same length. During energy minimization, *v*-junctions thus tend to shrink. In the cortical elasticity model (**b**), the mechanical anisotropy was introduced by weighting the length of *v*-junctions by a factor β, so that they would contribute more to cell mechanical energy than equally sized *t*-junctions and tend to shrink during energy minimization. For simplicity we consider in this figure that only one junction separating two hexagonal cells i and j has an increased contractility. Consequently the two cells are stretched and elastic tension builds up along their contour. Although both elastic tensions γ_v^{el} and γ_t^{el} are proportional to perimeter deviations of the cells, the elastic tension along the *v*-junction was β times larger, compared with *t*-junctions (Rauzi et al. [4], with permission)

cell deformation, such as stretching, for example. In this context we assume that local changes in the contractility of the actomyosin network along a junction, such as those resulting from the increase in Myosin II concentration, change the line tension value along this junction and/or modify its stretching state (Fig. 3a), thereby producing elastic tension (Fig. 3b). In both cases, if tensions at cell junctions are unbalanced, the network reconfigures until a new mechanical equilibrium is reached. During energy minimization, some junctions shrink and vanish. Local neighbour exchanges (T1 transitions) are then allowed, and new junctions may expand if this lowers the energy.

We explored in silico how the anisotropy, and in particular its amplitude, impacts on cell and tissue descriptors. By "descriptors" I mean that you could look at extension, you could look at disorder, topological disorder of the tissue, many things that you can measure experimentally. And then you compare the modelling results with the experimental data, because we start from the same lattice of cells for which we have time-lapse photography results, and make an in silico simulation, where all we "play around" with is this anisotropy factor. And then we test the consistency of the model through independent estimates of anisotropy that best match the in vivo and in silico data.

I will illustrate this approach, starting with the most simple, which is tissue elongation. So it is a real situation of cells in vivo, and you have a certain rate of elongation, resulting from intercalation. And of course what you can do now is to implement various amounts of anisotropy at the junctions, and then look at the outcome in terms of intercalation. And the first rhythm that we saw is that, as we increased the anisotropy factor, we increased the relative elongation up to a plateau value. In fact there seem to be many more values of anisotropy which result in maximum elongation; and these values, as we will come back to later, are not far from the real tension anisotropy that we measured in vivo.

Now, what you could do here as a proxy for time – because, as you realise, there's no time in this simulation, that's why it's gross simplification, it's pure thermodynamic equilibrium – we use a number of what we call T1 transitions, which is an elementary event of intercalation that can be used as a proxy for time. And, interestingly, the rate of elongation is highly dependent upon the anisotropy factor. If you increase the tension to 1.4, just a 40 % increase, i.e., not quite enough to elongate the tissue with the same relationship to the number of T1 transitions, basically all you do in this case is to elongate a maximum with not enough transitions. Now if you have a much greater anisotropy, you very rapidly transit through all the cells that you previously failed to elongate, and, in the end, the tissue will elongate to relax the stress which you have stored in it. And there is somewhat intermediate behaviour, not quite linear, but more linear than any of the other behaviours, where, as the T1 transitions occur, the tissue elongates step by step.

Passing through this configuration is in fact a T1 process, it's an exchange of neighbours – they are topological transitions. So, you have two steps (Fig. 2d), but mathematically it's one particular, unique, process, and this one process doesn't care about physics, it's a topological transition. I think that's the best way to describe it. But the point I want to make is that the elongation rate is highly dependent on the anisotropy factor using T1 numbers as a proxy for time. And because we also have in vivo data, we can examine what the behaviour actually is in vivo – and we find, interestingly, that the in vivo data match anisotropy values that are around 1.7–1.8.

And now you can use another descriptor, which is a measurement of pattern deformation, which you basically compute. The tool was developed by François Graner working on soap bubbles; it can be used with cells, or any system that you want, to measure local deformations. And you know that the ratio of these two, the "ellipsoidicity", I don't know how you say it, the ellipticity of these structures, you know, it's direct measurements of local deformations – and of course this evolves greatly during the course of intercalation. And actually you always go through an increase and then a relaxation of the deformation, and that again is very dependent on the anisotropic factor. If you have a great anisotropy of 2.4, you know you are going to get some extremely aberrant deformation, and then an eventual relax. If you have very low anisotropy, you have a very mild increase and relaxation, and you have some kind of intermediate behavior – I don't mean to say that the in silico data are a perfect fit with the in vivo data with respect to the actual shape, but with respect to the increase, the maximum of the deformation and its eventual relax, the in vivo data are, again, not far from this 1.8 intermediate value. So that gives us another independent estimate of the anisotropy, looking at the in vivo and in silico data. I don't have the time to go through all the measurements.

Initially we did not introduce fluctuations; it was deterministic. Then we introduced some fluctuations: we measured the mean square displacement of vertices in vivo, because it does fluctuate in vivo – and if I have time, I will come to that, it's a very, very important feature. And if you implement in silico these fluctuations that exist in vivo, they basically facilitate the search through the energy landscape, and thus greatly facilitate the rate at which you can do T1 processes, and intercalation and everything. So when you do introduce these fluctuations, then when you do the calculation twice, the results are not exactly the same, so the truth is that all these studies are extremely time-consuming, and we could not do statistical measurements of that. In fact, everything I've told you is based on two embryos. You can do simulations several times, but the fitting with in vivo data was very limited for practical reasons.

And tension anisotropy can be measured in vivo. So the other part of the study, probably the most important, was using a system to measure tension in vivo. As a parenthesis, I think that models are all interesting as a way to test a plausibility argument: can it work following a specific biological or physical assumption. And the most interesting situation is when it *doesn't* work. If it doesn't work, you have to scratch your head and think of something else. If it does work, we all agree that it's just because we were lucky, or because we were looking for parameters that best fit, and, from my perspective, that amounts to saying we have not learned anything about the system.

Matteo Rauzi, a Ph.D. student, developed the nanoscissors; thus one could measure the distance of the vertices after ablation, We used femtosecond laser ablations, because, for ablation, precisely what matters is not the total energy you put in, but the peak intensity. These are pulse lasers; and for a given peak intensity, a femtosecond laser will deliver less total energy than a picosecond laser – so there will be much less collateral damage, which is why we use them as nanoscissors in our system. And indeed we observed that we can ablate, and remove actomyosin without perforating the membrane. We can demonstrate that because we inject beforehand, in the embryo, a caged molecule, fluorescein, which has a fusion coefficient of 200 square microns per second: that means that it fills the embryo within a second. And, before or after ablation, if you encage it, it doesn't leak through the membrane. So any hole that is made in the membrane is so small and so rapidly sealed that it doesn't permit diffusion of this molecule. Hence we are very confident that we only break the actomyosin network, that it works really well. Breaking the actomyosin network leads to a relaxation of the vertices, *the speed of which can be measured.* And we can infer local tension based on the contribution of the actinomyosin network solely. Now that doesn't say that the membrane doesn't have a contribution, it just means that we don't touch it, we don't affect it.

You can think of the context as a bunch of actomyosin filaments, about 300 nm thick, and basically you cut it. So you have a spring (the actomyosin), you cut the spring. Subsequent to a 2–3 s laser pulse, the actomyosin pulls away and brings along whatever sticks to it, like cadherin complexes – cadherins are these molecules that anchor tension. They, too, are also repositioned away, because they are pulled away from the site by the actomyosin. And all of that is basically stripped away from the membrane. Again – we've done all controls – it's very selective for Actin-Myosin II, and whatever molecule binds Actin-Myosin II. And E-cadherin is also removed, and that is dependent on alpha-Catenin which is the main mechanical coupling between these two systems.

The purpose of this experiment (Fig. 4) was, first, to measure anisotropy. So I would like to just go into the two main results. We make the very strong but reasonable assumption, or we could reason whether it's reasonable, that the peak velocity is related to the local tension divided by a viscosity factor, and we say that the viscosity is isotropic. Maybe it's wrong, but if the viscosity is isotropic, we can explain differences in V_{max}, the peak velocity, solely in terms of local tension. And this peak velocity, or tension, if you want, is dependent on Myosin II. If you use an inhibitor of Myosin II, this peak velocity drops down to basal level.

Now, the most important experiment is to then measure the peak velocity in vertical junctions, and also in transverse junctions, which do not have high levels of Myosin II, just basal level. And then if you look at the ratio of these two speeds, you find that the tension anisotropy in vivo is about twofold. Which, interestingly, is not too far from this 1.8 estimate that we have from in silico predictions.

At this point we concluded that tissue elongation emerges from the collective effect of planar polarized (re)modelling of cell junctions. And the *cortical tension anisotropy*, which is controlled by Myosin II planar polarity, drives the whole process. That's the main conclusion we arrived at 2 years ago. Cadherins are removed as well. But I think that even if you remove Cadherins, the interpretation holds true, that (1) the tension we measure is Myosin II dependent, (2) it is anisotropic and relates to Myosin II concentration.

Now, there is a key question, namely, the symmetry-breaking mechanism. Conceptually, the symmetry-breaking mechanism has two parts: one, is what are

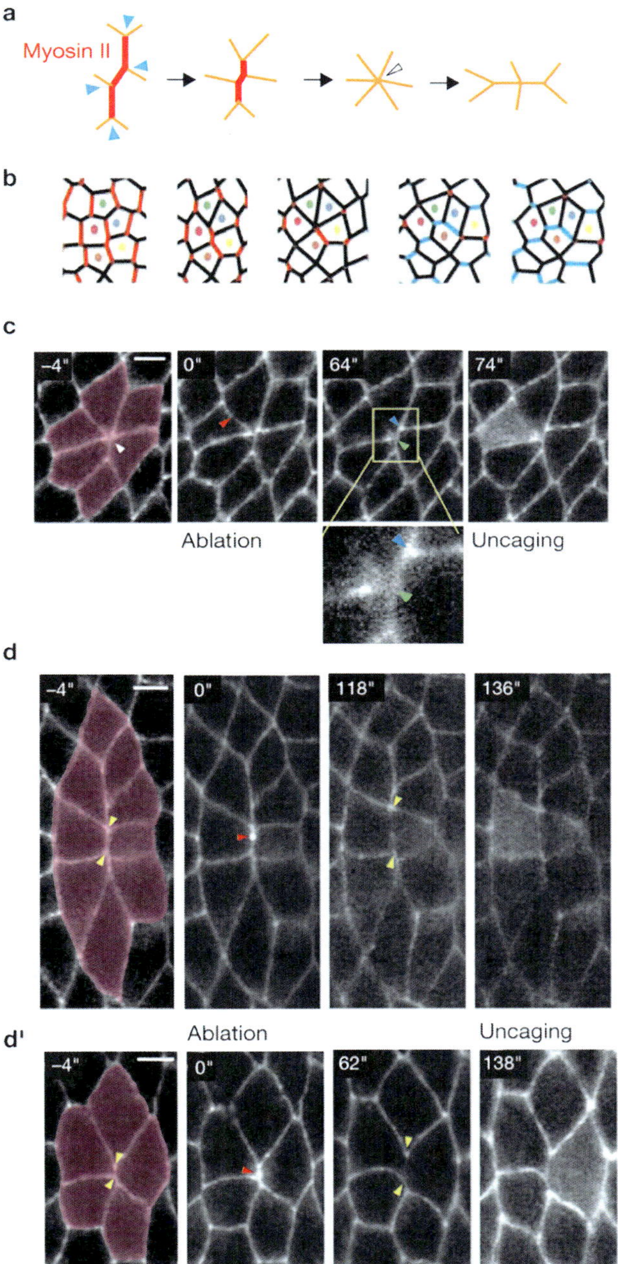

Fig. 4 Mechanics of high-order structures. (**a**) A rosette structure (*white arrowhead*) occasionally emerges from the coalescence of multiple (≥3) 3-way vertices (*blue arrowheads*) and resolves by the expansion of new junctions along the horizontal direction. (**b**) Formation and evolution of a

the cell contacts and signalling events that initiate this process of symmetry breaking such that neighbours along the anteroposterior axis talk to each other, saying, "we're going to put more Myosin II here," and not the dorsal or ventral cells; and the other, is, once you've done this signalling at the junctions, what are the downstream biological processes – and processes is broad, because it could be anything – that lead to Myosin II activation. We know nothing about it, because the system is genetically highly redundant; the only thing we know is that it is *not* dependent on the *frizzled/disheveled* system. Either flies have invented a new system in the early stages of development, or it's a stripped-down version of what we know in other systems.

So, the downstream cell biological processes? Well, the new phenomenology came from new imaging possibilities. We noticed that, during the process of cell polarisation, Myosin II, which you can visualise with a GFP label, is present, not just at the cortex, but, both before then and at the time you see it at the cortex, in local patches that span the junctional area apically. Now, these speckles are very annoying, because they are not in any way similar in position, it's highly random, apparently. And if you do time-lapse, it's even more amazing, because you can see speckles, you can see convergence of speckles to particular regions, and then you can see some movement towards the cortex, so it's very complex dynamics: you have Myosin II in two populations, and the two populations are in continuity. The first one is at the cortex (the cortical population is the one I talked to you about, it has planar polarity). And then you have another population, the medial population, 500 nm above, that is in continuity at the junction, it's like a little dome, and has this very complex dynamics. And these two systems are interconnected. Interestingly, if you look at the distribution of the temporal dynamics of these Myosin II profiles, you see that these populations have pulses of activity. It's not periodic, it's not even oscillatory, it's just pulsatile. We have not done a very detailed Fourier analysis of that, but there are some clear durations of pulses, in the order of a few tens of seconds.

So now for the third parameter, the junction dynamics – the shrinkage is not a smooth process. It also proceeds in steps, small steps then pause, steps, and so on and so forth. So we're interested in understanding what the relationships are between these progressive stepwise kinetics of shrinkage and the pulsatility of the actomyosin networks. The novelty is to take into account these very erratic, random fluctuating systems in the medial network. And what you can already notice if you

---

Fig. 4 (continued) five-cell rosette in silico. Three contiguous junctions with increased tension shrink and bring three 3-way vertices into contact. Cortical elasticity anisotropy: $\beta = 1.8$. Junctions with increased tension are shown in *red*, and new junctions after T1 transitions are shown in *blue*. (**c**) Relaxation of a rosette structure after nano-dissection (*red arrowhead*) along a junction emanating from the rosette centre (*white arrowhead*). After ablation, a 4-way vertex (*green arrowhead*) and a 3-way vertex (*blue arrowhead*) separate and decompose the rosette structure (zoomed-in *inset below*). (**d**, **d′**) Relaxation of a shrinking junction connected to other shrinking junctions, during rosette formation. Ablation point (*red arrowhead*). Scale bars are 5 μm (*white*) (Rauzi et al. [4], with permission)

look at all these pulses is that the cortical and medial pulses always segregate together: wherever you have one, the other follows, and so on and so forth. So they are not randomly distributed with respect to one another.

And the approach we used was to look at cross-correlation of raw data: you take *one* profile, you take the other one, you shift it in time, and you measure the correlation function, and you see a maximum. And when you have the maximum, you can also have the time delay, which tells you which is preceding the other, or which is lagging behind. This time delay could be positive or negative.

When we measured the cortical pulse maximum, we found that it consistently lagged about 9 s behind the medial pulses, and I think the experimental approach is basically the following: you have very complex fluctuating systems, and if you want to make specific hypotheses as to what lies upstream and may cause the other, you want to make this cross-correlation analysis to rule out specific hypotheses. If one is lagging behind, it cannot be the cause of the previous one, in this cluster where they are. In other words, if A lags behind B, then A cannot be the cause of B.

And the most important, if you want to look at force, is not just to look at the intensity but also to look at the first derivative. So if you look at the contraction rate of the cortical system and its first derivative, the very, very striking result is that if you cross-correlate the derivative of the cortical contraction, i.e., the contraction rate of the cortex, and the shrinkage rate of the junction, the latter precedes the former by a few seconds, i.e., the maximum of contraction of the cortex lags behind the shrinkage maximum of the junction. This means that cortical contraction does not cause junction shrinkage. So that raised the possibility that maybe the medial contraction is concurrent with the shrinkage of the junction, and indeed it was: we found that these two events were, statistically, not separable in time. So that suggests very strongly that the medial network, although very fluctuating and faint, mechanically contributes to the shrinkage step. And the cortical system, which lags behind, is not causing the shrinkage – maybe it does stabilise it.

We again used our nanoscissors, where we now decided, not just to ablate the cortex (as we had already done), but to ablate the medial pulses. And using this we have shown that if you ablate the medial network, you are going to affect the shrinkage of the junction. After you ablate it, the junction relaxes, and then you ablate again, and it relaxes, again and again. You can quantify this event and show that the medial pulse mechanically contributes to the junction shrinkage.

Now the point is that we have a division of labour, if you like. The medial network contributes the shrinkage step, and the cortical system is essential for the persistant shrinkage, because it does stabilise whatever steps of contraction have occurred. So it's like a ratchet, if you wish. You contract a bit, and you hold; a bit more, and you hold. And all of that is driven by, first, a pulse of contraction in the medial region. You have division of labour, but you have a very strong and robust hierarchical organisation of these two systems. Because the medial web contracts, and it has a flow towards the cortex, and it contributes to the polarity of the cortex. So the two systems are not independent. And this can be shown qualitatively in a time-lapse (film), where we can see the medial pulses and their migration, or flow, towards the vertical junctions. So the two systems are not independent – they are mechanically coupled, and they are hierarchically organised.

Now, we arrived at the notion that cortical anisotropy of tension drives the process, and I showed you the evidence to arrive at this. Taking into account this new network, its pulsatile activity, its major role in the shrinkage step, in bringing polarity to the cortex and its contribution to the persistent shrinkage, together with the Actin, which is all around, you arrive at the notion that there is a continuous viscoelastic sheet where two systems in the medial region just flow to the cortex, contribute to the process of shrinkage. Now, the key point is that it is continuous, but it's separated by membranes and contacts by Cadherins. So clearly the adequate way to think about it is not as something that is discontinuous with just boundaries, but as something more widespread, where cells maybe have this inherent contractility and maybe coordinate it.

Because the key point about planar polarity and symmetry breaking is that the pulses occur randomly, then they flow towards the vertical junctions. So there's an inherent polarity in the flow pattern. And the initial bias, the initial symmetry-breaking event, is the medial pulse dynamics. The pulse arises anywhere, but then flows to the cortex. Now that contributes progressive shrinkage.

This is the picture you have to have in mind. Polarity starts with polarity and dynamics in the medial Myosin II network, and what we showed is that the dynamics of this major network is controlled by the anisotropy of the Cadherin-anchoring points of the cortex. So clearly this continuous network is very important.

3 Open Questions and Problems

I would like to end with this sort of question and problem for us: it's clear that, if you look microscopically, you have something that is extremely robust, very deterministic, never fails, and elongates. And you can describe that at the cellular level by a very robust pattern of cell intercalations. The details are going to change, but the frequency of T1 processes is very conserved. Now if you look at the local mechanics, then you arrive at an increasingly complex fluctuating dynamic system, where *everything* is pulsatile. Whatever you measure in time, whether Myosin II in the medial network, Myosin II at the cortex, Cadherin anisotropy – all of that is never stable. You can take *averages* from images, but if you look in real time, you have fluctuations. We don't think these fluctuations are simple byproducts of an imperfect system. Or at least, if they are, they are also used for a good purpose, which progressively canalises the system into something which, as you go from supramolecular complexes to cellular scale and tissue scale, is more and more robustly assembled. And I think the real ambition of this kind of model is to try to integrate all these different levels of description, starting with the fluctuations of the system and trying to understand how you go to something which, at the tissue scale, is very robustly organised, and proceeds as we see it.

I think the major point of what we've found in the past 6 years can be summarized as follows. We started with the view that if you see this pattern of cell intercalation, your first thought is that the cells sneak their way and have some kind of protrusive activity, and migrate. They don't. The active process is not a protrusive activity, like those you see in other examples of intercalation such as mesenchymal cells in the Xenopus mesoderm, for instance. It results from mechanical properties of cell interfaces. So it's a change of a referential. Not to say that in other contexts migration is not important; maybe it is important along the lateral surface, but the key determinant is starting at the apical surface, and there you don't have to invoke any migration whatever – we don't see it, actually. What we see is a change in cortical tension controlled by medial pulses of contractile activity.

Acknowledgments I would like to thank Matteo Rauzi, who contributed all the work on nanoscissors and did all the work on the pulsatility of the actomyosin network, in collaboration with Pierre-François Lenne, physicist; and Ed Munro, mathematician and experimental biologist.

Associated Publications

Primary Articles

1. Bertet C, Sulak L, Lecuit T (2004) Myosin-dependent junction remodelling controls planar cell intercalation and axis elongation. Nature 429:667–671
2. Cavey M, Rauzi M, Lenne PF, Lecuit T (2008) A two-tiered mechanism for stabilization and immobilization of E-cadherin. Nature 453:751–756
3. Pilot F, Philippe JM, Lemmers C, Lecuit T (2006) Spatial control of actin organization at adherens junctions by the synaptotagmin like protein Btsz. Nature 442:580–584
4. Rauzi M, Verant P, Lecuit T, Lenne PF (2008) Nature and anisotropy of cortical forces orienting Drosophila tissue morphogenesis. Nat Cell Biol 10:1401–1410

Reviews

5. Lecuit T (2008) "Developmental mechanics": cellular patterns controlled by adhesion, cortical tension and cell division. HFSP J 2(2):72–78
6. Lecuit T, LeGoff L (2007) Orchestrating size and shape during morphogenesis. Nature 450:189–192
7. Lecuit T, Lenne PF (2007) Cell surface mechanics and the control of cell shape, tissue patterns and morphogenesis. Nat Rev Mol Cell Biol 8:633–644

Small Regulatory RNAs and Skeletal Muscle Cell Differentiation

Anna Polesskaya, Irina Naguibneva, Maya Ameyar-Zazoua, Cindy Degerny, Jeremie Kropp, Nora Nonne, Neri Mercatelli, Mouloud Souidi, Gueorgui Kratassiouk, Guillaume Pinna, Linda L. Pritchard, and Annick Harel-Bellan

Abstract Until recently, RNA was considered to be merely a downstream effector of the "noble" genome, the latter having all the information and therefore occupying a position at the very heart of gene regulation, according to the "central dogma" of DNA transcribed into RNA translated into protein. Although we all knew that RNAs also have accessory functions, and that non-coding RNAs intervene at all stages of gene expression, these essential functions were considered to be mere "housekeepers," and RNA was denied a regulatory role. This dogma was, however, "blown up" several years ago by the concomitant discovery of RNA interference and microRNAs in a model organism, the worm *Caenorhabditis elegans*. We now know that small regulatory RNAs are widely conserved in plants and animals, and that microRNAs and short interfering RNAs are not the only kinds of regulatory small RNAs that exist. Indeed, the variety of functions in which small non-coding RNAs have been shown to play essential roles has grown rapidly. Basically, they are involved in controlling large genetic programs or large regions of cell genomes, and they participate in determining what is called cell fate, or the balance between cell proliferation, differentiation and death. This equilibrium is strictly controlled under normal conditions, and its deregulation leads to oncogenesis. One of our main interests is the function of microRNAs in mammalian skeletal muscle. We describe here the high-throughput screening strategy used in our laboratory to identify and validate the microRNAs and their specific targets which are essential for muscle cell differentiation.

Adapted from the Sackler Talk given at the Institut des Hautes Etudes Scientifiques on 13 January 2010 by Dr. Harel-Bellan (corresponding author).

A. Polesskaya • I. Naguibneva • M. Ameyar-Zazoua • C. Degerny • J. Kropp • N. Nonne • N. Mercatelli • M. Souidi • G. Kratassiouk • G. Pinna • L.L. Pritchard • A. Harel-Bellan (✉)
Laboratoire Epigénétique et Cancer, Centre National de la Recherche Scientifique and Université Paris-Sud, CEA Saclay, Gif-sur-Yvette 91191, France
e-mail: Annick.Harel-Bellan@cea.fr

V. Capasso et al. (eds.), *Pattern Formation in Morphogenesis*, Springer Proceedings in Mathematics 15, DOI 10.1007/978-3-642-20164-6_7,

Abbreviations

AGO	Argonaute
ES cells	Embryonic stem cells
Hox	Homeobox
MCK	Muscle creatine kinase
MHC	Myosin heavy chain
miRNP	MicroRNA-containing ribonucleoprotein complex
MRF4	Myogenic regulatory factor 4 (also called Myf6, myogenic factor 6)
Myf5	Myogenic factor 5
MyoD	Myogenic differentiation 1
ncRNA	Non-coding RNA
Pax	Paired box
PIWI	AGO-family protein originally called "P-element induced wimpy testis" in *Drosophila*
piRNAs	PIWI-associated RNAs
Pol	Polymerase
Pol III	RNA polymerase III (in eukaryotes)
RasiRNAs	Repeat-associated small interfering RNAs
RISC	RNA-induced silencing complex
RNA	Ribonucleic acid
RNase	RNA endo- or exonuclease
rRNA	Ribosomal RNA
siRNA	Small interfering RNA
TAP	Tandem affinity purification
tRNA	Transfer RNA

Muscle differentiation (Fig. 1) basically occurs in two major steps: in the first, the totipotent embryonic stem cells (ES cells) acquire determination, i.e., they become committed to differentiate into muscle cells.[1] These committed cells, called myoblasts, continue to proliferate. They don't have a muscle phenotype, but they are predetermined to become muscle cells, so they may be considered as being muscle precursor cells.

[1] There is a discussion of the possibility of going back, in this first step, from determined cells to stem cells. That can be done artificially, as you may know, by introducing certain genes; and the question is, does it also occur naturally in a given cell population? This possibility is being addressed by using mathematical modelling (in collaboration with Dr. N. Morozova and Professor V. Capasso).

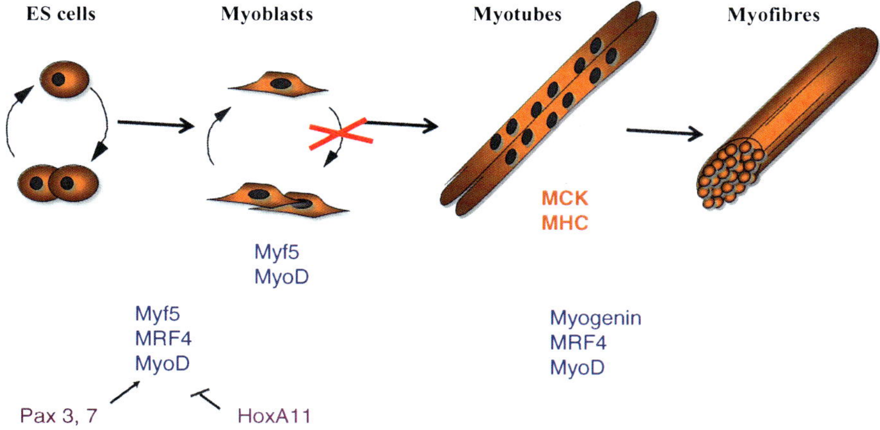

Fig. 1 Schematic representation of embryonic stem (*ES*) cell differentiation into myofibres, showing key regulatory factors that intervene in the process

At some point these myoblasts stop proliferating and enter the terminal differentiation programme, which can be schematically summarised as consisting of cell fusion into huge multinucleated cells called myotubes that will later mature into myofibres, and expression of a specific genetic programme that includes certain proteins considered as being markers for differentiation, such as muscle creatine kinase (MCK) or myosin heavy chain (MHC). This differentiation programme is, of course, orchestrated by transcription factors [1]. That said, and without going into the details, in the last few years it has become apparent that this control exercised at the transcriptional level is not the only one involved in differentiation – and in fact it is not the only one involved in gene expression in general, because it is now firmly established that RNA also has regulatory functions (reviewed in [2, 3]). This is a relatively recent discovery.

1 New Roles for RNA

The well-known "central dogma" of molecular biology states that, for gene expression, DNA is transcribed into RNA, this RNA is translated into protein,[2] and proteins basically "do everything" in the cells, including regulate gene expression. Although it has long been known that RNAs have other important functions in this process of gene expression, they are considered as being "housekeeping" functions, and consequently not very noble, in the mind of a biologist. In addition to the messenger RNAs, or coding RNAs, which contain the information necessary for

[2] The central dogma refers to information flow, and not to the subcellular location where the different processes occur. In eukaryotes, transcription takes place mainly in the nucleus, but translation occurs overwhelmingly or perhaps exclusively in the cytoplasm.

(protein) gene expression, there exist a multitude of large and small non-coding RNAs (ncRNAs) that are involved in this gene expression at several levels. For example, there are ncRNAs involved in the translation procedure itself, because there are RNAs in the ribosomes, the "machines" essential for translating the genetic code into protein molecules – and, of course, there are also the transfer RNAs (tRNAs), which guide the individual amino acids to the site on the ribosome where they will be incorporated into the growing protein chain. So RNA is involved at all levels, but these are housekeeping functions.

What is interesting and much more recent, in fact, is the finding that the genome also encodes ncRNA *genes*, and that their products can be rather small. Many of them are around 20–30 nucleotides in length – much shorter than the coding RNAs, the messenger RNAs, which are $\approx$1,000–6,000 nucleotides long. These very small ncRNAs are now known to regulate gene expression at several different levels, their best-characterised function being to prevent the translation of their messenger RNA targets. They are widely conserved in plants and animals, with the notable exception of the budding yeast *Saccharomyces cerevisiae* (though they are present in the fission yeast *Schizosaccharomyces pombe*).

2 Small Regulatory RNAs

Small RNAs are diverse and certainly not fully characterised to date. Some of these small RNAs act at the nuclear level, directly on genes or on transcription units. These nuclear small regulatory RNAs include, among others, siRNAs in plants and in the yeast *S. pombe*, rasiRNAs in drosophila, piRNAs in all animals, etc. Small RNAs also act at the cytoplasmic level on gene products, i.e., messenger RNAs, as outlined below. How these RNAs were discovered is an interesting story.

Studying development in the worm *Caenorhabditis elegans*, Victor Ambros did a genetic screen where he found a number of hits that, when mutated, impaired the timing of development. Among them were a gene coding for the Lin-28 protein and a hit called *lin-4* that produced a 700-nucleotide RNA with no open reading frame (hence unable to be translated into protein), but that nevertheless inhibited Lin-28 expression. The active portion of the 700-nucleotide transcript is the *lin-4* sequence, which is only about 20 nucleotides long and is partially complementary to the $3'$ untranslated region ($3'$UTR) of Lin-28 messenger RNA ([4]; see also [5]).

The mechanism responsible for this kind of inhibition was not reported by Victor Ambros, but was elucidated through the study of what is known as "RNA interference," a process discovered by Andrew Fire and Craig Mello [6], who likewise analysed *C. elegans* genetics, but in a different manner: by introducing what are called "antisense RNAs" into the worms to modulate gene expression. Fire and Mello were jointly awarded the Nobel Prize in Physiology or Medicine 2006 "for their discovery of RNA interference – gene silencing by double-stranded RNA."[3] It turns out that RNA interference in fact uses what we now call the microRNA pathway (reviewed in [7]). The difference between the exogenous compound and the natural molecule is that, in the natural microRNA, there are usually bulges and mismatches in the double-stranded sequence, whereas silencing RNAs (siRNAs) are perfectly matched.

2.1 Cytoplasmic Small Regulatory RNAs: The MicroRNA Pathway

RNA interference is a gene regulation mechanism that is highly conserved in eukaryotic cells. It involves small double-stranded RNAs (20–30 nucleotides in length), either siRNAs (short interfering RNAs) produced by the cleavage of longer double-stranded sequences or microRNAs transcribed from intergenic or intronic sequences. MicroRNAs are often transcribed as long precursors (priMiRNAs), which are processed in the nucleus into short hairpins (about 70 nucleotides in length) by the "microprocessor" (the RNAse Drosha in association with its co-factor Pasha/DGCR8). Hairpins are transported into the cytoplasm by exportin-5, where they are further processed into short double-stranded species by the RNAse Dicer. SiRNAs, on the other hand, are produced from long double-stranded sequences that can be either endogenous (from repeats and/or transposons) or exogenous (most often viral RNAs) by the RNAse Dicer. Dicer is a single protein in mammals, whereas there are four different Dicer isoforms in plants and two in *Drosophila*.

SiRNAs are essential in plants, but their importance in animals is not well documented. On the other hand, microRNAs are key elements in all organisms. In particular, the microRNA pathway is essential in all animals, from *C. elegans* to mammals. For example, in mouse, inactivation of the unique Dicer gene blocks development at a very early stage, due to the absence of embryonic stem cells [8]. The pathway is also essential in adult organisms.

In monocellular organisms (*S. pombe*), the best characterised function of the pathway takes place in the nucleus, where siRNAs/microRNAs induce chromatin condensation and gene silencing through covalent modification of chromatin components (in particular, the methylation of histones). In multicellular organisms, various functions have been ascribed to the pathway. In plants, as in *S. pombe*, siRNAs/microRNAs are involved in nuclear transcriptional gene silencing, and they also act at the cytoplasmic level. In animals, nuclear functions seem to be performed by specific categories of small RNAs (rasiRNAs, piRNAs), whereas microRNAs and siRNAs seem to be more specialized in cytoplasmic functions.

The microRNAs and siRNAs essentially have a common mode of action (Fig. 2): the short double-stranded RNA recruits a complex of proteins with

[3] http://www.nobelprize.org/nobel_prizes/medicine/laureates/2006/

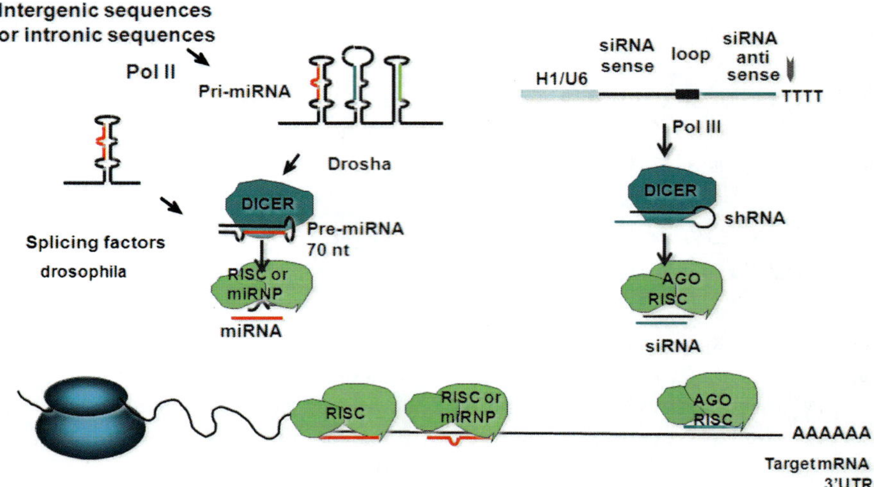

Fig. 2 RNA interference uses the microRNA (miRNA) pathway. Biogenesis of microRNAs is shown on the *left*, and siRNA production from RNA transcribed from an exogenous plasmid vector under the control of a Pol III promoter (H1/U6) is illustrated on the *right*

which it will form what is called the RISC complex, for "RNA-induced silencing complex." Among the proteins in the complex is a very important one called Argonaute. In both microRNA-mediated and siRNA-mediated silencing, the RISC complex contains an Argonaute family protein; but whether the two complexes are strictly identical or slightly different is still a matter of debate. In any case, the first thing that happens is that one of the strands in the small double-stranded RNA is removed from the complex and deleted – this deleted strand is called the "star" strand in the case of microRNAs – whereas the other strand remains associated with the protein complex, conferring on it a sequence specificity that guides the RISC complex toward target messenger RNAs.

Again, in the case of siRNAs, homology with the target sequence is complete, with the result that the Argonaute protein in the complex cleaves the messenger RNA, roughly at the centre of the recognition sequence, and the messenger RNA is subsequently degraded. Both protein-coding and regulatory sequences can be targeted.

In the case of microRNAs, the recognition sequences are usually in the 3′ UTR of the genes. Two things can happen: if the microRNA is fully homologous to the target sequence – and this usually happens in plants, but very little in mammals – the process will be exactly the same as for a siRNA: the messenger RNA will be cleaved near the centre of the sequence. But in most cases, in mammals, the microRNA is not fully homologous to the target sequence. In fact, if we start from the 5′ end and proceed toward the 3′ end of the microRNA, there is a sequence of seven to eight nucleotides that are fully homologous to the target – this is called the "seed" region – then there is a bulge, and then there is a region of more- or less-important homology between the microRNA and the target sequence. By analysing the x-ray crystallography structure of the complex between an Argonaute protein and a nucleic acid, it has been shown that this bulge prevents the cleaving of the target sequence [9]. Another reason why the target sequence is not always cleaved is that not all Argonaute family proteins have a cleaving activity. Mammals have four different Argonaute proteins, plus four others, called PIWI proteins, which are only expressed in germ cells, but only Argonaute 2 (AGO2) has a cleaving activity. So, for these two reasons, there is no cleavage of the target sequence when the homology is incomplete.

Usually there are several binding sites for microRNAs in the target messenger RNA – multiple sites for a given microRNA and/or sites for two or more different microRNAs. So there are several binding sites for microRNAs in the target messenger RNA, and each microRNA can target several different messenger RNAs – the exact number is not known – thus the system of interaction between microRNAs and messenger RNAs is highly complex. When the result is messenger RNA cleavage, the reaction is of course irreversible; but in absence of cleavage, microRNA-mediated translation inhibition is reversible, which may have biological significance. In any case, what happens is that there is inhibition of the translation of the messenger RNA. This can occur at the level of translation initiation or at the level of elongation of the translated protein – and in fact there are arguments for both ideas, both hypotheses. When a messenger RNA is translated, it is associated with several ribosomes, which are the translating entities, so when it migrates in a density gradient, it migrates with what are called the polysomes, RNAs that have several ribosomes attached to them. And in some studies, miRNAs have been shown to prevent the association of target messenger RNAs with polysomes, which is exactly what happens when you inhibit translation. This occurs in the cytoplasm, although there is a debate about the possibility of nuclear translation in some instances. However, in other studies there was no shift of the messenger RNAs (toward the light monosomal fraction, where only one ribosome is attached to the messenger), so some scientists think that inhibition occurs at the level of initiation, and some think that inhibition occurs at the level of elongation; other mechanisms have also been proposed (see below).

To make it a little bit more complicated, a fraction of the Argonaute proteins – in fact a fraction of the whole complex – is located in what are called the P-bodies, which are cytoplasmic entities in which are also located several enzymes used for RNA decay. For example, there are decapping enzymes (you know that messenger RNAs have a special structure at the 5′ end which is called the "cap"), and depolyadenylating enzymes, and when an RNA loses its poly-A tail and its cap, it's more prone to degradation. So in fact there is, in any case, a little decay of the messenger RNAs.

Thus there are different mechanisms for cytoplasmic gene inhibition by microRNAs which can be more-or-less summarised as: initiation block, elongation block, deadenylation and decay. There are also some studies suggesting proteolysis, but this is not totally clear.

To make things still more complicated, inhibition of translation or degradation of messenger RNAs in the cytoplasm is not the only mechanism by which short non-coding RNAs influence gene activity. In fact, in many species, and more-or-less now in all the species, they can act at the level of the nucleus, influencing gene transcription by changing the structure of chromatin.

2.2 Nuclear Small Regulatory RNAs: Gene Silencing

As you may know, DNA is not naked but, at least in eukaryotic cells, is in a structure which is called chromatin: there is a family of proteins, the histones, which form the core of what is called the nucleosome, and the DNA is wrapped around this core of histones. But this structure can be condensed (tightly packed), or it can be relaxed. And in fact the active parts of the genome are generally relaxed, and the repressed parts are generally condensed. Furthermore, there are specific biochemical "marks" on the chromatin components, either the DNA or the histones, which correspond to the different statuses of the chromatin, actively transcribing or not. When chromatin is condensed, DNA is methylated, histones are deacetylated, and specific histones are methylated. In contrast, when chromatin is relaxed, DNA is not methylated, histones are acetylated, and certain other histones are methylated; so there are very specific marks corresponding to the different statuses (reviewed in [10]).

In plants, DNA methylation, which is associated with gene repression, is triggered by RNA interference (reviewed in [11]), and it's usually the inverted repeats, or transposons, or sequences like that, which are regulated in this manner. Double-stranded sequences containing an inverted repeat are processed by the enzyme Dicer, which yields small RNAs of a given length – and it is a very specific Dicer which is responsible for that in plants. These short RNAs, about 27 nucleotides long, guide a complex of proteins that includes an Argonaute protein, but also DNA methylase, toward a given target. The end result of the process is chromatin condensation, with repression of the corresponding locus.

The same phenomenon occurs in the fission yeast *S. pombe* [11], although the modification is not exactly the same, because in that case it is not the DNA but a histone, histone H3, which is methylated on a specific amino acid; this particular histone modification constitutes a "mark" that is associated with chromatin condensation.

What is very striking is that the use of small RNAs to silence, or "turn off", unused portions of the genome also occurs in totally different organisms, such as paramecia or tetrahymenas (reviewed in [12]). In fact, in these organisms, the unused portions of the genome are not condensed, but deleted. Each paramecium has two nuclei, the macronucleus and the micronucleus. The micronucleus contains the entire genome, and is used to duplicate the genome for the daughter cells; but otherwise it is totally condensed, totally unused. The macronucleus, in fact, has all the active genomic sequences; but in the macronucleus, anything which is unused is simply deleted, and the sequences are lost.

Although in some species the mechanism used to silence the unused portion of the genome is only chromatin condensation, whereas in others it is gene deletion, in both of these very distinct mechanisms small RNAs are involved in choosing the sequence which will be unused. So it's a very highly conserved mechanism.

What about mammals? In mammals, the situation is a bit less clear. The microRNAs and siRNAs constitute only a small portion of the regulatory small non-coding RNAs that have now been characterised. In germ cells, there are specific RNAs called piRNAs, which are associated with Argonaute proteins called PIWIs that are specific to the germ cells. The piRNAs were first discovered in drosophila, and they are responsible for repeat and transposon silencing in mammals as well as in flies. What's going on in somatic cells is a matter of debate at the moment. What we are doing in this debate is trying to understand whether there is any nuclear Argonaute in somatic cells in mammals, and if there is, what its function is.

In experiments with nuclei purified on sucrose gradients, staining with antibodies against Argonaute proteins revealed that indeed there are Argonaute proteins inside these highly purified nuclei. Now, to understand the function of these Argonaute proteins, we have purified the Argonaute proteins biochemically using what is called a TAP-Tag[4] approach; we purified the Argonaute protein both from the chromatin-associated fraction, which is the Argonaute that is associated with DNA, and from the cytosolic fraction as a control. And what we are doing is characterising the proteins and the small RNAs that are associated with Argonautes in the chromatin fraction in the nucleus.

What we find in association with the nuclear Argonaute which is completely absent from the cytoplasmic fraction are proteins associated with chromatin modification and chromatin condensation: we find a histone deacetylase – again, deacetylated histones are associated with condensed chromatin – we find a histone methyltransferase, which is also specific for condensed chromatin, and we find a protein called HP1gamma (heterochromatin protein 1-gamma), which is also specific for condensed chromatin. So, it really looks like there is a mechanism at work in mammalian cells, even in somatic cells, that is similar to what is seen in the yeast *S. pombe*, or in plants.

3 Molecular Mechanisms

About the function of microRNAs at the molecular level, it is clear that this is a very important issue which has been largely overlooked until recently. The microRNA pathway is indeed a *very* conserved mechanism: microRNAs are in fact present in all organisms, from plants to mammals, with only one exception, which is the budding yeast *Saccharomyces cerevisiae*. They may represent 1–5% of the transcribed sequences, and more than a thousand have thus far been cloned in humans. The microRNAs themselves are also highly conserved, and there are more than 70 microRNAs known to be 100% conserved in sequence from zebra fish to human, which is a very striking degree of conservation. This is quite a small number compared to the tens of thousands of microRNAs that some predictions suggest, but more than 3% of the microRNAs that have actually been cloned to date in

[4] Tandem affinity purification (TAP) based on Argonaute proteins that carry two or more distinct biochemical modifications, or "tags," that bind to specific reagents immobilised on gel beads.

humans, for example. The discrepancy between predictions and "reality" is at least partly due to the difficulty in predicting a function for a sequence that should have only partial homology to its target, and certainly also to the fact that microRNAs represent only a portion of the small non-coding RNAs in a cell.

Another observation that really supports the idea that microRNAs have very important functions is that it is a highly redundant system: for many microRNAs there are several isoforms. And not only are there several isoforms, but the same isoform can also be encoded in several loci, so you have several copies of a given microRNA, exactly the same sequence, in the genomes.

3.1 MicroRNAs in Development

Concerning their physiological function, microRNAs are essential during development (reviewed in [13]). Indeed, as already mentioned, the first microRNA that was discovered regulates the timing of development in *C. elegans*. MicroRNAs are also essential for development in higher organisms. Mammals, in contrast to other species, have only one Dicer, which is the enzyme that cuts the precursor microRNA. Because there is only one, you can easily inactivate it completely; and the inactivation of Dicer is embryonic lethal at very early stages [8]. The problem the embryos have is that they do not have any embryonic stem cells. So microRNAs are necessary for embryonic stem cell formation. You can also, of course, inactivate Dicer in specific organs or during the generation of specific organs, which brings us to the subject of pattern formation, due to the participation of microRNAs in the formation of the limb bud.

You have heard about the Hox cluster of genes, which is important for patterning. Well, the Hox cluster of genes also encodes microRNAs; and one microRNA in particular, miR-196, regulates a Hox protein that is involved in limb formation: HoxB8 induces Sonic Hedgehog, which in turn controls the antero-posterior polarity of the limbs; and the interesting story is that HoxB8 is induced by retinoic acid, which is an important morphogen. In the wild-type situation, HoxB8 is induced in the forelimbs but not in the hindlimbs. However, if for some reason there is no Dicer in the limb, HoxB8 will in fact be induced in both the forelimbs and the hindlimbs. What normally prevents induction of HoxB8 in the hind limbs is a microRNA, miR-196, whose presence of course depends on the existence of Dicer [14]. So there is a strong relationship between microRNAs and development, of course. Now, if we examine what microRNAs actually do at the cellular level, we find that they influence cell proliferation, cell differentiation, and cell death, which is exactly what we are interested in our muscle model.

3.2 MicroRNAs in Cell Differentiation

We are addressing the function of microRNA during muscle differentiation, essentially during the terminal step of the process. The first thing that you have to do when you do that is what is called profiling – determining which microRNAs are expressed, because, of course, if they are not expressed they cannot have any function. By doing expression analysis at the genome-wide level, using different techniques, we found that there are actually quite a number of microRNAs that are differentially regulated during the terminal differentiation process. So what function do they have?

3.2.1 In Vivo Studies

We started to analyse the function with a specific microRNA that we found to be up-regulated during differentiation, miR-181 [15]. One of several observations proving this is the expression of miR-181 during muscle regeneration in vivo. Muscle normally regenerates at a very low level in adults. However, if you destroy muscle tissue by injecting a toxin, very locally, then it's regenerated in about two weeks in adult mice. Using what is called in situ hybridisation, which means looking for the microRNA inside the cell using a fluorescent probe, we only found miR-181 in the cells that also express a specific protein which is only present during regeneration.

So this microRNA is up-regulated during muscle regeneration, which involves differentiation. Does it have a function? To answer that, we need a loss-of-function assay, and the best approach, of course, would be to inactivate the gene in vivo, that is to say, in the mouse. Now, the problem is that miR-181 is one of those microRNAs that are highly redundant, and there are several isoforms. Indeed, there are four isoforms of this microRNA, two of which (miR-181a and b) are expressed in the muscle. But each of these two isoforms is also present in two copies, at two different loci, on two different chromosomes of the mouse genome. What we began by doing was to inactivate one copy of one isoform, and this knockout mouse so far has no obvious phenotype. We are now inactivating the other cluster, and we will see if there is any phenotype with that, then we will of course cross the two mouse strains to totally inactivate at least one of the isoforms, a very long and complicated procedure, and the results are not yet available.

3.2.2 In Vitro Experiments

It is, in fact, much easier for us to inactivate the microRNA in vitro, because there are in vitro differentiation systems that we can use very readily. So how do we do this? We designed a loss-of-function assay [16] for microRNA that involves replacing the complementary strand of the microRNA with an oligonucleotide that is fully complementary to the microRNA, so that there is no 3′ protruding sequence, thus preventing it from being recognised as a natural double-stranded small RNA. This oligonucleotide contains a modification which is called LNA (for "locked nucleic acid"); and, very briefly, what the LNA does is to maintain the sugar moiety in the proper conformation, the best conformation, for base stacking.

So it's a modification that dramatically strengthens the association between the two strands – you have two strands which more-or-less cannot be separated.

The association between this LNA-modified oligonucleotide and its target microRNA prevents the microRNA from functioning. In what is called a "northern blot" for the detection of RNAs, you can detect both the precursor and the mature forms of the microRNA. And when the cells receive an antisense sequence, you can no longer detect any of the microRNA by this method. So there is a way to monitor the inhibition. Now what happens to muscle cells if you inhibit miR-181? In cells that have received a control oligonucleotide, myotubes with several nuclei are clearly seen, and they also express myosin heavy chain (MHC), which is a differentiation marker. Now, when the cells receive the antisense against miR-181, there are no myotubes, there is very little expression of the marker, i.e., you see an inhibition of differentiation. We can also assess inhibition of differentiation by measuring the expression of another marker, muscle creatine kinase (MCK), which is likewise dramatically inhibited in the cells that receive the LNA antisense.

What we are doing at the moment is extending these observations to the genome-wide level.[5] This is not very complicated to do with microRNAs, because, again, there are only about 900 microRNAs[6] characterised in humans at this point. So we have an antisense library, which covers the microRNAs characterised in humans, and we are transfecting human myoblastic cells with this library. We routinely use, alternatively, mouse and human models with no problems. The myoblasts are incubated under differentiation conditions, and then differentiation is monitored by labelling with the marker MHC, which is of course done by a robot, because it still requires a large number of tissue-culture plates. Mock-transfected cells, that is, cells subjected to the entire testing process, including treatment with the transfection reagents but without any antisense oligonucleotide, are used as a negative control to determine the "background" level of differentiation into myotubes. The positive control is the antisense against miR-181a: very few myotubes form when miR-181 has been knocked down. Another negative control is a mutant antisense, which is based on the miR-181a sequence but does not recognise miR-181a; treatment with the mutant antisense does not inhibit myotube formation.

We scored the number of nuclei per cell, which is an index of the fusion of the cells, and the level of MHC fluorescence, which is a marker for differentiation. The analysis is done with a software package called Spotfire®, which assists us in analysing the distribution of the results and enables us to choose appropriate thresholds. Using this screen, we obtained 77 hits, which means that among the 900 microRNAs characterised in humans, 77 are required, absolutely required, for proper differentiation.

[5] "Genome-wide" with respect to the number of human miRNAs identified at the time the work was begun.

[6] miRBase 18, released in November 2011, contains 1,527 precursors, corresponding to 1921 mature microRNAs in humans.

Differentiation can be either inhibited or, on the contrary, increased, by the inhibition of a microRNA. In the first case, called hypodifferentiation, cells of course do not differentiate properly – but we also found a hyperdifferentiation phenotype, indicating that some of these microRNAs are required to *delay* the differentiation process. And indeed it really is a delay, because if you do the experiment in a shorter time frame, the hyperdifferentiating cells differentiate, but the controls do not. So some microRNAs control the timing of the human muscle differentiation process.

Obviously we have to validate these hits; and, for that, we first do what are known as rescue experiments. Introducing an oligonucleotide sequence into the cell can trigger many side effects, unrelated to the microRNA under investigation, but nevertheless interfering with the differentiation process. So to prove that the effect we see is *not* one of these off-target effects, but is in fact due to inhibition of the microRNA being studied, we have to do rescue experiments that reintroduce an excess of this microRNA into the cells. This is done very easily by using a synthetic double-stranded sequence that mimics the natural double-stranded compound – you put a very high level of the molecule into the cells. And of course we do not do the experiment by mixing the antisense and the microRNA, because the microRNA would bind to the antisense in the test tube, and so prevent the effect of the antisense, but that would not really be a rescue. Instead, we introduce the antisense into the cells on day 0, to "knock down" the endogenous microRNA, and then re-introduce the same microRNA (albeit a synthetic version), and induce differentiation, 1 day later. And if we do that with our positive control microRNA, miR-181, what happens is that we have a very good restoration of differentiation [15]: the antisense inhibits differentiation, but differentiation is restored by rescue. We are doing that now with our screening hits.

3.3 *Fishing for Targets*

The most important thing is to understand how the different microRNAs impact on differentiation, what genetic pathways they regulate – and for that, of course, we need to characterise the targets. Many bioinformatic programmes, software packages, have been designed to predict the targets; they are based on looking for the seed sequence mentioned earlier, and then taking into account different considerations – the homology in particular regions, different G:C contents, conservation, location in the 3′ UTR, and others. According to some of these programmes, each microRNA can have up to several hundred targets or more. But it's not quite clear at this point that these predictions are fully reliable.

We are developing two approaches to validate targets. The first is an unbiased biochemical way of identifying the targets. The idea is as follows: the microRNA guides the (Argonaute) complex to the messenger RNA target, so it's bound to the messenger RNA. Hence if we can pull down the microRNA, precipitate it from the cell extract, maybe we can pull down the messenger RNAs attached to it and characterise them. The microRNA of interest is "tagged" with a biotin molecule, and, to avoid a high background, we also tag the Argonaute proteins with a completely different marker. Then we do a tandem affinity purification (TAP) procedure, first pulling down Argonaute and then the microRNA. This procedure, which we call "TAP-Tar"[7] is working really well for validation of predicted targets [17]. What we would like to do now is identify new targets, or identify "the" targets, the real targets, of microRNAs of interest; but at this point the yield is very low, and needs to be improved.

Obviously, the most important thing is the actual function of a given target in the phenotype of interest. In fact, showing that the microRNA binds a given messenger RNA, or that the microRNA can induce the inhibition of a reporter gene containing its target sequence, or any other approach to validating the target, will only tell you that this messenger RNA *can* be a target of this microRNA, but it does not show that this target *is* important for your given phenotype. So our second approach to validating the targets, determining the physiological importance of their interaction with the microRNA, is addressed as follows. Let's say that, for example, the microRNA has three targets, one of which is known to be important in the phenotype of interest. In the case of miR-181, it's Hox-A11.

Hox-A11, actually, brings us back again to pattern formation, because it's a protein of the Hox cluster. It's mainly involved in kidney and urinal tract formation, but it is also involved in muscle differentiation, because it's expressed in limb bud pre-myoblasts in mouse. The limb bud muscle forms from the somites, and the precursor cells migrate to the somites. Hox-A11 is in fact expressed in the pre-myoblast that migrates, and is an inhibitor of differentiation. The protein, but not the messenger RNA, is down-regulated upon terminal differentiation, so it has a post-translational level of regulation. It represses MyoD expression, and it also represses terminal differentiation [18]. So it *looks* like a very good target. But is this in fact the case?

Let's begin by assuming that Hox-A11 is indeed an important target in our system. In that case, if we delete miR-181, the main consequence in the cell will be that we have too much Hox-A11; and since it's an inhibitor of differentiation, if we have too much Hox-A11, then we will not have any differentiation. If this is true, then we need only to reduce the amount of Hox-A11 to restore a normal phenotype. And that's the type of experiment we are doing. Indeed, when Hox-A11 is inhibited using an appropriate siRNA, both messenger RNA and protein levels in the cells decrease markedly, so this particular siRNA can be used to decrease Hox-A11 levels in this system. Now, when the cells receive the wild-type antisense LNA (against miR-181) but a control siRNA, which does not inhibit Hox-A11, differentiation is inhibited as expected. When, however, the cells receive the antisense against miR-181 plus the siRNA against Hox-A11, the differentiation is partially restored. Moreover, the level of differentiation observed is in fact inversely

[7] *TAP* procedure with *tar*get pull-down and identification.

correlated to the level of Hox-A11 present in the cells: in the cells with severely impaired differentiation there is a high level of Hox-A11, which is normal, because we inhibited the microRNA; and in the cells which received the antisense plus the siRNA there is an intermediate level of Hox-A11 and an intermediate level of differentiation. Together these data show that miR-181 inhibits myoblast differentiation through its effects on Hox-A11. So that is actually the proof of principle for the screen that we are implementing to identify the phenotypically important targets of the microRNAs.

The strategy is thus to use an LNA against one of our microRNA hits; this LNA will be introduced into cells together with a library of siRNAs, this time against genes chosen from among the predicted targets of the microRNA. Then the cells will be allowed to differentiate, we will monitor differentiation at the end, and all the hits that restore a normal differentiation phenotype will be investigated in greater depth.

To summarize, the questions that we are addressing are the following:

- Which messenger RNAs are targeted by the hits in our microRNA screen – can we design a methodology to characterise these targets in an unbiased manner, and thereby decipher the genetic pathways influenced by microRNAs? Also, can we develop an integrated view of these pathways, including the genes and the microRNAs which are involved in the differentiation process?
- By which mechanism (transcriptional gene silencing, messenger RNA decay, translational inhibition, others) do the microRNAs inhibit their target genes? [19] Can we explain, by a unifying theory, their mode of action?
- Do small non-coding RNAs have a nuclear function in mammalian cells?

Acknowledgements High-throughput data analysis was carried out in collaboration with Robert Young, and mathematical modelling, in ongoing collaborations with Andrei Zinovyev and Vincenzo Capasso. Financial support was provided by the Centre National de la Recherche Scientifique, Paris-Sud University, the Commissariat à l'Energie Atomique et aux Energies Alternatives (CEA), INSERM, the Ligue National contre le Cancer, the Ligue contre le Cancer in Île-de-France, the Association Française contre les Myopathies, the European Union's sixth framework programme (Integrated Projects RIGHT, contract number LSHB-CT-2004-005276, and SIROCCO, LSHG-CT-2006-037900), the Agence Nationale de la Recherche (project ANR-08-SYSC-003 CALAMAR), and the Cancéropole, Région Île de France.

References

1. Buckingham M (2006) Myogenic progenitor cells and skeletal myogenesis in vertebrates. Curr Opin Genet Dev 16(5):525–532
2. Mattick JS, Gagen MJ (2001) The evolution of controlled multitasked gene networks: the role of introns and other noncoding RNAs in the development of complex organisms. Mol Biol Evol 18(9):1611–1630
3. Suh N, Blelloch R (2011) Small RNAs in early mammalian development: from gametes to gastrulation. Development 138:1653–1661
4. Lee RC, Feinbaum RL, Ambros V (1993) The *C. elegans* heterochronic gene lin-4 encodes small RNAs with antisense complementarity to lin-14. Cell 75:843–854

5. Wightman B, Ha I, Ruvkun G (1993) Posttransctiptional regulation of the heterochronic gene lin-14 by lin-4 mediates temporal pattern formation in *C. elegans*. Cell 75:855–862
6. Fire A, Xu S, Montgomery MK, Kostas SA, Driver SE, Mello CC (1998) Potent and specific genetic interference by double-stranded RNA in *Caenorhabditis elegans*. Nature 391:806–811
7. Nelson P, Kiriakidou M, Sharma A, Maniataki E, Mourelatos Z (2003) The microRNA world: small is mighty. Trends Biochem Sci 28(10):534–540
8. Bernstein E, Kim SY, Carmell MA, Murchison EP, Alcorn H, Li MZ, Mills AA, Elledge SJ, Anderson KV, Hannon GJ (2003) Dicer is essential for mouse development. Nat Genet 35:215–217, Corrigendum in Nat Genet (2003) 35:287
9. Parker JS (2010) How to slice: snapshots of Argonaute in action. Silence 1(1):3
10. Gan Q, Yoshida T, McDonald OG, Owens GK (2007) Concise review: epigenetic mechanisms contribute to pluripotency and cell lineage determination of embryonic stem cells. Stem Cells 25:2–9
11. Verdel A, Vavasseur A, Le Gorrec M, Touat-Todeschini L (2009) Common themes in siRNA-mediated epigenetic silencing pathways. Int J Dev Biol 53:245–257
12. Coyne RS, Lhuillier-Akakpo M, Duharcourt S (2012) RNA-guided DNA rearrangements in ciliates: is the best genome defense a good offense? Biol Cell 104(6):309–325
13. Pauli A, Rinn JL, Schier AF (2011) Non-coding RNAs as regulators of embryogenesis. Nat Rev Genet 12:136–149
14. Hornstein E, Mansfield JH, Yekta S, Hu JK, Harfe BD, McManus MT, Baskerville S, Bartel DP, Tabin CJ (2005) The microRNA miR-196 acts upstream of Hoxb8 and Shh in limb development. Nature 438(7068):671–674
15. Naguibneva I, Ameyar-Zazoua M, Polesskaya A, Ait-Si-Ali S, Groisman R, Souidi M, Cuvellier S, Harel-Bellan A (2006) The microRNA miR-181 targets the homeobox protein Hox-A11 during mammalian myoblast differentiation. Nat Cell Biol 8(3):278–284
16. Naguibneva I, Ameyar-Zazoua M, Nonne N, Polesskaya A, Ait-Si-Ali S, Groisman R, Souidi M, Pritchard LL, Harel-Bellan A (2006) An LNA-based loss-of-function assay for micro-RNAs. Biomed Pharmacother 60(9):633–638
17. Nonne N, Ameyar-Zazoua M, Souidi M, Harel-Bellan A (2010) Tandem affinity purification of miRNA target mRNAs (TAP-Tar). Nucleic Acids Res 38(4):e20
18. Yamamoto M, Kuroiwa A (2003) Hoxa-11 and Hoxa-13 are involved in repression of MyoD during limb muscle development. Dev Growth Differ 45(5–6):485–498
19. Zinovyev A, Morozova N, Nonne N, Barillot E, Harel-Bellan A, Gorban AN (2010) Dynamical modeling of microRNA action on the protein translation process. BMC Syst Biol 4:13

Pattern Formation in Sea Urchin Endomesoderm as Instructed by Gene Regulatory Network Topologies

Isabelle S. Peter and Eric H. Davidson

Abstract Animals consist of body parts which are spatially discrete functional units. The spatial separation of these body parts and the diversification of their function and structure are developmentally controlled by gene regulatory networks. The transcription factors and signaling molecules which participate in the spatial organization of a developing organism are components of these networks. The causal linkages in the network consist of the regulatory interactions of each factor with its target genes. Interactions among different regulatory genes are responsible for forming specific spatial patterns of gene expression. The architecture of these regulatory interactions and how they instruct the formation of specific spatial domains is directly determined by the genomic sequence. In the sea urchin embryo, many such spatial domains are established early in development. A well-characterized gene regulatory network underlies the specification of endodermal and mesodermal regulatory domains in this embryo. We review multiple examples which reveal the causal logic underlying genomic control strategies for pattern formation during sea urchin embryogenesis.

1 Genomic Information Is Identical in All Cells and Underlies Functional Diversification

The genomic sequence encodes the developmental program which determines the progression from fertilized egg to organized body plan. Among individuals within a species, there is little variation of the body plan: the location, form and function of head, thorax, legs, antennae, eyes and so on are all about the same. These individuals share the genetic information required for the development of the body plan. Any changes in the composition of the genetic information leads to

I.S. Peter (✉) • E.H. Davidson (✉)
Division of Biology, MC 156-29, California Institute of Technology, Pasadena, CA 91125, USA
e-mail: ipeter@caltech.edu; davidson@caltech.edu

V. Capasso et al. (eds.), *Pattern Formation in Morphogenesis*, Springer Proceedings in Mathematics 15, DOI 10.1007/978-3-642-20164-6_8,

fatal errors in the progression of developmental process, as was systematically assessed in sea urchin embryos in the early twentieth century by Theodor Boveri [1, 2]. The same genetic information is present in each cell of an individual, yet these cells will acquire different functions and shapes during development. In development cells function with respect to the organization of the body part to which they contribute, and cellular specialization has to occur in the context of this organization. Functionally, specialization is a stepwise process. Initially, a group of cells is defined whose descendants will form an entire body part. But cells within such "progenitor fields" acquire more and more specialized functions until the entire functional body part is formed, consisting of muscle cells, nerve cells, pigment cells, bone forming cells etc. Where these individual types of cells develop in respect to each other, and in respect to the major body axes, is crucial for the function of the entire body part; and where the body part forms is crucial for the entire organism. In the spatial organization of the organism there is therefore no room for randomness, noise and stochasticity. The genome sequence encodes the map underlying the progressive functional diversification of cells during development which invariably results in the pre-defined body plan organization of the entire organism.

Here we will focus on some of the mechanisms which determine the spatial organization of the sea urchin embryo. Sea urchins develop indirectly. After fertilization, their embryos develop into swimming larvae which feed, providing a nourishing environment for the juvenile sea urchin developing inside of it. The relatively simple structure of sea urchin larvae and their relatively low number of different cell types are major advantages for unraveling the mechanisms by which their spatial organization is determined. The first signs of spatial organization are already apparent a few hours after fertilization. The gene regulatory networks which determine the spatial organization in the sea urchin embryo are among the best studied gene regulatory networks and will here be used to elucidate some of the spatial control mechanisms.

2 Regulatory State Domains

Very few morphological differences are recognizable in the first 30 h of sea urchin embryonic development. Morphologically, the blastula stage embryo appears to consist of a layer of cells forming a hollow ball, except for a few cells within the blastocoel, which will generate the larval skeleton. However, the next few hours will entirely change the shape of the embryo. Gastrulation starts at about 30 h of development and includes the invagination of the gut tube, the archenteron, and the formation of muscle cells and neuronal cells, the ingression of pigment cells and blastocoelar cells which have immune functions. Groups of cells fulfilling each of these functions are specified hours before any functional or morphological diversification can be recognized. Nevertheless, these differences can easily be detected by analyzing spatial gene expression patterns. As a first sign of diversification, cells start to transcribe different sets of regulatory genes. These transcripts in turn can be detected by in situ hybridization, indicating the spatial domain in which the analyzed gene is expressed. When gene expression is detected within a spatial domain but not in cells outside of the domain, it indicates that a specific regulatory state must be present in that spatial domain. **Regulatory states** are defined as the set of transcription factors and signaling molecules which are expressed in a given spatial domain and which distinguish this domain from other spatial domains. Regulatory genes expressed in all cells are not considered in the definition of spatial regulatory states, because they do not contribute any spatial information.

From an early stage on, all cells which will constitute the endoderm and mesoderm of sea urchin larvae are divided into several different regulatory states. The endomesodermal regulatory domains are formed in concentric rings around the vegetal pole even before gastrulation and give rise to specific parts of the larva, as shown in Fig. 1. The regulatory domains in successive radial positions from the vegetal pole are as follows:

- **Small micromeres** are set aside cells formed by fifth cleavage (6 h post fertilization) which will contribute to the formation of the rudiment, where the juvenile sea urchin develops inside the larva. Small micromeres constitute a separate regulatory domain, as indicated by the specific expression of *foxc* by 21 h [3].
- **Skeletogenic micromeres** develop from large micromeres, the sister cells of the small micromeres. All descendants of these large micromeres give rise to skeletogenic cells responsible for forming the larval skeleton. Specific transcription of *pmar1* begins in the progenitors of small and large micromeres and continues for a few hours exclusively in the large micromere descendants, as shown in Fig. 1. *Pmar1* is the earliest regulatory gene with micromere-specific zygotic expression [4, 5].
- **Oral and aboral mesoderm** are the two regulatory domains which constitute the remainder of the mesoderm: they give rise to pigment cells, blastocoelar cells, muscle cells and coelomic pouch cells. The progenitors of these mesoderm cells are marked by *gcm* expression at about 16 h, as shown in Fig. 1. A few hours later, the regulatory states within this domain differentiate, with *gcm* expression fading in about one quarter of the cells on the oral side (the side of the embryo where the mouth is going to form) and remaining only in the aboral mesoderm progenitors [6]. Different regulatory genes are at this point expressed in the oral and aboral mesoderm precursor cells, marking an initial step in the subdividsion of the mesoderm progenitor cells into the different cell fates [7, 8]. For example, *gcm* continues to be expressed in pigment cells up to larval stages and encodes a transcription factor with an important role in the pigment cell specification GRN [6].
- **Anterior and posterior endoderm** regulatory domains are distinguished long before a gut tube even starts to form. The anterior endoderm shares common progenitor cells with the mesodermal cell types and becomes a distinct domain by about 18 h [9]. Posterior parts of the gut on the other hand derive from a

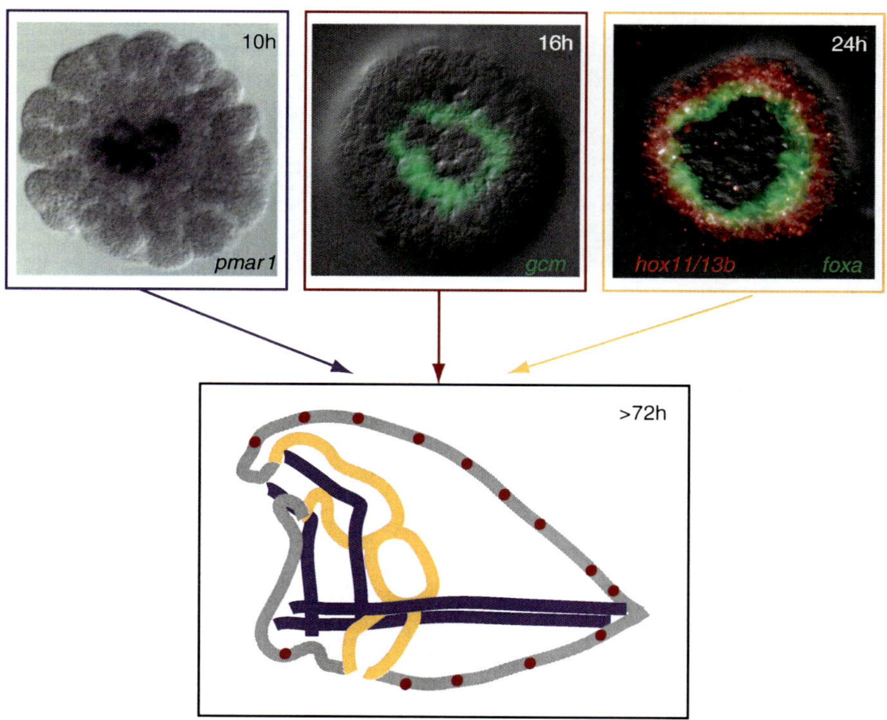

Fig. 1 Embryonic regulatory state domains and the larval cell types they give rise to. The *top panel* shows three embryos which were analyzed by in situ hybridization for the expression of indicated regulatory genes. In the 10 h embryo, four cells at the vegetal pole express *pmar1* [3]. All descendants of these cells will produce the larval skeleton (*purple*) as instructed by the skeletogenic GRN expressed downstream of *pmar1*. At 16 h, the *gcm* expression domain forms a ring surrounding the eight skeletogenic cells then present. All descendants of the *gcm* expressing cells will give rise to mesodermal cell types, although only those on the aboral side of the mesoderm will continue to express *gcm* and develop into pigment cells (*red*); these migrate from the vegetal plate and integrate into the aboral ectoderm. By 24 h, all endodermal precursor cells express either one of two regulatory states, as shown here by the expression of *foxa* in the inner ring (cells of the future anterior gut) and *hox11/13b* in the outer ring (cells of the future posterior gut). Both rings of cells surround the mesodermal domain. During gastrulation, these cells invaginate and form the tri-partite gut (*yellow*)

separate ring of cells located immediately adjacent to the ring of cells constituting the future anterior endoderm. By 24 h, the cells which will later form the entire gut [10] have separated into two separate regulatory domains, marked by *foxa* expression in the anterior progenitor cells, and *hox11/13b* expression in the posterior progenitor cells.

With the exception of the skeletogenic cells, these regulatory domains demarcate cells with related but different developmental programs. However, later in development, these regulatory domains will each further segregate into multiple regulatory subdomains, until individual cell types differentiate. The formation of different spatial regulatory state domains is the product of gene regulatory networks. These networks determine all regulatory interactions which control the expression of **regulatory genes**, genes encoding transcription factors and signaling molecules, that is, all the molecules which in turn comprise the spatially specific regulatory states. The regulatory interactions executed by previous regulatory states install each new regulatory state. Gene regulatory network architecture, which is based on the structure of regulatory interactions, determines spatial organization in development.

3 Gene Regulatory Networks

3.1 Gene Regulatory Network Structure

Gene regulatory networks consist of regulatory genes and the regulatory interactions among them. The function of GRNs is to control cell fate specification processes in time and space. The underlying computation which during development drives an organism from low to high complexity in respect to the numbers of different types of cells present is executed by combinatorial transcriptional regulation. Transcription factors individually recognize short specific DNA sequences (mostly <10 nucleotides long), which randomly occur throughout the genome. However, in DNA sequences that function to regulate genes, termed ***cis*-regulatory modules,** multiple transcription factors bind simultaneously within a short distance, and recruit additional co-factors. The *cis*-regulatory modules controlling a given gene may be located in its immediate proximity, or in its introns, or sometimes distantly upstream or downstream. In many *cis*-regulatory modules all required factors have to bind in order for RNA polymerase to initiate transcription. Other *cis*-regulatory DNA sequences bind transcription factors which act as repressors, by interfering with the activation of transcription. The expression of a single gene may be controlled by multiple *cis*-regulatory modules, each module including the binding sites for multiple transcription factors, and therefore active only in the spatial and temporal domain where these factors are present as a set. Diverse *cis*-regulatory modules thus control the diverse phases of expression of a gene. When and where each gene is expressed is therefore dependent on its *cis*-regulatory code, and on when and where the regulatory factors which control its expression are co-expressed. In turn, the co-expression of specific combinations of transcription factors depends on the *cis*-regulatory sequences which control their own expression. The function of each transcription factor and signaling molecule during development is determined by its expression pattern and by the target genes it controls. Since multiple regulatory factors are required to control expression of each gene, and since each regulatory factor controls the expression of multiple target genes, the total constellation of regulatory interactions is a network.

Gene regulatory networks connect all the regulatory genes in the genome but can be analyzed separately for each individual developmental process. The formation of regulatory state domains, the progressive specification of cell types, cell migration, the functional differentiation of cells and many other processes depend on the expression of specific genes and are thus subject to gene regulatory network control. Specific regulatory interactions are required to drive each process and their logic operation determines the causality of the process. For example, a new regulatory domain Y is apparent, marked by the expression of a gene A. No other gene is expressed in the same group of cells before gene A is expressed. One way to turn on gene A, which is frequently observed in developmental process, is if gene A expression depends on two or more transcription factors which are each expressed in domain Y, though not exclusively. Each of the upstream regulatory factors has an expression domain which is broader than Y. But their expression overlaps in domain Y which is where specific expression of gene A is activated. This is one example where the simultaneous requirement of two regulatory factors for the activation of gene expression results in the formation of a new spatial regulatory domain. Many such subcircuits have been found where the constellation of regulatory interactions among a few regulatory genes has a specific function in development. These subcircuits can be repeatedly observed in different aspects of development [11]. Each time, the architecture of regulatory interactions is the same even though the regulatory genes involved are different and specific to the developmental context.

An important feature which discriminates gene regulatory networks from other known networks like social interaction networks or protein-protein interaction networks is that regulatory interactions are obligatorily uni-directional. In a GRN model, each linkage between two genes represents the functional binding of a transcription factor to the *cis*-regulatory sequences of its target gene. Even though this target gene might encode a transcription factor which in turn regulates the upstream regulatory factor in a positive feedback constellation, regulatory interactions can never simply be reversed. In addition, the functionality of these regulatory interactions always depends on the regulatory context. Thus a given transcription factor can be required at a given time to drive the expression of its target gene, and loss of this transcription factor by experimental perturbation will result in a failure of target gene expression. However, the target gene will depend as well on additional transcription factors which must be present simultaneously to control expression of this gene. An important consequence of this mode of combinatorial gene regulation is that regulatory factors can be used in multiple different processes, and in each of these processes a given factor might fulfill a different function, i.e., drive the expression of a different set of target genes. This explains how the control of developmental process can rely on a limited number of molecules. Only a few signaling pathways are used over and over during development, in various different processes. Similarly, there are only a limited number of transcription factors encoded in the genome. Sea urchin embryos, for example, deploy more than 80 % of their transcription factors in the first stages of their embryonic development, and therefore have to re-deploy many of them again and again at later stages [12].

3.2 Gene Regulatory Network Model

The gene regulatory network model shown in Fig. 2 represents the regulatory genes and regulatory interactions responsible for driving the specification of all endodermal and mesodermal precursor cells in sea urchin embryos up to 30 h post fertilization. Each regulatory state domain present at the end of this time period is represented by a colored rectangle. Within the rectangle are shown all the specifically expressed regulatory genes plus regulatory interactions which could contribute to the distinct function of that domain. For modeling of GRNs it is of utmost importance to represent the spatial domains in which each part of the GRN is active. In this presentation regulatory genes involved in multiple processes are represented separately in each of these processes. The result can be seen in the GRN model in Fig. 2. Instead of showing all the potential target genes of, for example, Delta/Notch signaling which are present throughout the genome, this kind of representation focuses on specific developmental process and specifies the few target genes that the Delta/Notch signal affects within mesodermal precursor cells in a given time window ("Oral NSM" and "Aboral NSM"). The GRN model in Fig. 2 does not represent temporal aspects, but an online version of this model shows the regulatory interactions active in each 3 h time interval (http://sugp.caltech.edu/endomes/).

The number of regulatory logic transactions which are required for a given developmental process also affects the temporal dynamics of the process. Each regulatory state controls the expression of specific target genes. These target genes are first transcribed in the nucleus, and the transcripts then have to be transferred to the cytoplasm to be translated to protein molecules. Newly translated transcription factor proteins must then return to the nucleus to control their own target genes. This process requires a certain amount of time, which in sea urchin embryos has been calculated and measured [5, 13]. For sea urchin embryos developing at 15 °C, it takes approximately 2–3 h to activate target gene transcription after an upstream transcription factor has begun to be transcribed. Each step in a GRN therefore occurs at a given rate and the number of these steps, the depth of the GRN in a given process, correlates with the temporal requirement for that process.

3.3 Gene Regulatory Network Analyses

The goal of GRN analyses is to understand the mechanisms responsible for the spatial assignments of different body parts and for driving development of these body parts. The spatial definition of specific domains is a multi-step process which during development progressively installs new spatial regulatory state boundaries, ultimately so as to generate the numerous cell fate domains in the body. The scientific value of a GRN model increases with its degree of completeness. Thus GRN analyses are preferentially performed at a systems level, including at least the great majority of the regulatory factors involved in a given process.

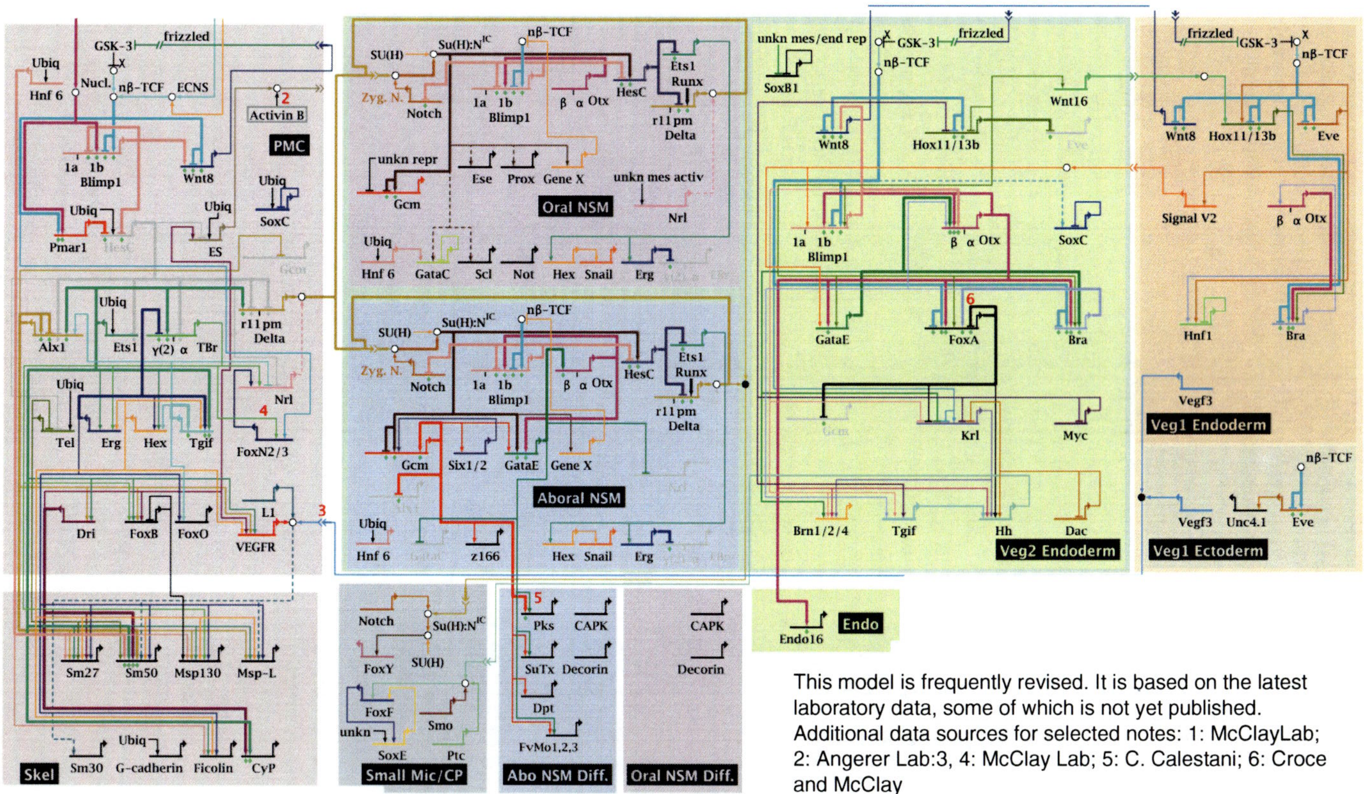

This model is frequently revised. It is based on the latest laboratory data, some of which is not yet published. Additional data sources for selected notes: 1: McClayLab; 2: Angerer Lab:3, 4: McClay Lab; 5: C. Calestani; 6: Croce and McClay

This approach has been successfully applied in sea urchin embryos [5, 9, 10]. Genes encoding transcription factors were identified in the entire sea urchin genome and spatial expression patterns were analyzed for all the regulatory factors expressed at sufficient levels before 24 h post fertilization [14–18]. The subsequent analyses of regulatory interactions between these factors are in turn greatly facilitated by knowing all the possible players. All transcription factors which show spatially specific expression within a particular domain are perturbed one by one, and the resulting consequences for expression of all other regulatory genes in this domain are examined. This approach identifies for each regulatory factor its set of potential target genes. The constellation of regulatory interactions which produce observed perturbation results, and the observed spatial and temporal gene expression pattern, are logically inferred, based on these data. Furthermore, the accuracy of such predictions can be proven by showing for each predicted regulatory interaction that the corresponding transcription factor binding sites for the upstream transcription factor are actually present and functional in the *cis*-regulatory control sequence of the target gene. The more regulatory factors have been involved in the perturbation analyses, the finer will be the resolution of the resulting GRN model. Because GRNs function hierarchically, perturbation of an early upstream transcription factor usually affects the entire process and thus most of the genes which are subsequently expressed in the same domain. However, when such an experiment includes all regulatory genes within a GRN, then it might become clear that the upstream factor directly controls only a few other regulatory genes, while these in turn control a number of other genes, and so on. The sophisticated constellations of regulatory interactions reveal the mechanisms which drive developmental processes, and they provide the causal explanation for the occurrence of these processes.

Fig. 2 The GRN model for specification of endodermal and mesodermal domains in sea urchin embryos. *Differently colored rectangles* indicate individual regulatory domains: *PMC* skeletogenic domain, *Skel* skeletogenic differentiation gene battery, *oral NSM* oral non-skeletogenic mesoderm, *aboral NSM* aboral non-skeletogenic mesoderm, *Small Mic./CP* small micromeres and later coelomic pouch domain, *Oral NSM Diff.* and *Aboral NSM Diff.* respective differentiation gene batteries, *Veg2 Endoderm* Veg2- derived endoderm, future anterior endoderm, *Endo* endodermal differentiation gene battery, *Veg1 Endoderm* Veg1-derived endoderm, future posterior endoderm. The regulatory genes involved in the specification of each domain are shown in the respective rectangles. Regulatory genes are indicated by *horizontal bars* which represent their *cis*-regulatory sequences, where upstream transcription factors bind to control their expression. The *outgoing arrows* from each regulatory gene represent the expressed transcription factor molecules. Regulatory linkages from one regulatory gene to its downstream target genes are represented by *color coded lines ending in arrows* (activation) or *small horizontal bars* (repression). Linkages between two domains are mediated by signaling interactions. Regulatory genes which as a result of a specific repression are not expressed within a domain are shown in *light grey*. For an updated version of this model and for the temporal period during which specific regulatory interactions are active, please visit http://sugp.caltech.edu/endomes/

4 Formation of Spatial Regulatory States in Sea Urchin Embryos

The GRN modeled in Fig. 2 is responsible for setting up the regulatory state domains described above in the developing sea urchin embryo. A few examples will now be discussed in more detail to illuminate how regulatory interactions mediate spatial specification.

4.1 *Function of the Skeletogenic Micromeres in Embryonic Development*

The specification of the skeletogenic cells represents an interesting problem for various reasons. These cells are the earliest cells in the sea urchin embryo to be specified. Specific gene expression in these cells starts at fourth cleavage stage (about 5 h after fertilization), just before these cells become separated from the small micromeres [4]. After that, the descendants of the large micromeres all give rise to skeletogenic cells, cells which will produce the larval skeleton. What is very unusual is that the descendants of these cells, which are formed at 6 h post fertilization, all acquire the same cell fate with no other cells acquiring this same fate in the sea urchin embryo. In addition, these large micromere descendants have a very important role in the specification of other nearby cells. Removing the large micromeres affects the specification of the nearby endodermal domain which in normal embryos will form the larval gut [19]. On the other hand, adding large micromeres to an ectopic site of the embryo results in the formation of a second gut [20, 21]. The large micromeres therefore produce signaling molecules which instruct these adjacent cells to acquire certain cell fates. The GRN which determines the specification of these skeletogenic cells was recently analyzed at the system level and the results provide the causality for each of these observations [5].

4.2 *Control of the Earliest Spatially Restricted Expression: The Double Negative Gate Subcircuit*

The earliest zygotic gene which is specifically expressed in the skeletogenic lineage is *pmar1* [4]. The timing of *pmar1* expression can be explained by the transcriptional control of this gene which does not rely on other zygotically expressed genes but on two maternal factors. Otx and Tcf/β-catenin are transcription factors which accumulate in the nuclei of micromere derived cells. Maternal determinants are responsible for the anisotropic distribution of these molecules [22, 23]. Both transcription factors together are required to activate the expression of *pmar1* in the micromeres [4, 24]. Once Pmar1 transcription factor is present, it does not simply activate additional regulatory genes in the skeletogenic domain. Instead, what Pmar1 does is to repress the expression of a transcription factor which is almost ubiquitously expressed in all cells of the early embryo, except in the skeletogenic cells [25]. This second transcription factor, HesC, is encoded by the only known target gene of Pmar1, and just like Pmar1 it functions as a transcriptional repressor. The multiple target genes of HesC, however, are required for the acquisition of the skeletogenic cell fate. These regulatory genes, which are activated by ubiquitous transcription factors, are expressed exclusively in cells which at this stage do not express HesC, restricting their expression to the Pmar1-expressing micromere descendants.

This subcircuit, which has been termed the double negative gate, is depicted in Fig. 3a. Even Here we see how this tandem repression ultimately leads to activation of the target genes of the second repressor in the spatial domain of the first repressor; this constellation of regulatory interactions has a specific spatial impact. In cells which do not express the first repressor Pmar1, crucial skeletogenic regulatory genes are being repressed by HesC thus preventing skeletogenic cell fate specification throughout all the rest of the embryo. The regulatory logic instructed by this subcircuit is shown in Fig. 3b. Its consequence is that the embryo is first of all divided into two regulatory domains: one in which the first repressor Pmar1 and the target regulatory genes of the subcircuit, the skeletogenic regulatory genes, are expressed, and one in which the second repressor HesC is expressed. Spatially, the double negative gate subcircuit mediates an X/(1-X) logic affecting the entire embryo. For the very early stage in which this subcircuit is active, before other cells are specified, such general regulation seems desirable and is in fact observed in multiple places in the early endomesoderm GRN. Other parts of this GRN are initially controlled by transcription factors which function in response to signaling interactions [6, 9]. Early response transcription factors like Tcf can function as activators in cells receiving a signaling input and as repressors in cells which are not exposed to the signal. Again, the target genes of these transcription factors are regulated in all cells of the embryo, either activated or repressed in an X/(1-X) logic.

4.3 Independent Progression of Skeletogenic Fate Specification in the Entire Skeletogenic Lineage

A few regulatory genes are directly controlled by the double negative gate as shown in Fig. 3a. The particular function of these transcription factors in the skeletogenic GRN is to turn on three additional regulatory genes, *erg*, *hex* and *tgif*. Multiple layers of positive feedback regulatory interactions between these three regulatory genes leads to the stable expression of the skeletogenic regulatory state [5]. Now activation from the upstream transcriptional regulators is no longer required and in fact, skeletogenic *pmar1* expression ceases after only a few hours. No signal from

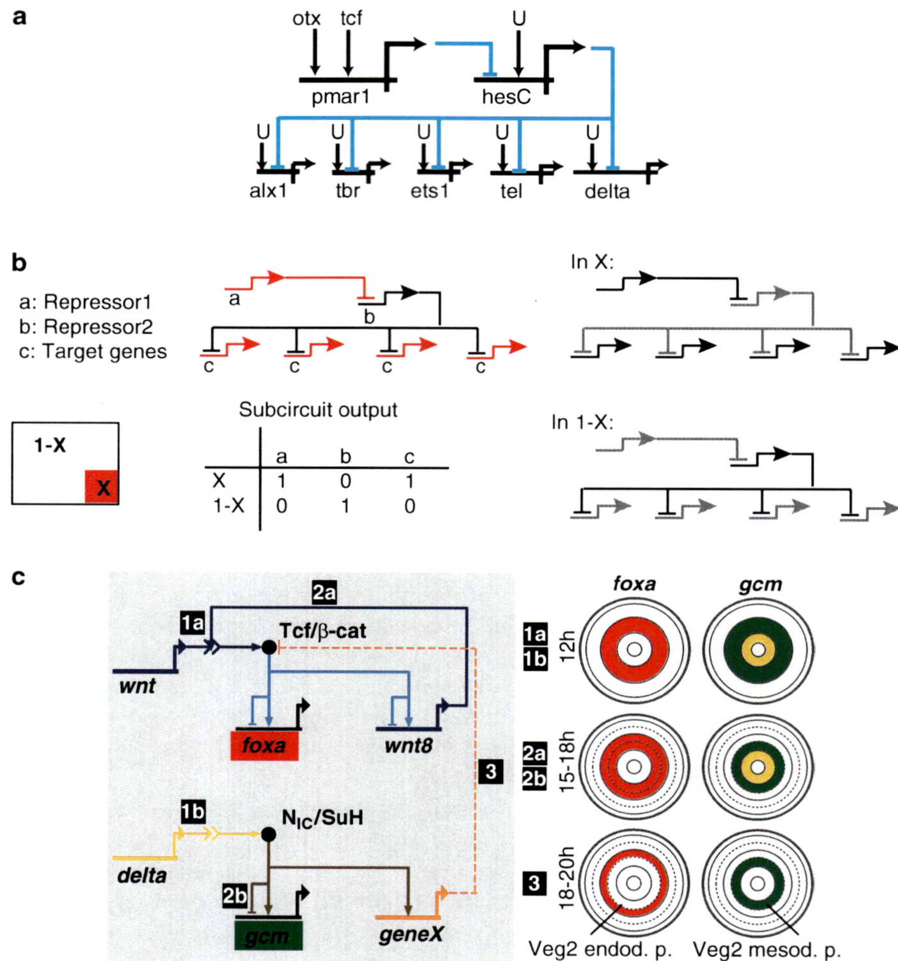

	a	b	c
X	1	0	1
1-X	0	1	0

Fig. 3 Spatial pattern formation by regulatory circuitry logic. (**a**) The double negative gate subcircuit. Expression of *pmar1* is activated by Otx and Tcf/β-catenin in the skeletogenic lineage. Pmar1 represses *hesC* which in turn encodes a repressor controlling the expression of multiple skeletogenic regulatory genes. (**b**) Regulatory logic mediated by the double negative gate: repressor1 (*a*) is expressed in domain X (*red*), in the same domain where the target genes (*c*) are expressed as well. In all other cells of the embryo (1-X; *white/black line*) repressor2 (*b*) is expressed. The specific parts of the subcircuit expressed in X or in 1-X are shown in *black* (expressed) or *grey* (not expressed). The subcircuit output is given in a Boolean activity matrix where gene expression is indicated as 1 (expressed) or 0 (not expressed) (Modified from [11]). (**c**) Subcircuit determining separation of endoderm and mesoderm regulatory domains and schemes of resulting gene expression patterns. Many endodermal regulatory genes, here represented by *foxa*, are controlled by Tcf/β -catenin which is activated by Wnt signaling. The mesodermal regulatory gene *gcm* is controlled by Delta/Notch signaling. Also under the control of Delta/Notch signaling is the predicted geneX which mediates the repression of endodermal regulatory genes, most likely by interfering with β-catenin activity. The resulting spatial

surrounding cells affects skeletogenic cell specification at this stage [5], and all cells expressing first *pmar1* and subsequently the positive feedback subcircuit will go on to become skeletogenic cells. Even if the micromeres are removed from the embryo and cultured in vitro, they continue to function as skeletogenic cells all the way up to differentiation [26].

4.4 Signaling Functions in the Skeletogenic GRN

Removing or transplanting the micromere descendants has dramatic consequences for the rest of the embryo. Cells which will form the mesodermal and endodermal cell types require specific signals which are emitted from the skeletogenic cells. One of these signals is mediated by the signaling ligand Delta. Since it is also controlled by the double negative gate, *delta* expression is specific to the skeletogenic cells between 9 and 18 h [27, 28]. Once the protein is produced, Delta ligand is located on the cellular membrane. Functional interaction with its receptor Notch on the cell surface of adjacent cells leads to the cleavage of the Notch protein, whereupon the intracellular fraction of the protein (N_{IC}) transfers to the nucleus and interacts with the transcription factor suppressor of hairless (Su (H)). Together, N_{IC} and Su(H) activate the transcription of their target genes. If N_{IC} is not available, in the absence of Delta/Notch signaling, Su(H) interacts with a co-repressor molecule, Groucho, and mediates the repression of the very same target genes: again an X/(1-X) control system. Another signaling molecule which is expressed in skeletogenic cells at an early stage is *wnt8*. Its control is independent of the double negative gate and instead is mediated by the transcription factor Tcf [29]. If the Wnt signaling pathway is active, a co-activator molecule, β-catenin, which is usually captured in the cytoplasm and degraded, will instead transfer to the nucleus and interact with Tcf to activate the transcription of their target genes. Again, as we have seen for Su(H), Tcf will form a complex with Groucho if β-catenin is not available and now represses the same target genes that it would have activated in the presence of β-catenin [30]. *Wnt8* gene expression is therefore subject to a positive feedback regulation, encoding a ligand protein which activates the Wnt signaling pathway which in turn activates the Wnt8 gene. Unlike Delta, Wnt8 is a secreted ligand which can diffuse to cells not in immediate contact with the signal producing cells. Similar to *pmar1*, also a Tcf target gene, *wnt8* expression is initially exclusive to the micromere lineage, but its expression spreads out to

Fig. 3 (continued) expression patterns for *foxa* (*red*), *gcm* (*green*) and *delta* (*yellow*) are shown in the embryo schemes in the *right panel*. At 12 h, *gcm* and *foxa* are co-expressed in the 16 veg2 derived cells surrounding the *delta* expressing skeletogenic cells. However, by about 16 h these cells have divided radially and the outer ring of cells have ceased expressing *gcm*. A few hours later, those mesodermal precursor cells expressing *gcm* have ceased to transcribe endodermal regulatory genes, by a mechanism which depends on Delta/Notch signaling (Modified from [10])

adjacent cells by 8 h post fertilization [29]. Both signals affect the specification of nearby cells.

4.5 Induction of Mesodermal Cell Fate

All mesodermal cell types besides the skeletogenic cells derive from cells which at 15–18 h are located right adjacent to the skeletogenic cells. At this stage, all future non-skeletogenic mesodermal cells form one regulatory domain and express one of the important early nodes of the pigment cell specification GRN, the regulatory gene *gcm*. A few hours later, the mesodermal regulatory domain is further split, and *gcm* is now only expressed in a fraction of future mesodermal cells, the ones giving rise to pigment cells [6]. The gcm gene becomes active at about 10 h, in a ring of about 16 cells immediately surrounding the skeletogenic cells. However, at this point, not all descendants of these 16 cells give rise to mesoderm. About half of the cells deriving from these veg2 lineage cells give rise to anterior parts of the gut, which is an endodermal organ. The regulatory state domain marked by *gcm* expression is therefore more and more restricted, at first encompassing cells which form mesoderm and anterior endoderm, then encompassing all non-skeletogenic mesoderm cells, and later encompassing only cells which give rise to pigment cells. The crucial division between cells giving rise to mesodermal cell types and cells acquiring anterior endodermal cell fates occurs as the 16 *gcm* expressing cells divide radially between 12 and 15 h, now generating two rings of about 16 cells each. Only the inner ring of cells continues to express *gcm* and give rise to mesodermal cell types. The reason for this spatial resolution of cell fates lies in the regulatory code of the endodermal and mesodermal GRNs. The regulatory control of the *gcm* gene depends on Delta/Notch signaling. The *cis*-regulatory sequences of *gcm* encode several Su(H) binding sites which mediate both, the activation of *gcm* in cells where the Delta signal is received and repression of *gcm* in cells lacking the signal [6]. The Delta/Notch signal is activated only in cells that are in direct contact with cells expressing the Delta ligand protein on their membranes. Since Notch is provided maternally and is initially present in all cells of the embryo, the spatial activation of Delta/Notch signal depends exclusively on the localization of Delta expressing cells. In the 10 h old sea urchin embryo, the only source of Delta ligand is the skeletogenic lineage. Therefore, when the veg2 lineage divides to form the two rings of cells, only cells remaining in contact with the central skeletogenic cells will maintain *gcm* expression [6, 9, 31]. At this stage, *gcm* definitely marks the regulatory domain which gives rise to all mesodermal cell types.

4.6 Induction of Endodermal Cell Fate

Two additional regulatory state domains form concentric rings around the mesodermal regulatory state domain at 18 h [9]. The cells which descend from the inner ring cells will form the anterior parts of the gut, the foregut (esophagus) and some of the midgut (stomach). The outer ring of cells on the other hand will give rise to cells forming the remainder of the gut, the rest of the midgut and the hindgut (intestine). Even though these gut compartments cannot really be distinguished morphologically before 60 h of development, the two rings of cells in the 18 h embryo already express different regulatory states. The mechanisms distinguishing anterior and posterior endoderm have been discussed elsewhere [9, 10], and we here focus on the mechanisms which initially determine the formation of the anterior or veg2 endoderm regulatory state domain. Most nodes of the anterior endoderm GRN (the "veg2 endoderm" GRN in Fig. 2) contain in their *cis*-regulatory sequence the information to respond to Tcf, the early response transcription factor for Wnt signaling [9, 32, 33]. When the 16 veg2 cell descendants form a ring around the skeletogenic domain, levels of nuclear β-catenin are high in these cells and in consequence, they turn on the expression of endodermal regulatory genes. The high levels of β-catenin are partly a consequence of maternal factors and partly a consequence of Wnt signaling ligands expressed in the skeletogenic lineage. As the veg2 derived cells divide to form the two rings of cells, expression of regulatory genes of the veg2 endoderm GRN initially continues in all veg2 descendants [9].

4.7 Spatial Separation of Endodermal and Mesodermal Cell Fates

Signals emitted from the skeletogenic lineage initially affect adjacent cells simultaneously. Thus by about 10–12 h the single ring of 16 veg2 lineage cells surrounding the skeletogenic cells is exposed to both Wnt and Delta signaling ligands, and due to the wiring of the endodermal and mesodermal GRNs, these cells activate regulatory genes of both specification GRNs. The distinction of endoderm and mesoderm cell fates in cells descending from a common endomesoderm progenitor cell represents a classical problem in developmental biology. At least in the sea urchin embryo, the endomesoderm regulatory state is simply the result of overlapping expression of at least two specification GRNs. No regulatory interactions were observed between transcription factors which participate in these simultaneously expressed but distinct GRNs. However, by 18 h the endodermal and mesodermal cell fates are completely separate. The first step in this process is the radial division of veg2 derived cells into two rings of cells. Whereas β-catenin levels remain high in both rings of veg2 cell descendants, Delta/Notch signaling only continues in the inner ring of cells which are in direct contact with the Delta expressing skeletogenic cells. All veg2 lineage cells not immediately adjacent to skeletogenic cells therefore lose *gcm* expression, and exclusively express the veg2 endodermal regulatory state. The cells which remain in contact with the skeletogenic cells, on the other hand, continue expression of both specification GRNs, until the endoderm GRN becomes repressed in these mesoderm precursor cells. Though the precise mechanism for repression of endoderm fate in mesoderm precursor cells is not yet known, it is clear that this repression depends on Delta/Notch signaling, and that it is mediated by the Tcf sites in the endoderm regulatory genes. The most plausible explanation is that an unknown gene ("Gene X" in Fig. 3c) which is controlled by Delta/Notch signaling, encodes a protein which interferes with nuclear β-catenin. In the absence of β-catenin, as we have seen, Tcf associates with Groucho and acts as a repressor of all the genes which would otherwise be activated, in this case the endodermal regulatory genes. As a result, two completely exclusive regulatory states have arisen in the veg2 lineage by 18 h: one forming the anterior endoderm, the other defining the future non-skeletogenic mesodermal lineages.

5 Concluding Remarks

The spatial organization of the body plan is the outcome of a series of logic operations which are encoded in the genome. Mostly, these logic operations are executed at the level of transcriptional regulation such that different parts of a developing organism display different transcriptional activity. The goal of GRN analysis is to be able to solve these genomic logic operations in molecular terms, to causally relate each regulatory state to its preceding and following regulatory states. Though this field is still fairly young, its insights and promises are likely to become key to the study of development, evolution, genomics, genetics and many other areas in Biology.

Acknowledgement We are pleased to acknowledge support from NIH Grant HD037105.

References

1. Boveri T (1907) Zellenstudien VI. Die Entwicklung dispermer Seeigeleier. Ein Beitrag zur Befruchtungslehre und zur Theorie des Kerns. Gustav Fischer, Jena
2. Laubichler MD, Davidson EH (2008) Boveri's long experiment: sea urchin merogones and the establishment of the role of nuclear chromosomes in development. Dev Biol 314:1–11
3. Ransick A, Rast JP, Minokawa T, Calestani C, Davidson EH (2002) New early zygotic regulators expressed in endomesoderm of sea urchin embryos discovered by differential array hybridization. Dev Biol 246:132–147
4. Oliveri P, Carrick DM, Davidson EH (2002) A regulatory gene network that directs micromere specification in the sea urchin embryo. Dev Biol 246:209–228
5. Oliveri P, Tu Q, Davidson EH (2008) Global regulatory logic for specification of an embryonic cell lineage. Proc Natl Acad Sci U S A 105:5955–5962
6. Ransick A, Davidson EH (2006) cis-regulatory processing of Notch signaling input to the sea urchin glial cells missing gene during mesoderm specification. Dev Biol 297:587–602
7. Cameron RA, Fraser SE, Britten RJ, Davidson EH (1991) Macromere cell fates during sea urchin development. Development 113:1085–1091

8. Ruffins SW, Ettensohn CA (1996) A fate map of the vegetal plate of the sea urchin (*Lytechinus variegatus*) mesenchyme blastula. Development 122:253–263
9. Peter IS, Davidson EH (2010) The endoderm gene regulatory network in sea urchin embryos up to mid-blastula stage. Dev Biol 340:188–199
10. Peter IS, Davidson EH (2011) A gene regulatory network controlling the embryonic specification of endoderm. Nature 474:635–639
11. Peter IS, Davidson EH (2009) Modularity and design principles in the sea urchin embryo gene regulatory network. FEBS Lett 583:3948–3958
12. Howard-Ashby M, Materna SC, Brown CT, Tu Q, Oliveri P, Cameron RA, Davidson EH (2006a) High regulatory gene use in sea urchin embryogenesis: implications for bilaterian development and evolution. Dev Biol 300:27–34
13. Bolouri H, Davidson EH (2003) Transcriptional regulatory cascades in development: initial rates, not steady state, determine network kinetics. Proc Natl Acad Sci U S A 100:9371–9376
14. Howard-Ashby M, Materna SC, Brown CT, Chen L, Cameron RA, Davidson EH (2006b) Identification and characterization of homeobox transcription factor genes in *Strongylocentrotus purpuratus*, and their expression in embryonic development. Dev Biol 300:74–89
15. Howard-Ashby M, Materna SC, Brown CT, Chen L, Cameron RA, Davidson EH (2006c) Gene families encoding transcription factors expressed in early development of *Strongylocentrotus purpuratus*. Dev Biol 300:90–107
16. Materna SC, Howard-Ashby M, Gray RF, Davidson EH (2006) The C2H2 zinc finger genes of *Strongylocentrotus purpuratus* and their expression in embryonic development. Dev Biol 300:108–120
17. Rizzo F, Fernandez-Serra M, Squarzoni P, Archimandritis A, Arnone MI (2006) Identification and developmental expression of the ets gene family in the sea urchin (*Strongylocentrotus purpuratus*). Dev Biol 300:35–48
18. Tu Q, Brown CT, Davidson EH, Oliveri P (2006) Sea urchin Forkhead gene family: phylogeny and embryonic expression. Dev Biol 300:49–62
19. Ransick A, Davidson EH (1995) Micromeres are required for normal vegetal plate specification in sea urchin embryos. Development 121:3215–3222
20. Hoerstadius S (1939) The mechanics of sea urchin development studied by operative methods. Biol Rev Camb Philos Soc 14:132–179
21. Ransick A, Davidson EH (1993) A complete second gut induced by transplanted micromeres in the sea urchin embryo. Science 259:1134–1138
22. Chuang CK, Wikramanayake AH, Mao CA, Li X, Klein WH (1996) Transient appearance of *Strongylocentrotus purpuratus* Otx in micromere nuclei: cytoplasmic retention of SpOtx possibly mediated through an alpha-actinin interaction. Dev Genet 19:231–237
23. Weitzel HE, Illies MR, Byrum CA, Xu R, Wikramanayake AH, Ettensohn CA (2004) Differential stability of beta-catenin along the animal-vegetal axis of the sea urchin embryo mediated by dishevelled. Development 131:2947–2956
24. Smith J, Davidson EH (2009) Regulative recovery in the sea urchin embryo and the stabilizing role of fail-safe gene network wiring. Proc Natl Acad Sci U S A 106:18291–18296
25. Revilla-i-Domingo R, Oliveri P, Davidson EH (2007) A missing link in the sea urchin embryo gene regulatory network: hesC and the double-negative specification of micromeres. Proc Natl Acad Sci U S A 104:12383–12388
26. Okazaki K (1975) Spicule formation by isolated micromeres of the sea urchin embryo. Am Zool 15:567–581
27. Revilla-i-Domingo R, Minokawa T, Davidson EH (2004) R11: a cis-regulatory node of the sea urchin embryo gene network that controls early expression of SpDelta in micromeres. Dev Biol 274:438–451
28. Sweet HC, Gehring M, Ettensohn CA (2002) LvDelta is a mesoderm-inducing signal in the sea urchin embryo and can endow blastomeres with organizer-like properties. Development 129:1945–1955

29. Minokawa T, Wikramanayake AH, Davidson EH (2005) cis-regulatory inputs of the wnt8 gene in the sea urchin endomesoderm network. Dev Biol 288:545–558
30. Range RC, Venuti JM, McClay DR (2005) LvGroucho and nuclear beta-catenin functionally compete for Tcf binding to influence activation of the endomesoderm gene regulatory network in the sea urchin embryo. Dev Biol 279:252–267
31. Croce JC, McClay DR (2010) Dynamics of Delta/Notch signaling on endomesoderm segregation in the sea urchin embryo. Development 137:83–91
32. Ben-Tabou de-Leon SB, Davidson EH (2010) Information processing at the foxa node of the sea urchin endomesoderm specification network. Proc Natl Acad Sci U S A 107:10103–10108
33. Smith J, Kraemer E, Liu H, Theodoris C, Davidson E (2008) A spatially dynamic cohort of regulatory genes in the endomesodermal gene network of the sea urchin embryo. Dev Biol 313:863–875

Part II
Mathematical Models

Modelling Oscillator Synchronisation During Vertebrate Axis Segmentation

Philip J. Murray Philip K. Maini, and Ruth E. Baker

Abstract The somitogenesis clock regulates the periodicity with which somites form in the posterior pre-somitic mesoderm. Whilst cell heterogeneity results in noisy oscillation rates amongst constituent cells, synchrony within the population is maintained as oscillators are entrained via juxtracine signalling mechanisms. Here we consider a population of phase-coupled oscillators and investigate how biologically motivated perturbations to the entrained state can perturb synchrony within the population. We find that the ratio of mitosis length to clock period can influence levels of desynchronisation. Moreover, we observe that random cell movement, and hence change of local neighbourhoods, increases synchronisation.

1 Introduction

Somitogenesis is the process by which the pre-somitic mesoderm (PSM) segments at periodic time intervals into regularly spaced blocks of epithelial cells. The frequency at which new pairs of somites are formed is regulated by the somitogenesis clock, a population of coupled molecular oscillators that is present in the posterior PSM, while a travelling front determines the position of somite formation.

Notch-Delta signalling is a key cell-cell communication mechanism in the PSM [1, 2] and it is thought that one of its roles is to synchronise the oscillations of

P.J. Murray • R.E. Baker
Centre for Mathematical Biology, Mathematical Institute, University of Oxford, 24-29 St Giles', Oxford OX1 3LB, UK
e-mail: murrayp@maths.ox.ac.uk; ruth.baker@maths.ox.ac.uk

P.K. Maini (✉)
Centre for Mathematical Biology, Mathematical Institute, University of Oxford, 24-29 St Giles', Oxford OX1 3LB, UK
e-mail: maini@maths.ox.ac.uk

V. Capasso et al. (eds.), *Pattern Formation in Morphogenesis*, Springer Proceedings in Mathematics 15, DOI 10.1007/978-3-642-20164-6_9,

neighbouring oscillators [3]. However, mutants for different Delta ligands are observed to affect oscillation patterns in the PSM in different ways, suggesting more than the single role of synchronisation in the PSM for Notch-Delta signalling [4, 5], or at the very least a complex interaction between Notch-Delta signalling components and the somitogenesis clock.

Cell movement in the mouse PSM has recently been quantified: there is a posteriorly increasing motility gradient along the anterior-posterior (AP) axis [6]. This random cell motion results in cells changing their neighbours, at least in the posterior tip of the PSM which, in turn, could influence the synchronisation of neighbouring cells. Random cell movement has also been quantified in zebrafish [1], although spatial variation in motility rates along the AP axis has not, to the best of our knowledge, been documented. A recent computational study has indicated that random cell movement should increase synchronisation in a population of locally coupled oscillators [7].

Synchronisation within the PSM is also perturbed by mitosis: when a cell enters M phase of the cell cycle it ceases the transcription necessary to progress its somitogenesis clock [3, 8]. Thus, when the cell divides both mother and daughter cells are out of phase with their neighbours. To the best of our knowledge, the importance of mitotic perturbations in the maintenance of oscillator synchrony has not been quantified.

There is a long history of modelling of somitogenesis (see [9], for a review). Recent models have tended to utilise the increasing knowledge of molecular regulation of the clock (e.g. [10–14]). However, given incomplete knowledge of the governing molecular networks, the seemingly intricate relationship between oscillator coupling and the molecular clocks, and the difficulty in unambiguously parameterising the current models, coarse-grained descriptions of molecular oscillators have recently been considered in which progression through the somitogenesis clock cycle is described by a single variable – the oscillator phase [15–17].

In this study we consider a phase description of a population of locally coupled oscillators and investigate how population synchrony is perturbed by both cell movement and mitosis. The layout is as follows: in Sect. 2 we introduce a spatially distributed population of coupled oscillators and describe how synchrony can be measured in the population; in Sect. 3 we introduce a perturbation representing cell proliferation and examine the effects on oscillator synchrony; in Sect. 4 we allow random movement within the population and again consider the effects on oscillator synchrony; and, finally, in Sect. 5 we conclude with a discussion on how these results are relevant to our understanding of somitogenesis.

2 A Simple Model for Cell Synchronisation in the PSM

As we are primarily interested in studying the synchronisation of oscillators in the context of the PSM, we consider a cellular automation (CA) of size 30×6 cells, which is approximately of the same proportion as the zebrafish PSM. We define

$\theta_i(t)$ to be the phase of the oscillator at time t and assume that the governing phase dynamics are given by

$$\dot{\theta}_i = \omega + \sum_j A\sin(\theta_j - \theta_i); \; i = 1 : N, \tag{1}$$

where A is the strength of diffusive coupling, ω is the oscillator frequency, the sum is taken over nearest neighbours on the CA lattice and N is the number of cells in the simulation. We note that the phase description of a population of coupled oscillators can be derived from an underlying molecular description with the sinusoidal coupling term representing the leading order term in a more general coupling function [18].

In order to quantify synchronisation in the oscillator population we define the global order parameters $r(t)$ and $\Psi(t)$ [18] such that

$$re^{i\Psi} = \frac{1}{N}\sum_{j=1}^{N} e^{i\theta_j}. \tag{2}$$

When $r = 0$ the oscillators are completely out of phase while $r = 1$ corresponds to global synchronisation. The variable Ψ represents the mean phase of the population so when $r = 1$, $\Psi = \omega$.

2.1 Results

In order to investigate perturbations from the spatially homogeneous phase distribution typically observed in the posterior PSM, we simulated populations of oscillators with phase dynamics given by (1) in which the initial phase of each oscillator was uniformly selected from the range $[0, \chi]$. Typically we choose $\chi \sim 1$ to represent a small initial perturbation from global synchrony. In Fig. 1 we present snapshots of phase distributions at $t = 0$ and $t = 3$ min, respectively.

The well-studied effect of nearest-neighbour sinusoidal coupling is to synchronise initially out-of-phase oscillators. As $t \to \infty$, the phase differences between oscillators tends to zero, and each oscillator oscillates with natural frequency, ω. Calculating the order parameter (see Fig. 2) allows us to quantitatively measure global synchrony in the population as time evolves. As an aside we note that it may be feasible in the future to track in real time the phase of individual oscillators using reporter genes. If so, the order parameter could be measured by relating the activity of the reporter to a nominally defined 'phase' and using (2) to measure global synchrony. The effects of perturbations to global synchrony, such as the qualitative experiments performed by Horikawa et al. [3], could then be measured experimentally and the resulting order parameter data could be directly compared with theoretical models, such as (1).

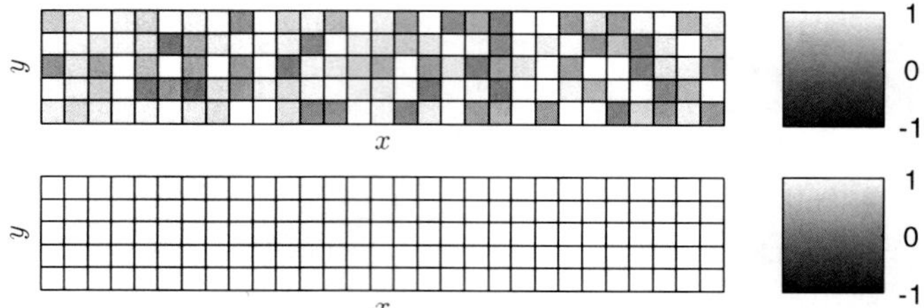

Fig. 1 $\text{Sin}(\theta_i)$ is plotted for each of the N cells on the CA lattice. *Top*: the initial phase distribution in a typical simulation ($\chi = 1$). *Bottom*: the spatially uniform phase distribution at $t = 3$. Cell phases on the CA lattice were updated using (1). Parameter values as in Table 1

Table 1 Parameters used in the simulations

Parameter	Value	Unit	Source
A	2.2	min^{-1}	Murray et al. [16][a]
ω	0.2	min^{-1}	Giudicelli et al. [19]
T	240	min	Kane [20]
M	>15	min	Horikawa et al. [3]

[a] Parameter estimated from previous model

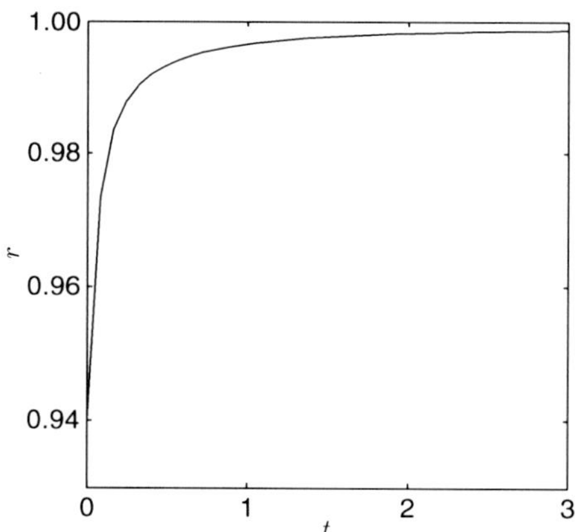

Fig. 2 The order parameter, r, is plotted against time, t. Cell phases on the CA lattice were updated using (1) and the order parameter was calculated using (2)

3 Investigating the Perturbation to Oscillator Synchronisation Arising from Cell Mitosis

Our goal in this study is to investigate how biologically-motivated perturbations to a population of sinusoidally coupled phase oscillators can influence phase synchronisation. We now examine if perturbations arising from mitosis can significantly destabilise oscillator synchrony within physiologically relevant parameter ranges.

3.1 *Model Development*

Using zebrafish, Horikawa et al. [3] observed that cells which were either in, or had just recently exited, M phase of the cell cycle tended to be out of phase with neighbouring oscillators in the posterior PSM. They concluded that during M phase of the cell cycle the transcription necessary to progress the somitogenesis clock was paused as the cells prepared for division, thus dividing cells drifted out of phase with their neighbours. After M phase the cells re-initiated the transcription necessary for somitogenesis clock progression, and, via cell-cell coupling, resynchronised with their neighbours. It was estimated that over a time observation window of one somitogenesis clock cycle (~30 min), 10–15% of cells had either undergone or were undergoing M phase of the cell cycle, which was thought to last for a minimum of 15 min.

In order to study the effect of this phenomenon on a two-dimensional lattice, we consider a modified equation for the phase dynamics given by

$$\dot{\theta}_i = \left(\omega + \sum_j A\sin(\theta_j - \theta_i)\right) H(T - M - a_i), \tag{3}$$

where a_i is the age of the i th cell, T the cell cycle period, $H(.)$ represents the Heaviside function and M represents the time spent in M phase of the cell cycle. Cell age increases proportionally with time and is reset to zero when $t = T$. When a cell's age lies in the range $a_i \in [T - M, T]$ it is in M phase of the cell cycle and does not update its somitogenesis clock. We note that when a cell reaches the end of the cell cycle we do *not* introduce a daughter cell into the simulation. This simplification allows us to investigate computationally the effect of mitosis on phase synchronisation, without having to account for processes such as growth and apoptosis and the many more parameters they would introduce into our model.

3.2 *Results*

We performed a range of simulations with different values of the parameters A and M. The initial phase distributions were as described in Sect. 2.1 while the initial cell cycle position for a given cell was chosen randomly from the uniform distribution $[0, T]$. In Fig. 3a the phase distribution is plotted at $t = 240$ while in Fig. 3b cells in M phase of the cell cycle are highlighted. Note the correlation between a cell being in M phase and being out of phase with its neighbours. The result of the mitosis-

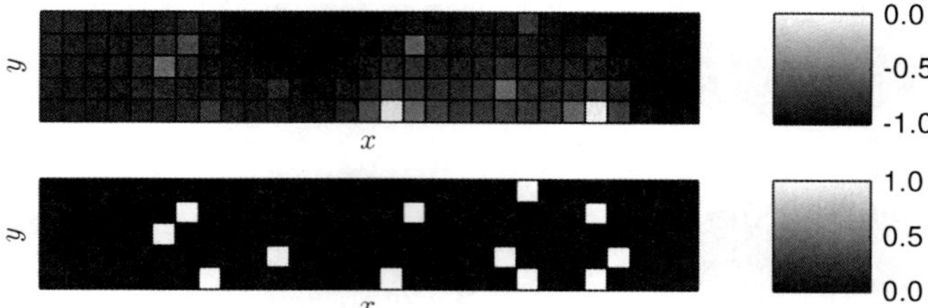

Fig. 3 Phase (*top*) and cell cycle (*bottom*) distributions at $t = 4$ h. Cell phases on the CA lattice were updated using (3)

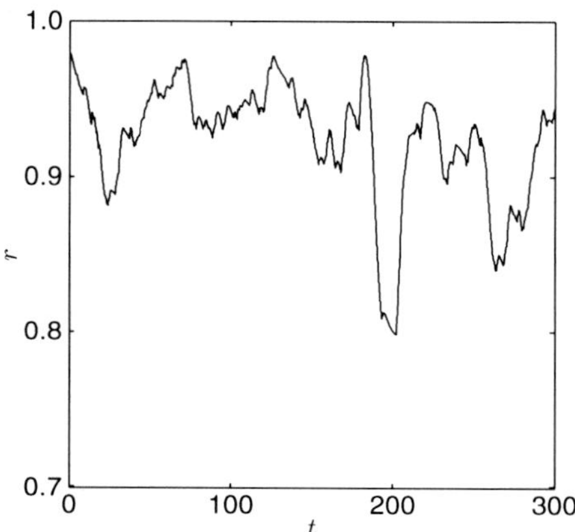

Fig. 4 The order parameter, r is plotted against time. Cell phases on the CA lattice were updated using (3) and the order parameter was calculated using (2). Parameter values as in Table 1

induced clock-stopping perturbation is that the cell population does not attain global synchrony (see Fig. 4).

In Fig. 5 we plot the time-averaged global order parameter for a range of different values of the parameters A and M. For a fixed value of the parameter A, the global order parameter firstly decreases and subsequently increases with increasing M. This biphasic behaviour can be explained by consideration of the period of the somitogenesis clock (~30 min) and the length of time spent in M phase (~15 min). As $M \to 0$ the somitogenesis clock is paused for increasingly shorter amounts of time and the perturbation to global synchrony is negligible. As M increases, dividing cells pause their somitogenesis clocks for longer times, and thus move out of phase with neighbouring cells; hence global synchrony decreases. However, as M increases further neighbouring non-dividing cells are almost a full somitogenesis clock cycle ahead of the cell that entered M phase when it returns to updating its somitogenesis clock. Thus the order parameter increases for larger values of M. The simulation results suggest that there is an optimal (in terms of phase desynchronisation) ratio of the parameters M and ω. One might

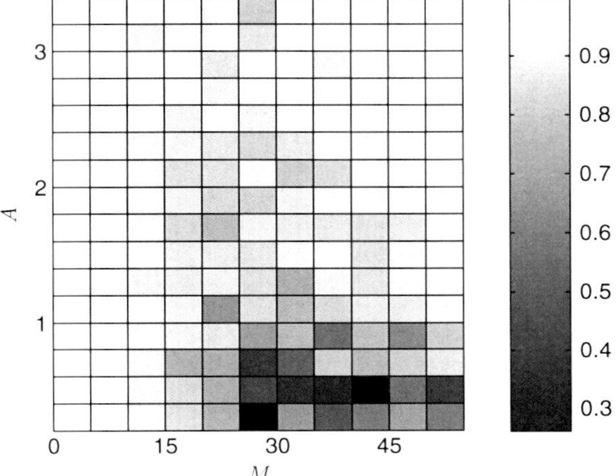

Fig. 5 The time-averaged order parameter plotted against A and M. For different parameter values cell phases on the CA lattice were updated using (3) and the time-averaged order parameter was calculated by averaging over (2)

expect the optimal ratio would be such that the time spent in mitosis was half the somitogenesis clock period. However, the simulation results suggest that the optimal length of M phase is longer than half the somitogenesis period. This occurs because when an oscillator pauses its somitogenesis clock for M phase of the cycle, its neighbours, which are updating their phase, continue to be influenced by it and the effect of the growing phase difference is to slow their effective oscillation rate.

4 Investigating the Perturbation to Oscillator Synchronisation Arising from Random Cell Movement

Bénazéraf et al. [6] have recently quantified random cell motion in the mouse PSM while, in a theoretical study, Uriu et al. [7] have demonstrated that random cell movement enhances synchronisation of neighbouring molecular oscillators. We now investigate the effects of random cell movement on the mitosis-perturbed phase dynamics introduced in Sect. 3.

4.1 Model Development

In order to simulate random cell movement on the CA lattice, at each time step of a simulation we make N random selections of a cell from the population and perform a swap with a randomly chosen neighbour with probability p. The phase dynamics are updated using (3). Thus the cellular automaton simulations are used to investigate the interaction between three key processes in the PSM: (a) phase oscillations and synchronisation; (b) perturbation to synchronisation due to mitosis; and (c) random cell movement.

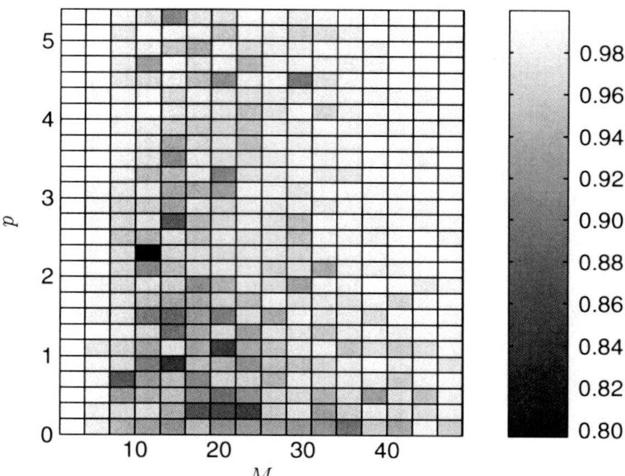

Fig. 6 The time-averaged order parameter plotted against p and M. For different parameter values cell phases on the CA lattice were updated using (3) and the time-averaged order parameter was calculated by averaging over (2). Cells were allowed to randomly swap positions with their neighbours with probability p

4.2 Results

In order to quantify the effects of random cell movement on global synchrony, we performed simulations in biologically relevant regions of parameter space and measured the time averaged global order parameter. Firstly, we held the coupling strength, A, fixed and varied the random cell movement and the mitosis-induced perturbation considered in the previous section. For small values of p, and hence little random movement, the order parameter has a minimum, as in Fig. 5, for some intermediate value of M (see Fig. 6). However, as p increases and cells interchange neighbours more frequently, global synchronisation in the simulations increases. Thus random cell movement can counteract the reduction in order parameter resulting from the perturbation due to mitosis.

Next we held the time spent in mitosis fixed at $M = 15$ min, the lower bound reported by Horikawa et al. [3], and examined how global synchrony in the simulation varied with the coupling strength and random movement (see Fig. 7). The order parameter is clearly, on average, lower in the bottom left corner of the figure, i.e. when the coupling strength and random cell movement rate are low. A low coupling strength can be compensated for by increasing the rate of random movement and *vice-versa*. This result is of particular interest as the two mechanisms enhancing synchrony in the simulations represent two completely different biological mechanisms: the coupling strength A represents the effect of biochemical cell-cell signalling while the parameter p represents a mechanical interaction. In zebrafish cell-cell coupling is strongly dependent on the Notch-Delta signalling pathway while in chick it has been shown that cell motility is strongly influenced by an Fgf gradient along the AP axis [6].

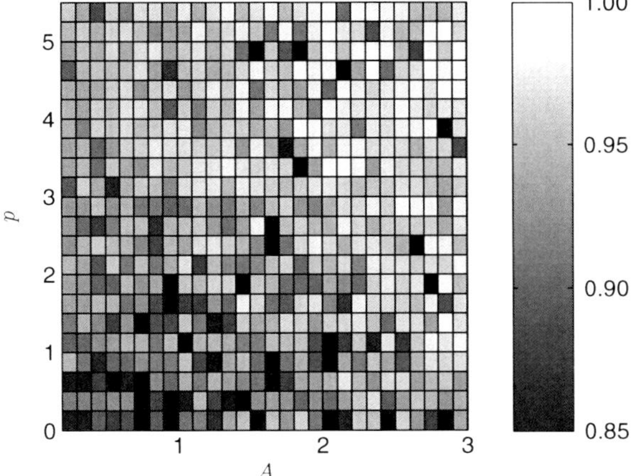

Fig. 7 The time-averaged order parameter plotted against A and p. For different parameter values cell phases on the CA lattice were updated using (3) and the time-averaged order parameter was calculated by averaging over (2). Cells were allowed to randomly swap positions with their neighbours with probability p

5 Discussion

The phenomenon of oscillator synchronisation plays an important role in the posterior PSM during somitogenesis: as a result of cell heterogeneity neighbouring cells have different natural frequencies and without some coupling mechanism the oscillators would drift out of phase. Our knowledge of the molecular interactions governing the coupling interactions is ever increasing, however, with so many molecular components at play, and additional complications, such as the influence of the coupling on the clock frequency itself, it is challenging to construct a predictive molecular model. As such, in this study we have considered a coarse-grained model of the coupling mechanism in which it is assumed that oscillators can be adequately described by their relative phases.

By considering a relatively simple mechanism of coupling we can probe how biologically motivated perturbations can influence synchrony in an environment such as the PSM. For example, we consider the influence of mitosis, where cells temporarily pause their somitogenesis clock, on synchrony and demonstrate that the ratio of time spent in mitosis to the somitogenesis period can influence global synchrony in a physiologically relevant parameter regime.

There is a posteriorly increasing cell motility gradient along the AP axis in chick and it has previously been suggested that random cell motion, and hence exchange of local neighbours, could play an important role in maintaining synchronisation in the posterior PSM. We introduced cell motion into the posterior PSM by allowing cells to swap positions with their nearest neighbours with some probability, p, which can be related to measurements of mean-squared displacements of cells in the posterior PSM. Our simulation results indicate that neighbour swapping does indeed promote synchrony in the PSM.

In this study the local coupling of neighbouring oscillators is modelled using an on-lattice CA framework. Whilst the CA was chosen as it allows us to trivially simulate random cell movements via neighbour exchanges, with the degree of mixing within the population quantified by a single parameter, we note that alternative frameworks, such as the cellular Potts, vertex or off-lattice models, could be used to investigate the effects of mitosis block and random movement on oscillator synchronisation. While we expect that the simulation results are independent of the particular cell-level model under consideration, this touches on the general issues of what is the appropriate cell-based framework in which to model such phenomena and do different frameworks give different results? This problem is difficult to investigate as one cannot always relate parameters in one model to those in another [21]. However, we note that in certain cases it is possible to compare coarsened versions of different models (e.g. [22]).

The theory of phase-coupled oscillators has been applied to a range of physical phenomena. A ubiquitous feature of spatially coupled phase oscillators is the existence of rich dynamical behaviours, such as spiral waves. In this study, motivated by the observation of spatially synchronous oscillations in the posterior PSM, we have considered perturbations about the spatially synchronous solutions and examined how various biologically motivated perturbations can perturb the steady state. This approach allows us to focus on the effect of the perturbations rather than concern ourselves with the more complex, but not observed in the PSM, dynamical solutions to the phase equations.

Acknowledgement PJM and REB acknowledge the support of the Engineering and Physical Sciences Research Council through an EPSRC First Grant to REB. PKM was partially supported by a Royal Society-Wolfson Research Merit Award.

References

1. Jiang YJ, Aerne BL, Smithers L, Haddon C, Ish-Horowicz D, Lewis J (2000) Notch signalling and the synchronization of the somite segmentation clock. Nature 408(6811):475–479
2. Rida PCG, Le Minh N, Jiang YJ (2004) A Notch feeling of somite segmentation and beyond. Dev Biol 265(1):2–22
3. Horikawa K, Ishimatsu K, Yoshimoto E, Kondo S, Takeda H (2006) Noise-resistant and synchronized oscillation of the segmentation clock. Nature 441(7094):719–723
4. Mara A, Schroeder J, Chalouni C, Holley SA (2007) Priming, initiation and synchronization of the segmentation clock by delta D and delta C. Nat Cell Biol 9(5):523–530
5. Mara A, Schroeder J, Holley SA (2008) Two delta C splice-variants have distinct signaling abilities during somitogenesis and midline patterning. Dev Biol 318(1):126–132
6. Bénazéraf B, Francois P, Baker RE, Denans N, Little CD, Pourquié O (2010) A random cell motility gradient downstream of FGF controls elongation of an amniote embryo. Nature 466(7303):248–252

7. Uriu K, Morishita Y, Iwasa Y (2010) Random cell movement promotes synchronization of the segmentation clock. Proc Natl Acad Sci U S A 107(11):4979
8. Mara A, Holley SA (2007) Oscillators and the emergence of tissue organization during zebrafish somitogenesis. Trends Cell Biol 17(12):593–599
9. Baker RE, Schnell S, Maini PK (2008) Mathematical models for somite formation. Curr Top Dev Biol Vol. 81 Chapter 6, 183–203
10. Goldbeter A, Pourquié O (2008) Modeling the segmentation clock as a network of coupled oscillations in the Notch. Wnt and FGF signaling pathways. J Theor Biol 252(3):574–585
11. Hirata H, Yoshiura S, Ohtsuka T, Bessho Y, Harada T, Yoshikawa K, Kageyama R (2002) Oscillatory expression of the bHLH factor Hes1 regulated by a negative feedback loop. Science 298(5594):840–843
12. Lewis J (2003) Autoinhibition with transcriptional delay: a simple mechanism for the zebrafish somitogenesis oscillator. Curr Biol 13(16):1398–1408
13. Rodríguez-González JG, Santillán M, Fowler AC, Mackey MC (2007) The segmentation clock in mice: interaction between the Wnt and Notch signaling pathways. J Theor Biol 248(1):37–47
14. Santillan M, Mackey MC (2008) A proposed mechanism for the interaction of the segmentation clock and the determination front in somitogenesis. PLoS One 3(2):e1561
15. Morelli LG, Ares S, Herrgen L, Schröter C, Jülicher F, Oates AC (2009) Delayed coupling theory of vertebrate segmentation. HFSP J 1(3):55–66
16. Murray PJ, Maini PK, Baker RE (2011) The clock and wavefront model revisited. J Theor Biol 283(1):227–238
17. Riedel-Kruse IH, Muller C, Oates AC (2007) Synchrony dynamics during initiation, failure, and rescue of the segmentation clock. Science 317:1911–1915
18. Kuramoto Y (1981) Rhythms and turbulence in populations of chemical oscillators. Physica A 106(1–2):128–143
19. Giudicelli F, Özbudak EM, Wright GJ, Lewis J (2007) Setting the tempo in development: an investigation of the zebrafish somite clock mechanism. PLoS Biol 5(6):e150
20. Kane DA (1998) Cell cycles and development in the embryonic zebrafish. Methods Cell Biol 59:11–26
21. Osborne JM, Walter A, Kershaw SK, Mirams GR, Fletcher AG, Pathmanathan P, Gavaghan D, Jensen OE, Maini PK, Byrne HM (2010) A hybrid approach to multiscale modelling of cancer. Phil Trans 368, 5013–5028
22. Murray PJ, Edwards CM, Tindall MJ, Maini PK (2009) From a discrete to a continuum model of cell dynamics in one dimension. Phys Rev E 80(3):031912-1–031912-10

Pattern Formation in Hybrid Models of Cell Populations

N. Bessonov, P. Kurbatova, and V. Volpert

Abstract The paper is devoted to hybrid discrete-continuous models of cell populations dynamics. Cells are considered as individual objects which can divide, die by apoptosis, differentiate and move under external forces. Intra-cellular regulatory networks are described by ordinary differential equations while extracellular species by partial differential equations. We illustrate the application of this approach to some model examples and to the problem of tumor growth. Hybrid models of cell populations present an interesting nonlinear dynamics which is not observed for the conventional continuous models.

1 Discrete-Continuous Models of Cell Populations

Pattern formation in biology attracts much attention since D'Arcy Thompson [1]. An important development was related to Turing structures and to their applications to various biological problems (see, e.g., [2, 3]). Other mechanisms of patterns formation, such as morphogenetic gradients, chemotaxis, or mechano-chemistry are also widely discussed [2, 4–7].

In this work we do not intend to describe a specific biological mechanism of pattern formation but rather to present a class of mathematical models which can be used to describe various mechanisms of pattern formation. These can be the mechanisms shortly indicated above but also some other mechanisms specific for the models under consideration.

N. Bessonov (✉)
Institute of Mechanical Engineering Problems, Saint Petersburg 199178, Russia
e-mail: nickbessonov@yahoo.com

P. Kurbatova • V. Volpert
Institut Camille Jordan, University Lyon 1, UMR 5208 CNRS, Villeurbanne 69622, France
e-mail: kurbatova@math.univ-lyon1.fr; volpert@math.univ-lyon1.fr

V. Capasso et al. (eds.), *Pattern Formation in Morphogenesis*, Springer Proceedings in Mathematics 15, DOI 10.1007/978-3-642-20164-6_10,

We will study cell population dynamics with so-called hybrid models where cells are considered as discrete or individual based objects while intra-cellular and extra-cellular biochemical substances are described with continuous models. It is in this sense that we understand here hybrid models. Hybrid models can be based on cellular automata or various lattice or off-lattice approaches (see [8–11] and the references therein).

Cells can interact with each other and with the surrounding medium mechanically and biochemically, they can divide, differentiate and die due to apoptosis. Cell behavior is determined by intra-cellular regulatory networks described by ordinary differential equations and by extra-cellular bio-chemical substances described by partial differential equations.

In order to describe mechanical interaction between cells, we restrict ourselves here to the simplest model where cells are represented as elastic balls. Consider two elastic balls with the centers at the points x_1 and x_2 and with the radii, respectively, r_1 and r_2. If the distance d_{12} between the centers is less then the sum of the radii, $r_1 + r_2$, then there is a repulsive force f_{12} between them which depends on the distance. If a particle with the center at x_i is surrounded by several other particles with the centers at the points x_j, $j = 1, \ldots, k$, then we consider the pairwise forces f_{ij} assuming that they are independent of each other. This assumption corresponds to small deformation of the particles. Hence, we find the total force F_i acting on the i-th particle from all other particles, $F_i = \sum_{j \neq i} f_{ij}$. The motion of the particles can now be described as the motion of their centers. By Newton's second law

$$m\ddot{x}_i + \mu m \dot{x}_i - \sum_{j \neq i} f\left(d_{ij}\right) = 0, \tag{1}$$

where m is the mass of the particle, the second term in the left-hand side describes the friction by the surrounding medium. Dissipative forces can also be written in a different form. This is related to dissipative particle dynamics [12].

Intra-cellular regulatory networks for the i-th cell are described by a system of ordinary differential equations

$$\frac{du_i}{dt} = F(u_i, u), \tag{2}$$

where u_i is a vector of intra-cellular concentrations, u is a vector of extra-cellular concentrations, F is the vector of reaction rates which should be specified for each particular application. The concentrations of the species in the extra-cellular matrix is described by the diffusion equation

$$\frac{\partial u}{\partial t} = D\ \Delta u + G\ (u, c), \tag{3}$$

where c is the local cell density, G is the rate of consumption or production of these substances by cells. These species can be either nutrients coming from outside and consumed by cells or some other bio-chemical products consumed or produced by cells. In particular, these can be hormones or other signaling molecules that can influence intra-cellular regulatory networks. In some cases, convective motion of the medium should be taken into account. We do not discuss here various details of this model related to cell division, the force f_{ji} and cell displacement, relation between discrete and continuous models of cell population, more complex cell geometry, influence of stochasticity and so on (see [13–15] for more details).

In this work we restrict ourselves to some model examples in order to illustrate nonlinear dynamics of hybrid models. In the next section, we consider the 1D case. We present three model examples which describe the competition between cell proliferation and apoptosis, between proliferation and differentiation, and between cell cycling and quiescent state. These examples are important from the biological point of view. In Sect. 3, we return to one of them in the 2D case in order to show that behavior of the system can depend on the space dimension. We also consider a model of tumor growth where cells compete for nutrients. Three different regimes of tumor growth are observed. Application of hybrid models to some particular biological processes with their specific intracellular and extracellular regulation will be published in the subsequent works.

2 1D Models

2.1 Model Examples

2.1.1 Self-Renewal and Apoptosis

We begin with the 1D model example where cells can move along the line. The coordinates x_i in (1) are real numbers. Each cell can divide or die by apoptosis. After division a cell gives two cells identical to itself. We suppose that cell division and death are influenced by some bio-chemical substances produced by the cells themselves. We consider the case where there are two such substances, u and v. We come to the following system of equations:

$$\begin{cases} \dfrac{du}{dt} = d_1 \dfrac{\partial^2 u}{\partial x^2} + b_1 c - q_1 u, \\ \dfrac{dv}{dt} = d_2 \dfrac{\partial^2 v}{\partial x^2} + b_2 c - q_2 v. \end{cases} \tag{4}$$

These equations describe the evolution of the extracellular concentrations u and v with their diffusion, production terms proportional to the concentration of cells c and with the degradation terms. We note that cells are considered here as point sources with a given rate of production of u and v. The cell concentration is understood as a number of such sources in a unit volume. In numerical simulations, where cells have a finite size, we consider them as distributed sources and specify the production rate for each node of the numerical mesh.

Intra-cellular concentrations u_i and v_i in the i-th cell are described by the equations:

$$\begin{cases} \dfrac{du_i}{dt} = k_1^{(1)} u(x_i, t) - k_2^{(1)} u_i(t) + H_1, \\ \dfrac{dv_i}{dt} = k_1^{(2)} v(x_i, t) - k_2^{(2)} v_i(t) + H_2. \end{cases} \tag{5}$$

Here and in what follows we write equations for intra-cellular concentrations neglecting the change of the cell volume. This approximation is justified since the volume changes only twice before cell division and this change is relatively slow. The first term in the right-hand side of the first equation shows that the intra-cellular concentration grows proportionally to the value of the extra-cellular concentration $u(x, t)$ at the space point x_i where the cell is located. It is similar for the second equation. These equations contain the degradation terms and constant production terms, H_1 and H_2. When a new cell appears, we put the concentrations u_i and v_i equal zero.

If the concentration u_i attains some critical value u_c, then the cell divides. If v_i becomes equal v_c, the cell dies. Consider first the case where $k_1^{(1)} = k_1^{(2)} = k_2^{(1)} = k_2^{(1)} = 0$. Then u_i and v_i are linear functions of time which reach their critical values at some times $t = \tau_u$ and $t = \tau_v$, respectively. If $\tau_u < \tau_v$, then all cells will divide with a given frequency, if the inequality is opposite, then all cells will die.

Next, consider the case where $k_1^{(1)}$ is different from zero. If it is positive, then cells stimulate proliferation of the surrounding cells, if it is negative, they suppress it. Both cases can be observed experimentally. We restrict ourselves here by the example of negative $k_1^{(1)}$. All other coefficients remain zero. Therefore, cells have a fixed life time τ_v. If they do not divide during this time, they die.

We carry out the 1D simulation where cells can move along the straight line. Initially, there are two cells in the middle of the interval. Figure 1 shows the evolution of this population in time. For each moment of time (*vertical axis*) we have the positions of cells (*horizontal axis*) indicated with blue points.

The evolution of the cell population in Fig. 1 (*left*) can be characterized by two main properties. First of all, it expands to the left and to the right with approximately constant speed. Second, the total population consists of relatively small sub-populations. Each of them starts from a small number of cells. Usually, these are two cells at the right and at the left of the previous sub-population. During some time, the sub-population grows, reaches certain size and disappears giving birth to new sub-populations.

This behavior can be explained as follows. The characteristic time of cell division is less than of cell death. When the cell sub-population is small, the quantity of u is also small, and its influence on cell division is not significant. When the sub-population becomes larger, it slows down cell division because of growth of u. As a result the sub-population disappears. The outer cells can survive because the level of u there is less.

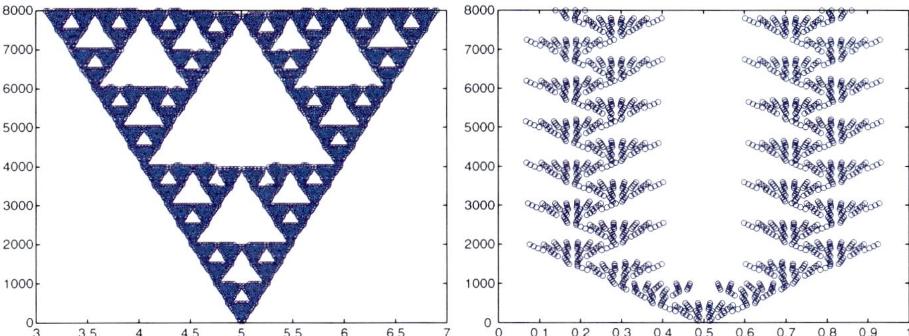

Fig. 1 Dynamics of cell population in the case where cells either self-renew or die by apoptosis. Cells are shown with *blue dots*. *Horizontal axis* shows cell position, *vertical axis* – time

The geometrical pattern of cell distribution for these values of parameters reminds Serpinsky carpet (Fig. 1 *left*), an example of fractal sets. It is obtained from an equilateral triangle by consecutive removing its parts. At the first step, the central triangle is removed. At the second step, the central part of each remaining triangle is removed and so on. In the case of the cellular pattern, the minimal triangle is determined by the cell size. Since this patterns grows with time, we can scale it to a triangle of a constant size. In this case, we obtain a decreasing sequence of removed (*white*) triangles, as it is the case for Serpinsky carpet. Thus, we observe here propagation of chaotic (*fractal*) waves.

The pattern of cell distribution depends on the parameters. Another example is shown in Fig. 1. The cell populations in Fig. 1 (*right*) remain bounded, and the pattern is time periodic.

The simulations presented here do not use the extra-cellular variable v. Instead of the variable u, which decelerate cell proliferation, we can consider assuming that it accelerates cell apoptosis. In this case, qualitative behavior of cell population is similar.

2.1.2 Self-Renewal and Differentiation

We consider the same systems (4) and (5) for the extra-cellular and intra-cellular concentrations. Instead of apoptosis, where dead cells are removed from the computational domain we consider here cell differentiation. When v_i becomes equal to v_c, cells differentiate into another type of cells. New cells remain in the computational domain, they interact mechanically with each other and with other cells but they do not grow or divide.

Evolution of the cell population depending on the critical value of v_c is observed. The simulations begin with two cells at the center of the interval. If v_c is sufficiently small, the most part of cells differentiate. For larger values of v_c, there is a central part of the domain filled by differentiated cells (Fig. 2, *left*) and additional groups of

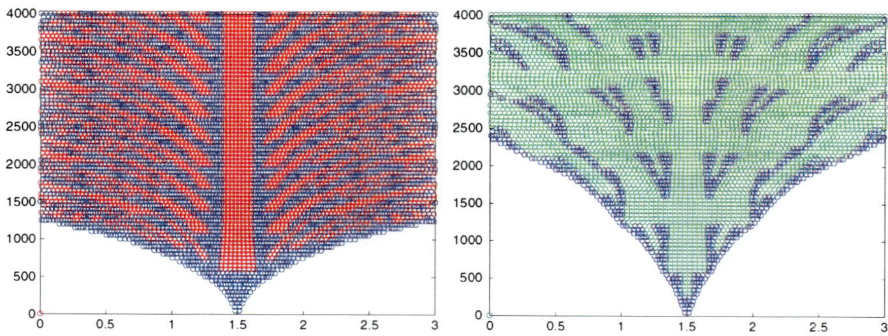

Fig. 2 *Left*: cell population dynamics in the case of self-renewal and differentiation. Self-renewing cells are shown in *blue*, differentiated cells in *red*. *Right*: cell population dynamics with dividing cells (*blue*) and quiescent cells (*green*)

differentiated cells which periodically appear from the left and from the right. They move to the borders of the interval and leave it. It should be noted that the size of the domain filled by differentiated cells in the center of the interval depends on v_c. Moreover, this dependence is not monotone. For v_c sufficiently large differentiated cells are located only at the center of the interval.

The mechanisms of pattern formation considered in this and in the previous sections seem to be biologically realistic. Indeed, cell proliferation, differentiation and apoptosis are determined by concentrations of intracellular proteins. Among many examples of extracellular regulation, we can cite density dependence of cell proliferation (see [16] and the references therein) where cells downregulate proliferation of surrounding cells. The choice between differentiation and proliferation in bacteria filament [17] is another interesting example. It corresponds to 1D modeling discussed in this section. Model problems considered here illustrate some basic properties of hybrid models. We use these approaches with more specific mechanisms of intracellular and extracellular regulation in order to study normal and leukemic hematopoiesis. These results will be published elsewhere.

The mechanisms of patterns formation in the hybrid models are different in comparison with Turing structures, morphogenetic gradients or mechano-chemistry. Formation of the domain filled by differentiated cells described above can serve, at the subsequent stages of morphogenesis, as a source of morphogenetic gradients.

2.1.3 Cell Cycle and Quiescent Cells

Let us consider the model where cells can be either in quiescent state or in cell cycle. When the cell is in cell cycle it grows and then divides into two identical to itself cells. Cells can leave cell cycle and remain quiescent and can return to cell cycle. We suppose that cells which are in the process of division produce some substance denoted by u such that it delays proliferation of the surrounding cells. At the same time, it influences quiescent cells delaying its return to cell cycle.

Thus, the role of this substance is to control the number of dividing cells. Its distribution is described by the equation

$$\frac{du}{dt} = d\frac{\partial^2 u}{\partial x^2} + bc_1 - qu, \tag{6}$$

where c_1 is concentration of cells in division. Intra-cellular regulatory networks for the i-th cell in cell cycle are described by the following system

$$\begin{cases} \dfrac{du_i}{dt} = k_1^{(u)} u(x_i, t) - k_2^{(u)} u_i(t) + H_u, \\ \dfrac{dv_i}{dt} = k_1^{(v)} v(x_i, t) - k_2^{(v)} v_i(t) + H_v. \end{cases} \tag{7}$$

When the value of v overcomes v_c, critical value of v_i, the cell enters the quiescent state.

Denote by p_i the intra-cellular substance which determines cell transition from quiescent state into cell cycle,

$$\frac{dp_i}{dt} = k_1^{(p)} u(x_i, t) - k_2^{(p)} p_i(t) + H_p \tag{8}$$

When the value of p_i overcomes p_{cr}, the critical value of p_i, the cell returns to cell cycle. Figure 2 (*right*) presents one of the simulations for different values of the parameter $k_1^{(p)}$.

3 2D Patterns

In this section we consider the same model as above but in the 2D case. System (4) for extra-cellular concentrations is replaced by the system

$$\begin{cases} \dfrac{du}{dt} = d_1 \Delta u + b_1 c - q_1 u \\ \dfrac{dv}{dt} = d_2 \Delta v + b_2 c - q_2 v \end{cases} \tag{9}$$

while system (5) remains the same. We consider the model example where cells stimulate apoptosis of the surrounding cells producing some substance which diffuses in the extra-cellular matrix. In this case, the coefficients $k_1^{(1)}, k_2^{(1)}, k_2^{(2)}$ of system (5) equal zero while the coefficient $k_1^{(2)}$ is positive.

The difference of the 2D case in comparison with the 1D case is that we need to specify the direction of cell division. This means that when we replace a mother cell by two daughter cells, we specify the positions of their centers. The distance

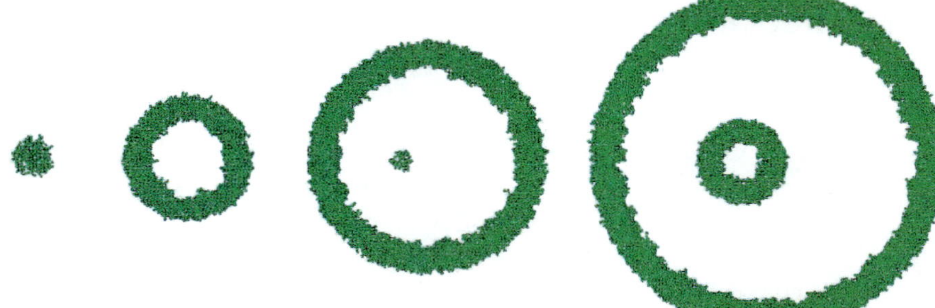

Fig. 3 Evolution of cell population in consecutive moments of time. Cells are shown by *green dots*. For the same values of parameters, the small annulus inside the big one can appear or not because of randomness in the direction of cell division (see the explanation in the text)

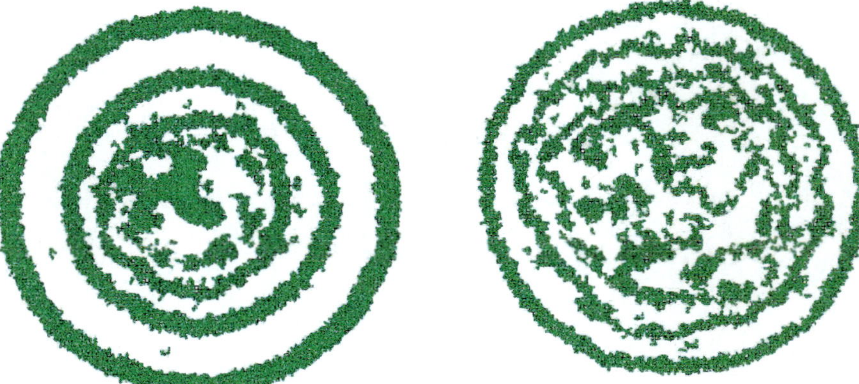

Fig. 4 Snapshots of cell populations for different values of parameters: $k_1^{(2)} = 6$ (*left*) and $k_1^{(2)} = 7$ (*right*)

between the centers equals two radii and the position of the center of mass does not change after the division. However, the direction of the line connecting the centers is not a priori given and should be specified. It can depend on the biological applications. In this model, we impose a random direction of cell division.

Figure 3 shows the evolution of cell population starting with a single cell at the center of the domain. The growing cell population forms an annulus. When it becomes sufficiently large, another one can appear inside the first one. In other simulations with the same values of the parameters the internal annulus may not appear. The difference in the simulations is related to the randomness in the direction of cell division. The pattern of growing cell populations becomes more complex if we increase the value of the parameter $k_1^{(2)}$ which determines the intensity of apoptosis (Fig. 4).

The model examples considered in this section can be used in modelling morphogenesis but more detailed mechanism of intracellular and extracellular regulation should be considered for specific biological processes.

4 Tumor Growth

In this section we consider the complete model (1), (2) and (3) assuming that is a scalar variable. It describes the concentration of nutrients which diffuse from the boundary of the domain and which are consumed by cells inside the domain. We consider the equation

$$\frac{\partial u}{\partial t} = D\Delta u - kcu, \tag{10}$$

where c is cell concentration and k is a positive parameter. The rate of nutrient consumption is proportional to the product of the concentrations. The form of the domain and of the boundary conditions will be specified below.

Next, we consider the scalar intra-cellular variable u_i where the subscript i corresponds to the cell number. It is described by the equation

$$\frac{du_i}{dt} = k_1 u(x_i, t) - k_2 u_i \tag{11}$$

The first term in the right-hand side of this equation shows that the intra-cellular concentration u_i grows proportionally to the value of the extra-cellular concentration $u(x, t)$ at the space point x_i where the cell is located. The second term describes consumption or destruction of u_i inside the cell. We suppose that the cell radius r_i grows proportionally to the increase of u_i:

$$\frac{dr_i}{dt} = \max\left(\frac{du_i}{dt}, 0\right). \tag{12}$$

The initial value of the radius for each new cell is r_0, the maximal radius r_m. When it is reached, the cell divides. If the cell does not divide before its maximal age, then it dies. The maximal cell age is a parameter of the problem.

Consider a circular domain Ω and complete equation (10) by the boundary condition $u = 1$ at the boundary $\partial\Omega$. We put a single cell in the center of the domain and begin numerical solution of system (10), (11) and (12). The results of the simulations are shown in Fig. 5 for several consecutive moments of time. The grey 2D surface shows the spatial distribution of the concentration $u(x, t)$. In the beginning it equals 1 everywhere in the domain. The constants k_1 and k_2, $k_1 > k_2$, are chosen in such a way that the intra-cellular concentration u_i growth. Consequently, the radius of the cell also grows and after some time the cell divides.

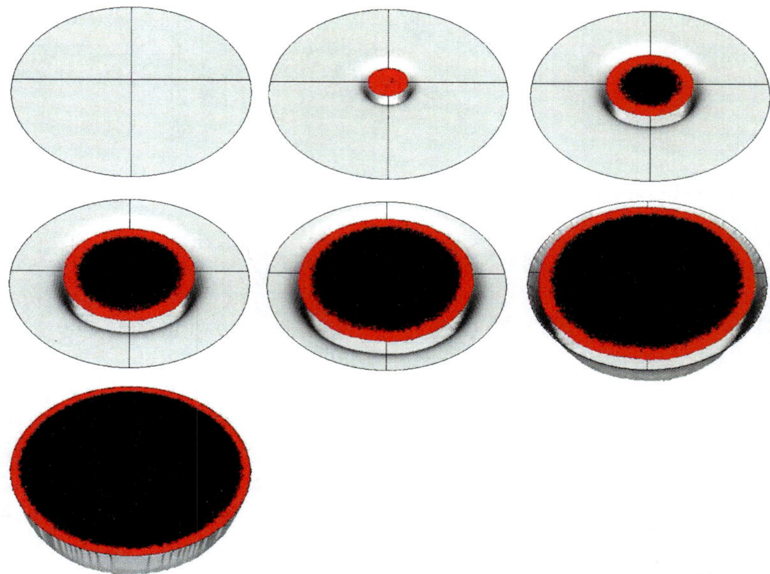

Fig. 5 Consecutive moments of tumor growth. It starts with a single cell at the center of the circle. Living cells are shown in *red*, dead cells in *black*. *2D grey* surface shows the level of nutrients

The new cells also consume nutrients, grow and divide. The part of the domain filled by cells forms a disk while the concentration $u(x, t)$ decreases in the center of the domain (second figure in the upper row). Hence, the right-hand side of (12) also decreases, the intra-cellular concentration stops growing or even decreases, and cells cannot divide before their maximal age τ. As a result, they die and form the black region in the center. Living cells shown in red form a narrow external layer. The region filled by cells grows in time and finally approaches to the boundary of the domain.

Dynamics of the cell population can be more complex if we decrease the maximal life time τ. Cells now have less time to accumulate enough nutrients for division. In this case, even a small decrease in nutrient concentration can become crucial from the point of view of the choice between proliferation and apoptosis. In the beginning cells form, as before, a circular region with living cells outside and the core formed by dead cells. The layer of living cells is narrower than in the previous example. Rather rapidly the layer of living cells becomes disconnected (Fig. 6, third in the upper row). After that the domain loses its radial symmetry which can be related to an instability with a certain wavenumber. Further growth of the region filled by cells makes the outgrowing parts more pronounced and this region less regular. Finally it reaches the boundary of the domain.

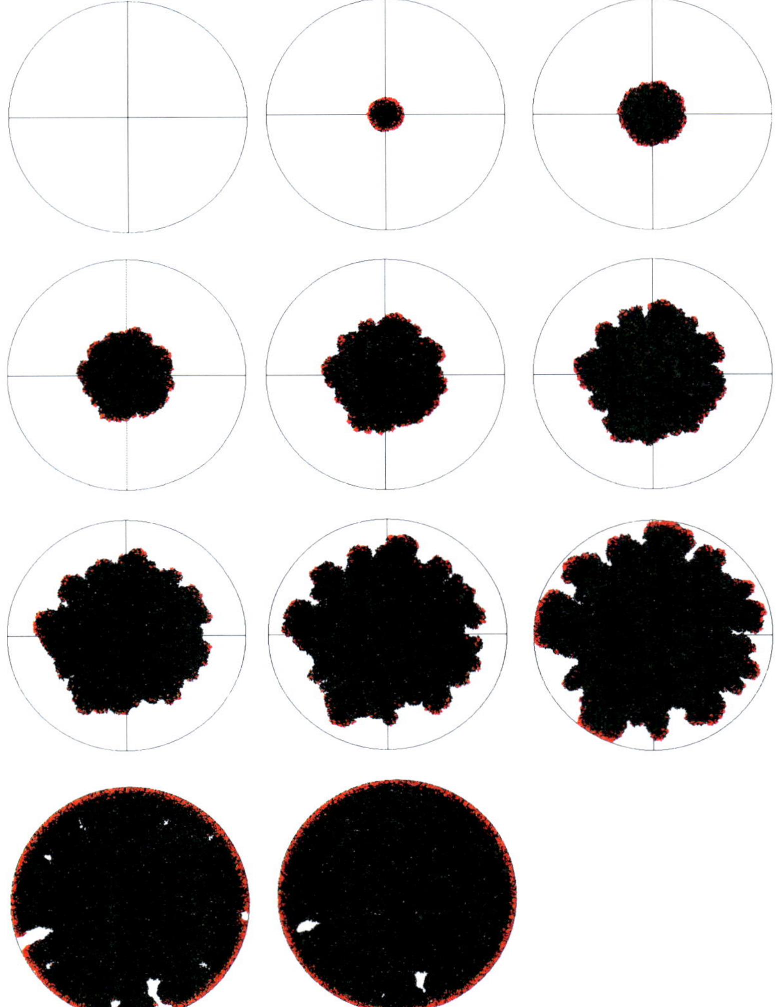

Fig. 6 Consecutive moments of tumor growth. If the life time of cells is short, they become more sensitive to the lack of nutrient. The region filled by cells loses its radial symmetry

If we increase the diffusion coefficient for nutrients, then another pattern can also be observed (Fig. 7). Dividing and apoptotic (or dormant) cells are mixed and do not form clearly separated populations as before. Moreover, detailed analysis of the results shows the presence of time oscillations in cell distributions.

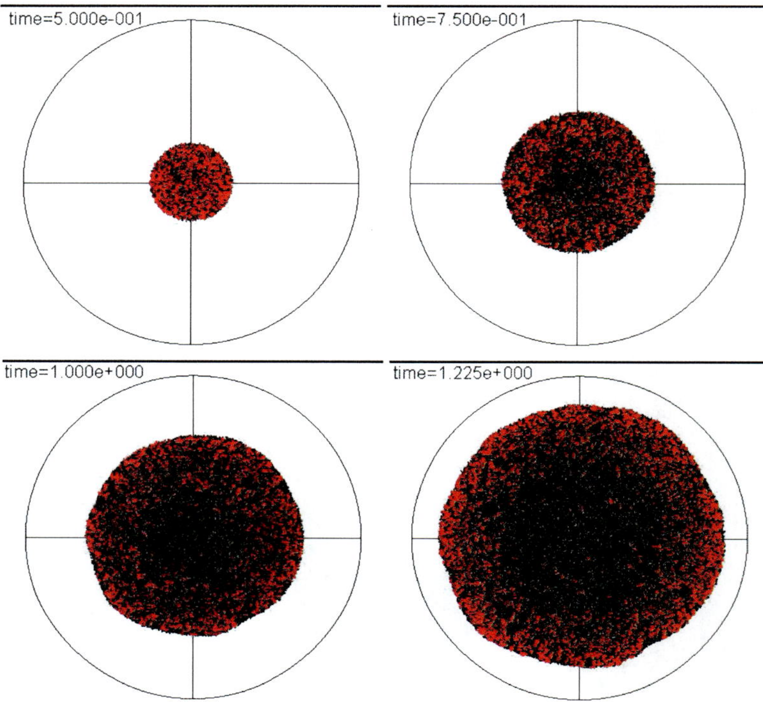

Fig. 7 Consecutive moments of tumor growth where dividing and apoptotic (or dormant) cells are mixed

References

1. Thompson DW(1992) On growth and form, 2nd edn, Dover reprint of 1942, Dover, New York
2. Murray JD Mathematical biology, 3rd edn. In: 2 vols: Mathematical biology: I. An introduction (2002), Mathematical biology: II. Spatial models and biomedical applications (2003). Springer, New York
3. Othmer HG, Painter K, Umulis D, Xue C (2009) The intersection of theory and application in elucidating pattern formation in developmental biology. Math Model Nat Phenom 4(4):3–82
4. Alber M, Chen N, Glimm T, Lushnikov P (2007) Two-dimensional multiscale model of cell motion in a chemotactic field. 53–76
5. Dallon JC (2007) Models with lattice-free center-based cells interacting with continuum environment variables. Single Cell Based Model Biol Med 197–219
6. Deutsch A (2007) Lattice-gas cellular automaton modeling of developing cell systems. 29–51
7. Meinhardt H (2008) Models of biological pattern formation: from elementary steps to the organization of embryonic axes. Curr Top Dev Biol 81:1–63
8. Anderson ARA (2007) A hybrid multiscale model of solid tumour growth and invasion: evolution and the microenvironment 3–28
9. Anderson ARA, Chaplain M, Rejniak KA *Eds.* (2007) Single cell based models in biology and medicine. Birkhauser, Basel
10. Anderson ARA, Rejniak KA, Gerlee P, Quaranta V (2007) Modelling of cancer growth. Evolution and invasion: bridging scales and models. Math Model Nat Phenom 2(3):1–29

11. Drasdo D (2007) Center-based single-cell models: an approach to multi-cellular organization based on a conceptual analogy to colloidal particles. 171–196
12. Karttunen M, Vattulainen I, Lukkarinen A (2004) A novel methods in soft matter simulations. Springer, Berlin
13. Bessonov N, Pujo-Menjouet L, Volpert V (2006) Cell modelling of hematopoiesis. Math Model Nat Phenom 1(2):81–103
14. Bessonov N, Demin I, Pujo-Menjouet L, Volpert V (2009) A multi-agent model describing self-renewal or differentiation effect of blood cell population. Math Comput Model 49:2116–2127
15. Bessonov N, Kurbatova P, Volpert V (2010) Particle dynamics modelling of cell populations. Math Model Nat Phenom 5(7):42–47
16. Swat A, Dolado I, Rojas JM, Nebreda AR (2009) Cell density-dependent inhibition of epidermal growth factor receptor signaling by p38 alpha mitogen-activated protein kinase via Sprouty2 downregulation. Mol Cell Biol 29(12):3332–3343
17. Sark S, Jeanjean R, Zhang CC, Arcondeguy T (2006) Inhibition of cell division suppresses heterocyst development in Anabaena sp. Strain PCC 7120. J Bacteriol 188(4):1396–1404

An Integrative Approach to the Analysis of Pattern Formation in Butterfly Wings: Experiments and Models

Toshio Sekimura

Abstract Color patterning in butterfly wings is one of the most beautiful and spectacular examples of pattern formation in biology. We see, at first glance, a complex and rich diversity in wing color patterns. However, if we look at these patterns in details, we can find principles, or rules governing how the patterns are formed, and we can also find some specific features, or aspects to allow us to understand the underlying mechanisms of pattern formation, such as the "genetic", "ecological", "developmental", "biochemical", "molecular evolutionary" and "food plant" aspects. It is clear that we cannot not ignore any of these aspects to realize the diversity and evolution of wing color patterns. Rather, we must integrate these aspects if we are to fully understand the patterning process. Recently, in order to elucidate the underlying evolutionary strategies of living organisms, new insights have emerged in non-model organisms such as butterflies more than in the so-called model organisms.

In this paper, I briefly review the above-mentioned aspects of pattern formation of wing color patterns in relation to their evolution. I will also stress the integrative approaches required to understand the diversity in wing patterns and rules governing the pattern formation.

1 What Is Color Pattern in Butterfly Wings

1.1 Parallel Row of Scale Cells and Color Pattern in Butterfly Wings

A butterfly wing is covered with thousands of monochromatic scale cells, which form parallel rows in the proximal-distal direction of the wing. Each scale cell in the wing

T. Sekimura (✉)
Department of Biological Chemistry, Graduate School of Bioscience and Biotechnology, Chubu University, Kasugai, Aichi 487-8501, Japan
e-mail: sekimura@isc.chubu.ac.jp

V. Capasso et al. (eds.), *Pattern Formation in Morphogenesis*, Springer Proceedings in Mathematics 15, DOI 10.1007/978-3-642-20164-6_11,

Fig. 1 What is the color pattern in butterfly wings? Color pattern in the wings is a kind of mosaic pattern of dozens to hundreds of finely-tiled monochromatic scale cells

has only one specific color, i.e., it is monochromatic. It is known that there exist two different types of sources of scale color in butterfly wings (1) pigmentation color by chemical pigments, and (2) structural color by physical reflection of light. Many butterflies use chemical pigments for scale colors, except for a few examples such as *Morpho*-butterflies, which use only structural color. In this paper, I deal with butterflies which use chemical pigments for scale colors, and these will be described in detail in Sect. 5. We assume that color pattern in butterfly wings is a kind of mosaic pattern of dozens to hundreds of finely-tiled monochromatic scale cells (Fig. 1).

1.2 Basic Features of the Color Pattern Formation

The butterfly wing consists of two cell layers, a surface layer and a back layer. It is known that the two layers are completely separated by the middle tissues, and developmental phenomena in each cell layer occur independently of the other. Parallel row formation of scale cells in each layer occurs through two different kinds of developmental events, (1) cell differentiation, and (2) cell rearrangement. Cell differentiation means differentiation of scale precursor cells (SPCs) from undifferentiated epithelial cells in the larval wing disc. SPCs change finally to scale cells in the adult wing, and cell rearrangement occurs to form parallel rows of SPCs in the wing disc (e.g., see [1]). These two phenomena are known to occur independently of color pattern formation in butterfly wings. This means that parallel row formation of scale cells is independent of the color pattern formation. Color pattern formation occurs in a monolayer of the wing disc and it is essentially two-dimensional in space without cell movement. Therefore, the problem of color pattern formation is how colors of scale cells are settled at specific positions in a two-dimensional entire wing surface.

Fig. 2 Caterpillars of the butterfly *Papilio machaon* (*left*) and the wing disc within (*right*). Four color-less wing discs, which become four wings of adult butterfly, are located near the head in the caterpillar body

2 Developmental Aspect: Development of Wing Disc

2.1 Pre-pattern and Color Pattern

The adult butterfly wing originates from the wing disc in the larva, which appears first in the late larval stage and continues to develop during the pupal stage. The wing disc is transparent and colorless throughout the larval and pupal stages of development (see Fig. 2). We see colors of the wing first through the shell of the pupa in the latest stage, i.e., just 1 or 2 days before emergence of the adult butterfly. However, antibody fluorescent techniques for the corresponding genes have revealed existence of the so-called "pre-pattern" on the wing disc, which overlaps completely with the color pattern of the adult butterfly wing (e.g., [2]). It is known in general that the pre-pattern on the wing disc is the original feature of the wing color pattern. Therefore, we assume that color pattern formation in butterfly wings starts in the larval wing disc. We also assume that the problem of color pattern formation is essentially that of pre-pattern formation in the wing disc, but not simple color setting by pigmentation in the latest pupal stage. Since the wing disc grows in size and shape, we have to understand the time-course of development of pre-pattern formation during the time period of larval and pupal stages.

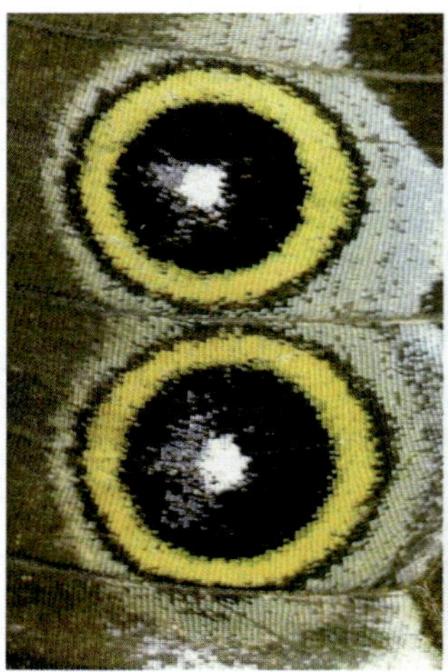

Fig. 3 Local pattern and global pattern. Eyespot patterns (local patterns) (*left*) and the global pattern of *Papilio dardanus* (♂) (*right*)

3 Local and Global Wing Color Patterns

Local pattern means localized pattern in a restricted area of the entire wing surface such as the eyespot pattern. Global pattern, on the other hand, means the whole pattern covering the entire wing surface and it is sometimes used for identification of butterfly species (Fig. 3). Global pattern includes local patterns as part of all pattern elements composing the global wing pattern. We should have in mind that research on local patterns is not independent of research on the global wing pattern. Recent research on the genetics of color patterns has been mostly concentrated on the local color pattern and genes such as *Distal-less* and *engrailed* controlling the eyespot pre-pattern have already been identified (e.g., [2]). On the other hand, genetic research on the global pattern has not been so far advanced. Recently, however, great progresses has been made using mimetic butterflies such as *Papilio dardanus*, *Heliconius* butterflies, and *Papilio polytes*. In the following sections, I review recent progresses in research for both the local and global color pattern formation.

4 Progress in Research for Local Color Pattern Formation

4.1 Eyespot Pattern and Basic Ideas on the Pattern Formation

Among local color patterns, the eyespot pattern has been studied most at both molecular and genetic levels. It is well known that there exist three concrete steps of development to establish eyespot patterns as follows,

(1) Determination of the eyespot focus, or signaling center in the wing disc, from which some signaling chemicals, or morphogens originate. The focus point on the disc is determined by complex genetic regulatory networks.
(2) Spreading out of morphogens into the surroundings of the focus point through diffusion, and activation of corresponding genes (e.g., *Distal-less, engrailed*) to establish concentric eyespot pre-patterns on the disc.
(3) Activation of pigmentation genes (e.g., *DDC, GTP-CH1, cinebar*) depending on pre-patterns, and then, color differentiation of scale cells in the adult wing.

4.2 Genetic Research for Eye-Spot Pattern Formation

Carroll et al. [3] found eye-spot foci in the wing disc of a species *Precis coenia* by using antibody fluorescent techniques for the gene *Distal-less*. The positions of the foci on the wing disc were found to correspond completely to eyespot centers on the adult butterfly [4]. On the other hand, Nijhout [5] found that computer simulations by using a mathematical model (e.g., Meinhardt model) could reproduce a focal point on the wing cell as a stable point-like solution of the model equation. The wing cell, in this case, means a localized area composed of scale cells between two adjacent veins near the wing margin, but not a biological single cell. These findings made clear existence and determination of the eyespot focus as the first step of eyespot pattern formation mentioned above.

Brunetti et al. [2] found that two different transcription factors, *spalt* and *engrailed* are expressed in the wing disc of the African species *Bicyclus anynana*. The expression patterns of concentric rings, i.e., eyespot pre-patterns correspond completely to the eyespot patterns on the adult butterfly. Some genes associated with pigmentation, that is, the third or final stage of color pattern formation, have been investigated and already identified so far, and will be discussed in Sect. 5.

Current researching trends for eyespot pattern formation are in the direction of detailed genetic analyses, which have taken into account newly founded gene regulatory networks. Mathematical and theoretical researching trends are also in the same direction (e.g., see [6]).

4.3 Genetic Research for Local Patterns with Respect to Evolution

Reed and Serfas [7] analyzed changes in the *Notch/Distal-less* intervenous expression patterns in wing discs of 8 different butterfly species such as *Vanessa cardui, Precis coenia and Bicyclus anynana*, with respect to evolution of pattern elements. They concluded that the combination of the *Notch/Distal-less* expression patterns is a good index or tool for understanding the evolutionary pathway of butterfly species.

5 Biochemical Aspect: Chemical Pigments Used for Wing Coloration

In this section, we concentrate our attention on the biochemical aspect of color pattern formation, i.e., pigmentation by chemical pigments, which are synthesized in cells by biochemical reaction networks. Genes associated with pigmentation have been already identified for some butterfly species.

The color setting (or pigmentation) of scale cells occurs in the final pupa stage just 1 or 2 days before emergence of the adult butterfly. The pigmentation process itself is independent of the essential part of pre-pattern formation, and the color setting is done after pre-pattern formation in the same positions already established by the colorless pre-patterns. The pigmentation process is, however, responsible for formation of a variety of colorful wing patterns. The major chemical pigments involved in the coloration of butterfly wings are [8]:

1. Melanin
   The color of melanin molecules is black or dark brown. Melanin is the most widely used molecule for coloration of scale cells of many butterfly species. One of the genes associated with melanin molecules is *DDC*.
2. Papiliochrome
   The color of papiliochrome is yellow, and it is used for coloration of the scale cells of the genus *Papilio*.
3. Bilichrome
   The color of bilichrome is blue, and it is used for coloration of the scale cells of some butterflies of the genus *Papilio* such as *Graphium sarpedon* and *Papilio phorcas*.
4. Ommochrome
   The color of ommochrome is red, and it is used for coloration in the genus *Heliconius*. Candidates for genes associated with ommochrome molecules are *cinebar* and *vermilion* [9].
5. Pteridine
   The color of pteridine is white for the genus *Pieris*, and yellow for the genus *Colias*. One of the genes associated with Pteridine molecules is *GTP-CH1*.

Some of the above major pigments are, of course, used in various combinations for the coloration of many butterfly wing patterns.

6 Progress in Research for Global Pattern Formation

6.1 Ground Plan and Basic Ideas on the Global Pattern Formation

Owing to the pioneering analyses of Schwanwitsch [10] and Süffert [11], diverse wing color patterns of Nymphalid butterflies could be reproduced from a hypothetical global wing pattern, i.e., the Nymphalid ground plan, by the effects of major

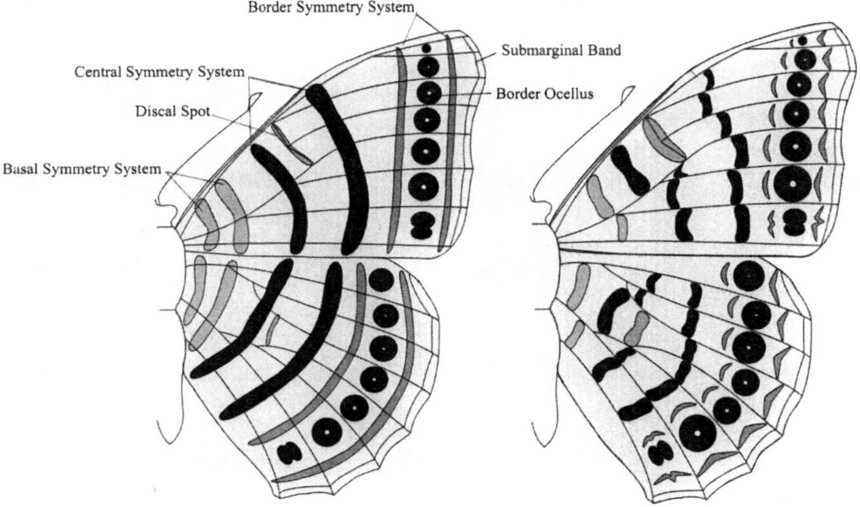

Fig. 4 The Nymphalid ground plan according to Nijhout [12]. The ground plan consists of a system of pattern elements such as bands and spots that run from the anterior to the posterior margin of each wing. The pattern elements are classified and named as symmetry systems, submarginal bands, border ocelli and discal spot

factors or functions such as disruption and dislocation (see below for details). Following on from this, Nijhout [12] improved the ground plan and proposed a new one. The ground plan consists of a number of patterning elements (Fig. 4):

(a) Symmetry systems
The ground plan is well systemized and mainly composed of three sets of symmetry bands called symmetry systems, which run in parallel along the anterior-posterior direction of the wing. Three symmetry bands, that is, the basal symmetry system, the central symmetry system, and the border symmetry system, make up the fundamental structures of butterfly wing patterns.
(b) Submarginal bands
We can see one or two bands running along the outer edge of the wing.
(c) Border ocelli
Color patterns in the border symmetry system are usually circular, and sometimes consist of concentric rings.
(d) Discal spot
In the central symmetry system, a large colored mark along the central line of the system can be seen.

It is generally assumed that diverse global wing patterns originate from the specific ground plan of each species through morphogenetic changes caused by some rules or factors. We next describe major key factors that produce diverse global wing patterns.

6.2 Major Factors or Functions for the Diversification of Global Wing Patterns

In order to understand the diversity in global wing patterns, we must know the vein system and vein positions in the wing, which play significant roles in generating the diversity. Precursors of wing veins, i.e., pre-veins, appear first in the larval wing disc and develop throughout the larval and pupal stages until the appearance of adult butterfly. We focus our attentions on the relationship between the pre-veins and the pre-pattern formation in relation to the diversification of global wing patterns (Fig. 5).

(1) Disruption
Disruption means that pre-veins divide pattern elements such as marginal bands running parallel to the wing margin into fragmental elements such as the border ocelli mentioned above. As an example, we can see in the species *Bicyclus anynana* a mutant without a wing vein showing that two adjacent eyes-pots merge into a single continuous pattern element [4].

(2) Dislocation
Dislocation means that pre-patterns running parallel to the wing margin lose their continuity over pre-veins by a kind of sliding movement along the pre-veins. For example, Koch and Nijhout [13] found a naturally occurring veinless mutant of the swallowtail butterfly *Papilio xuthus* showing extremely aberrant patterns, where the loss of veins changes normal dislocated marginal bands into continuous bands running parallel to the wing margin.

(3) Venous and intervenous patterns
The venous pattern appears precisely in the line of the wing veins, and intervenous pattern, by contrast, consists of a pigmented stripe that lies on the middle of two adjacent veins.

(4) Vein-independent patterns
This pattern is completely different from the above vein-dependent ones, and is usually based on pattern elements such as the central symmetry system running in the central part of the wing far from the wing margin. For example, in *Heliconius* butterflies, the major pattern elements, i.e., stripe-like patterns running through the middle of wing, appear to be vein-independent.

(5) Expansion, shrinking, and elimination of pattern elements

6.3 Research for Global Pre-patterns

There exist few butterflies that have been studied genetically to understand the formation of global wing patterns in relation to their evolution. Among them, we select here two mimetic butterfly species, *Papilio dardanus* and *Heliconius erato*, which have been of special interest to biologists for over a century because of the adaptive variation in their wing patterns.

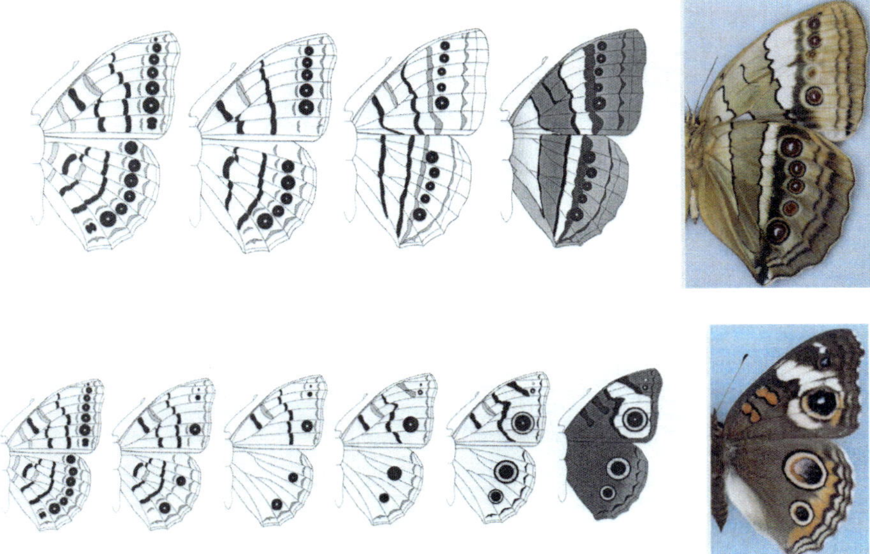

Fig. 5 Hypothetical evolutionary pathways from the ground plan to real butterflies [27]. The most complicated wing patterns can be derived from the ground plan by applications or effects of some key functions such as disruption, dislocation and so on. Two examples of butterfly: *Stichophthalma camadeva* (*top*) and *Precis coenia* (*bottom*)

6.3.1 *Papilio dardanus*

Papilio dardanus is a species of swallowtail butterfly widely distributed across sub-Saharan Africa. *P. dardanus* is well known for the spectacular phenotypic polymorphism in females. The females of different geographic races have evolved more than a dozen different wing color patterns and have come to mimic different species of unpalatable butterflies and moths in their specific regions.

Sekimura et al. [14] analyzed the evolution of pattern polymorphism in females by using a reaction-diffusion model. Mathematical and computational analyses were done on the basis of the ground plan of *P. dardanus*, which consist of three regions on the forewing and one region on the hindwing, where (1) the black pattern elements are major parts of the wing patterns, even though the background color attracts our attention most, and (2) the black pattern elements can increase or decrease in size, depending on the phenotypic form of each female. One of the main results is that the wing patterns may be due to underlying proximal-distal stripe-like patterns of pigment-inducing morphogen. This result is consistent with experimental results on the *Heliconius* butterfly [15]. Furthermore, our results suggest that different patterns in females are similar to each other and they could be generated essentially by a fixed sets of kinetic and diffusion parameter values in a reaction-diffusion system (Fig. 6). This result could be important from the genetic point of view, because it agrees with the finding that different forms of the females are controlled by a single genetic locus [16–18].

Fig. 6 A reaction-diffusion model for prepattern formation. Polymorphism in mimetic females of *Papilio dardanus* (*left*) [f. *trophonius* (*top left*), f. *cenea* (*top right*), f. *planemoides* (*bottom left*), and f. *hippocoonideas* (*bottom right*)]. Numerical results from the mathematical model (*right*) illustrating f. *trophonius*, f. *cenea*, f. *planemoides*, and f. *hippocoonideas*, respectively

6.3.2 *Heliconius* Butterfly

It is known that the species *Heliconius erato* is a mimetic butterfly with more than 20 different wing color pattern variants adapted to different regions across Central and South America. Recently, genetic analyses of *Heliconius* have started to unearth the characteristics of global wing color patterns. Reed and Gilbert [15] studied the relationship between wing venation and *Distal-less* expression in *Heliconius* displaying a mutant phenotype with a severe vein deficiency, and found that major color pattern elements develop independently of wing venation. This finding lends support to previous theoretical models suggesting that proximal-distal pre-patterns can be sufficient for establishing complex whole-wing color patterns in butterflies [14]. Reed et al. [9] investigated expression of candidate genes potentially involved with a red/yellow forewing band polymorphism in *H. erato*. They found that a transcript *cinnabar* expression was associated with the forewing band regardless of pigment color, and another *vermilion* expression changed spatially over time in red-banded butterflies, but was not expressed at detectable levels in yellow-banded butterflies, suggesting that regulation of *vermilion* may be involved with the red/yellow polymorphism.

7 Ecological Aspect: Population Dynamics

Global wing patterns are sometimes used for identification of butterfly species, because it is considered that diverse wing patterns have been caused by or derived from various types of adaptation to surrounding environments such as mimicry and behavior of females to choose food plants on which to lay eggs, and so on. Among ecological and adaptive aspects associated with evolutionary strategies of butterflies, I next describe changes in the number of butterfly individuals, that is, population dynamics. As an example of butterfly population dynamics, I take up a case study of the mimetic butterfly *Papilio polytes* as follows.

7.1 Population Dynamics of *Papilio polytes*

Papilio polytes is a species of swallowtail butterfly widely distributed across the Oriental region including the Sakishima Islands (southern islands group of Japan). It is known that the female is polymorphic with four forms (f. *cyrus*, f. *polytes*, f. *romulus*, and f. *theseus*), whereas the male is always monomorphic. In the Sakishima Islands, however, there exist only two female forms, f. *cyrus* and f. *polytes* (mimic). The form f. *cyrus* resembles the male, and the other form f. *polytes* resembles an unpalatable butterfly *Pachliopta aristolochiae* (model). Interestingly, in their distribution range, particular islands do not have *P. aristolochiae* and f. *polytes*, or *P. aristolochiae*.

Based on published records, Uesugi [19] reviewed temporal changes in the occurrence of *P. polytes* f. *polytes* and its model *P. aristolochiae* in the Sakishima Islands. He found that there had not been any records of *P. aristolochiae* (model) in the Sakishima Islands from 1907 to 1967, but it was first recorded in 1968 in the Yaeyama Islands, the southernmost islands group of the Sakishima Islands and the records of *P. polytes* f. *polytes* (mimic) increased rapidly after 1972 there. He also found that on Miyako-jima Island, a northern island of the Sakishima Islands, the number of the mimic butterflies has increased after establishment of the model butterfly in 1975 (Fig. 7). The increase in the population of *P. polytes* f. *polytes* was discussed and explained in the context of adaptive significance of the female-limited Batesian mimicry [20].

8 Molecular Evolutionary Aspect: Molecular Phylogenetics of Butterflies

Diversification of color pattern in butterfly wings is the result of evolution. In order to understand how diverse wing patterns have been generated during the life history, it is useful for us to know the evolutionary relationship of corresponding butterflies. One way to do this is to construct the molecular phylogenetic tree of butterflies, which is based on current molecular or genetic technologies such as PCR amplification and DNA sequencing (Fig. 8). Molecular phylogenetics is more objective and unambiguous than previous methods such as morphological studies. As an example, I show next the result of molecular phylogenetic research on swallowtail butterflies of the genus *Papilio*.

Vane-Wright et al. [21] determined sister group relationships of the mimetic butterfly *Papilio dardanus* by using four different DNA regions, i.e., mitochondrial 16SrRNA and cytochrome B regions, and nuclear Elongation Factor EF1 α and the ITS1 region of the ribosomal RNA locus. They concluded that all four data sets placed *P. dardanus* as the sister species of *P. phorcas* and *P. constantinus* as the next closest relative, and quite distant from other *Papilio* such as *P. nobilis*. This information is important for conclusions about the evolution of femalelimited mimicry in swallowtail butterflies and their wing patterns. On the other hand,

Fig. 7 Population dynamics of the butterfly *Papilio polytes*. The graph shows change in the ratio(%) (vertical axis) of the mimetic form (f. *polytes*) to the two forms (f. *cyrus* and f. *polytes*) of *P. polytes* for 15 years from 1975 to 1989 in the Miyakojima Island [20]. The ratio increased in a sigmoidal fashion from under 20% up to 45% since 1976, when the model butterfly *Pachliopta aristolochiae* has settled down in the Island. *Papilio polytes* (*photos*) in Miyakojima Island: (f. *cyrus* (*top left*), f. *polytes* (*top right*), male (*bottom left*), and *Pachliopta aristolochiae* (*bottom right*))

Caterino and Sperling [22] used mitochondrial cytochrome oxidase I and II genes (COI and COII) for 23 *Papilio* taxa to construct *Papilio* phylogenetics and to examine patterns of gene evolution across a broad taxonomic range. One of their most noteworthy findings is that neither *P. alexanor* nor *P. xuthus* belongs in the *P. machaon* group.

These days, however, people have noticed that it is necessary to combine and integrate various different methods to obtain reliable results, because each method has respective deficiencies even if the technology is very new.

9 Food/Host Plants

There exists another important factor to understand the butterfly and the evolution of butterflies, that is, food plants or host plants. Females of butterflies lay eggs selectively on leaves of specific plants (host plants), on which larvae feed from just after hatching until the beginning of the prepupal stage of development. It is known that the choice of host plants by mother butterflies is very precise and the leaves include specific chemical ingredients. The specific chemicals in leaves serve as different signstimuli, e.g. oviposition stimulus for the host seeking females and feeding stimulus for larvae, respectively. The specific relationship between butterfly and host plants is the result of strategic interactions for survival between butterflies and plants in the evolutionary history of life. For examples, the butterfly *Papilio*

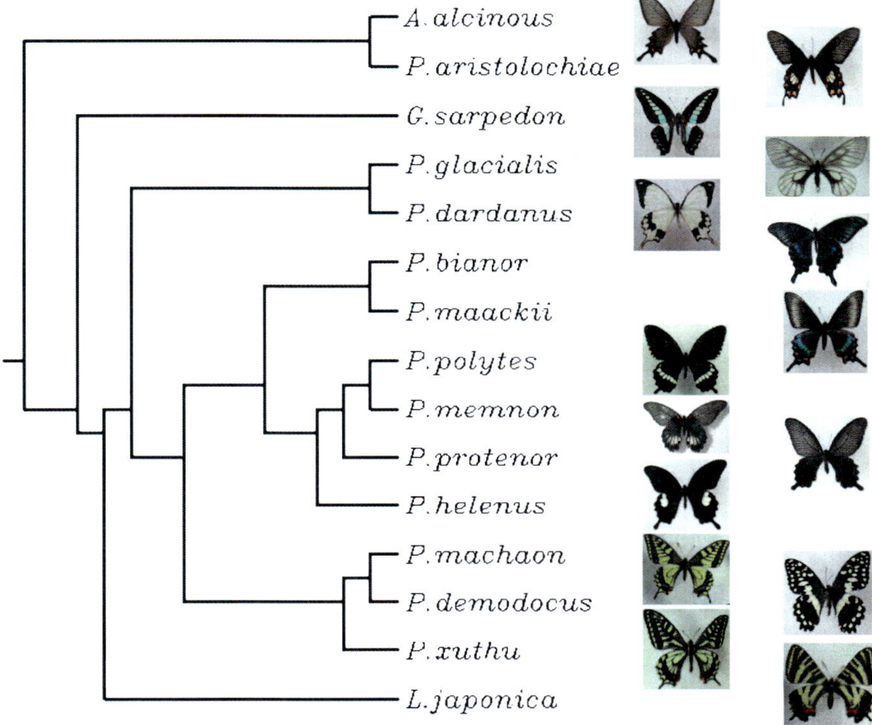

Fig. 8 Molecular phylogenetics. Molecular phylogenetic tree of the family *Papilionidae* through the analysis of the ND5 region of mtDNA

dardanus lays eggs on the host plant *mexican orange*, and the *Papilio polytes* does so on *citrus trifoliate* and *lemon*.

It would be possible to construct the molecular phylogenetics of food plants as well as that of butterflies, even though there exist additional and difficult problems, e.g., multi-component system of oviposition stimulants in *Papilio xuthus* [23]. However, it is exciting to understand the evolutionary pathway or the phylogenetic relationship among corresponding butterflies through molecular phylogenetic studies of both butterflies and food plants.

10 Discussions and Summary

I briefly reviewed current progress in the study of color pattern formation in butterfly wings. In order to understand how the diversity in wing patterns has been produced, we first have to know and analyze specific aspects (genetic, ecological, developmental, biochemical, molecular evolutionary, and food plant) underlying the pattern formation mechanisms and evolution. We then have to integrate them by new comprehensive methods including mathematical and computational analyses as one of the core methods. A hint how to do this is for us to remember or realize that a butterfly wing is an organ of a living organism, the butterfly, which is living under various types of environmental restrictions such as natural selection, interspecific and intraspecific competition, and so on. Since the matter of utmost concern for a butterfly would be to live safely or survive in a given environment, most of its characteristics including wing color pattern would have evolved to fit or adapt to the surrounding environment. This means that to the butterfly such characteristics represent some kinds of strategies for survival, which must be related ultimately to changes in the fitness in population biology. Here, the fitness roughly means the number of offspring per adult in the next generation, a useful quantity to discuss temporal changes in the population of an organism.

Wing color patterns, for example, have some specific biological functions, such as mutual recognition, protection from predators, and mimicry, which play important roles for butterfly survival. On the other hand, wing colors are mainly produced by chemical pigments, which are synthesized in cells by biochemical reaction networks. By use of the cost-benefit theory, Ohsaki linked the chemical pigment to mimicry [24, 25]. He examined a female-limited mimetic species *Papilio polytes*, whose females are known to mimic the wing-color patterning of unpalatable butterflies by means of the chemical pigment carotenoid, and he finally concluded that owing to their consumption of the chemical, the physiological life span of mimetic females is shorter than that of non-mimetic females. This conclusion may indicate a reduction in the fitness of the mimetic females. The above example shows that changes in wing color patterning could affect the fitness or the future population of a butterfly species, and at the same time, this shows a deep linkage of a biochemical process to a different ecological event.

Regarding mathematical models using PDE and ODE systems, there exists a problem of parameter values in the model system and they usually are given appropriately from outside and used in mathematical analyses. Another hint to connect different aspects is in parameter values, which indeed represent concrete specific biological functions, such as gene activities, and they are not values or tools intended for mathematical analysis. I would like to show the reader another example, even though the animal in question is not a butterfly, but a fish. This is an experiment-based theoretical paper that links parametric values in an activator-inhibitor system to activities of the *leopard* gene, which controls changes in Zebrafish skin color patterns [26]. This is a nice case study to connect a biochemical process directly to genetic activity.

We are now at the beginning stage in the whole course of an integrative approach to the analysis of pattern formation in butterfly wings. I think that as the first step we could connect at least a few of all the different aspects that are necessary to gain a full understanding of the reality and diversity in butterfly wing patterns and their evolution. All of these step by step processes will be integrated into a realistic and biological feature of wing patterns under the concept of strategies for survival of a beautiful creature, the butterfly. A variety of mathematical tools have contributed to our understanding of the problem so far and, in the future, they will continue to play an important role in the analysis of pattern formation and integration of the different aspects in question.

Acknowledgements This paper is based on my talks at Department of Mathematics, University of Sussex, Centre for Mathematical Biology, University of Oxford, and Department of Mathematics, University of Strathclyde in UK, where I stayed for 2 months in autumn, 2009, with supports from the LMS and the EPSRC through grants awarded to Dr. Anotida Madzvamuse of University of Sussex and Prof. Philip Maini of University of Oxford.

References

1. Sekimura T, Zhu M, Cook J, Maini PK, Murray JD (1999) Pattern formation of scale cells in lepidoptera by differential origin-dependent cell adhesion. Bull Math Biol 61:807–827
2. Brunetti CR, Selegue JE, Monteiro A, French V, Brakefield PB, Carroll SB (2001) The generation and diversification of butterfly eyespot color patterns. Curr Biol 11:1578–1585
3. Carroll SB, Gates J, Keys D, Paddock SW, Panganiban GF, Selegue J, Williams TA (1994) Pattern formation and eyespot determination in butterfly wings. Science 265:109–114
4. Brakefield PB, Gates J, Keys D, Kesbeke F, Wijingaarden PJ, Monteiro A, French V, Carroll SB (1996) Development, plasticity and evolution of butterfly eyespot patterns. Nature 384:236–242
5. Nijhout HF (1994) Genes on the wing. Science 265:44–45
6. Evans TM, Marcus JM (2006) A simulation study of the genetic regulatory hierarchy for butterfly eyespot focus determination. Evol Dev 8(3):273–283
7. Reed RD, Serfas MS (2004) Butterfly wing pattern evolution is associated with changes in a Notch/Distal-less temporal pattern formation process. Curr Biol 14:1159–1166
8. Umebachi Y (2000) The pigment of animals (in Japanese). Uchida-Rohkakuho, Tokyo
9. Reed RD, Macmillan WO, Nagy LM (2008) Gene expression underlying adaptive variation in Heliconius wing patterns: non-modular regulation of overlapping cinnabar and vermilion prepatterns. Proc R Soc Lond B Biol Sci 275:37–45
10. Schwanwitsch BN (1924) On the ground plan of wing-pattern in nymphalids and certain other families of rhopalocerous Lepidoptera. Proc Zool Soc Lond B 34:509–528
11. Süffert F (1927) Zur vergleichende analyse der schmetterlingszeichung. Biologisches zentralblatt 47:385–413
12. Nijhout HF (1991) The development and evolution of butterfly wing patterns. Smithsonian Institution Press, Washington/London
13. Koch PB, Nijhout HF (2002) The role of wing veins in colour pattern in the butterfly *Papilio xuthus* (Lepidopteran: Papilionidae). Eur J Entomol 99:67–72
14. Sekimura T, Madzvamuse A, Wathen AJ, Maini PK (2000) A model for colour pattern formation in the butterfly wing of *Papilio dardanus*. Proc R Soc Lond B Biol Sci 267:851–859
15. Reed RD, Gilbert LE (2004) Wing venation and Distal-less expression in *Heliconius* butterfly wing pattern development. Dev Genes Evol 214:628–634
16. Clark R, Brown SM, Collins SC, Jiggins CD, Heckel GD, Vogler AP (2008) Colour pattern specification in the Mocker swallowtail *Papilio dardanus*: the transcription factor invected is a candidate for the mimicry locus H. Proc R Soc Lond B Biol Sci 275:1181–1188
17. Clarke CA, Sheppard PM (1959) The genetics of some mimetic forms of *Papilio dardanus*, Brown, and *Papilio dardanus*, Linn. J Genet 56:237–259
18. Clarke CA, Sheppard PM (1960) The evolution of mimicry in the butterfly *Papilio dardanus*. Heredity 14:163–173

19. Uesugi K (1991) Temporal changes in records of the mimetic butterfly *Papilio polytes* with establishment of its model *Pachliopta aristolochiae*. Jpn J Entomol 59(1):183–198
20. Uesugi K (1996) The adaptive significance of Batesian mimicry in the swallowtail butterfly, *Papilio polytes* (Insecta, Papilionidae): associative learning in a predator. Ethology 102:762–775
21. Vane-Wright RI, Raheem DC, Cieslak A, Vogler AP (1999) Evolution of the mimetic swallowtail butterfly *Papilio dardanus*: molecular data confirm relationship with *P. phorcas* and *P. constantinus*. Biol J Linn Soc 66:215–229
22. Caterino MS, Sperling FAH (1999) *Papilio* phylogeny based on mitochondrial cytochrome oxidase I and II genes. Mol Phylogenet Evol 11(1):122–137
23. Ohsugi T, Nishida R, Fukami H (1991) Multi-component system of oviposition stimulants for a rutaceae-feeding swallowtail butterfly, *Papilio xuthus* (Lepidoptera: Papilionidae). Appl Entomol Zool 26(1):29–40
24. Ohsaki N (1995) Preferential predation of female butterflies and the evolution of Batesian mimicry. Nature 373:173–175
25. Ohsaki N (2005) A common mechanism explaining the evolution of female-limited and both-sex Batesian mimicry in butterflies. J Anim Ecol 74:728–734
26. Asai R, Taguchi E, Kume Y, Saito M, Kondo S (1999) Zebrafish leopard gene as a component of the putative reaction–diffusion system. Mech Dev 89:87–92
27. Nijhout HF (2003) The development and evolution of butterfly wing patterns, Chap 4. In: Sekimura T, Noji S, Morita R (eds) Diversity in pattern and form of biological systems – from DNA to ecological systems. Shokabo, Tokyo, pp 46–57

Modeling Morphogenesis in Multicellular Structures with Cell Complexes and L-systems

Przemyslaw Prusinkiewicz and Brendan Lane

Abstract We consider computational modeling of biological systems that consist of discrete components arranged into linear structures. As time advances, these components may process information, communicate and divide. We show that: (1) the topological notion of cell complexes provides a useful framework for simulating information processing and flow between components; (2) an index-free notation exploiting topological adjacencies in the structure is needed to conveniently model structures in which the number of components changes (for example, due to cell division); and (3) Lindenmayer systems operating on cell complexes combine the above elements in the case of linear structures. These observations provide guidance for constructing L-systems and explain their modeling power. L-systems operating on cell complexes are illustrated by revisiting models of heterocyst formation in *Anabaena* and by presenting a simple model of leaf development focused on the morphogenetic role of the leaf margin.

1 Introduction

There is a feedback between mathematics and studies of nature. On one hand, mathematical concepts—even though they may eventually be formalized in an axiomatic way—are often inspired and motivated by studies of nature. On the other hand, they facilitate these studies by providing proper mathematical tools (Fig. 1).

In this context, we consider computational methods needed to model the development of multicellular structures, in particular plants. We show that these methods are not merely a new application of partial differential equations, traditionally used to model spatio-temporal phenomena in mathematical physics. Instead, developmental modeling of multicellular structures requires an integration of

P. Prusinkiewicz (✉) • B. Lane
Department of Computer Science, University of Calgary, Calgary, AB T2N 1N4, Canada
e-mail: pwp@cpsc.ucalgary.ca

V. Capasso et al. (eds.), *Pattern Formation in Morphogenesis*, Springer Proceedings in Mathematics 15, DOI 10.1007/978-3-642-20164-6_12,

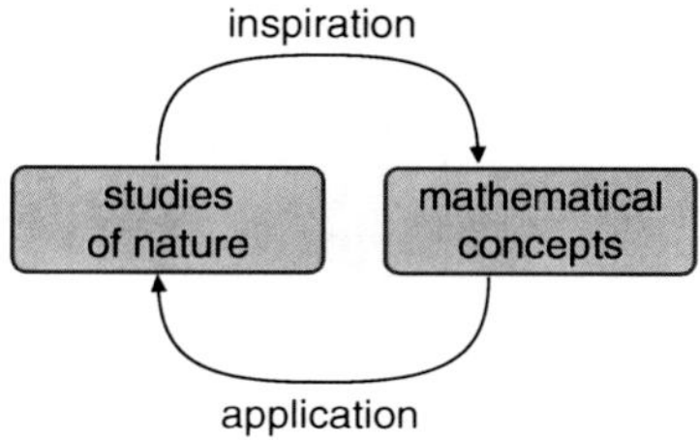

Fig. 1 A conceptual model of relations between natural science and mathematics

tools rooted in different branches of mathematics and computer science. This combination includes L-systems [1], ordinary differential equations, and the topological notion of cell complexes [2].

The structures we consider are the *spatial arrangements* of *discrete components* that *process information* and *communicate*. These *structures are dynamic*, which means that not only the state of the components, but also their number can change over time. The development is *symplastic*: the neighborhood relations can only be changed as a result of the addition or removal of components (in contrast to animal cells, plant cells do not move with respect to each other). We limit our examples to linear structures consisting of sequences of cells, although similar problems occur in the modeling of branching plant structures at the larger scale of architectural modules: branch and root segments, buds, leaves, flowers and fruits. The insights we obtain also extend to models of two- and three-dimensional structures.

2 Computation in Cell Complexes

Let us consider the fundamental process of diffusion in a filament as a running example. At any point in time, the distribution of the diffusing substance can be visualized by plotting concentration c as a function of position x along the filament (Fig. 2a).

How can we model changes in concentration due to diffusion over time? The first impulse may be to apply the well known partial differential equation for diffusion:

$$\frac{\partial c}{\partial t} = D\frac{\partial^2 c}{\partial x^2}. \tag{1}$$

Unfortunately, there is a problem with this approach. To derive partial differential equation (1), one starts with a discrete description of diffusion, then passes to the limit in space and time ([3], Chap. 9). When ascribing this equation to a multicellular structure, we go back to the discrete version. Such circular thinking should be avoided [4].

One step towards a solution is to ignore (1) and directly write the set of ordinary differential equations that describe the changes in concentration in each cell without going to the spatial limit (Fig. 2b):

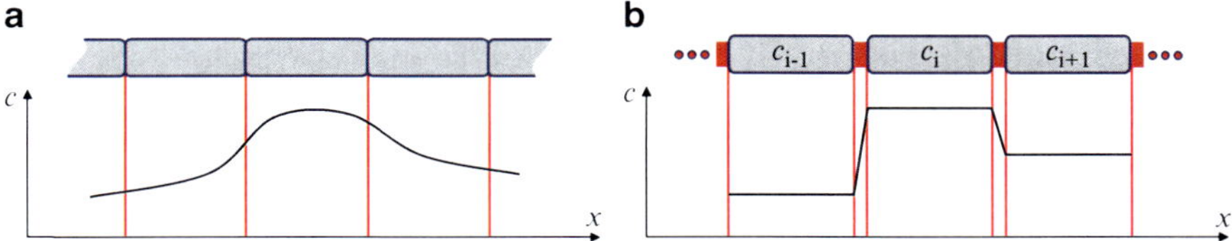

Fig. 2 Sample distribution of a diffusing substance in a spatially continuous (**a**) and spatially discrete (**b**) model of a multicellular filament

$$\frac{dc_i}{dt} = k_i(c_{i-1} - c_i) - k_{i+1}(c_i - c_{i+1}), \quad i = 2, 3, \ldots, n-1. \tag{2}$$

According to this system of equations, the concentration of the diffusing substance in each of the $n-2$ cell interior to the filament changes proportionally to the difference in concentrations across cell walls (a version of Fick's law [3]). For simplicity, we do not consider here the boundary cells 1 and n. We assume that the concentration of the diffusing substance is approximately uniform within each cell. This is a reasonable assumption as it is the cell walls, rather than the cells themselves, that present a significant obstacle to diffusion. The system of equations (2) highlights, however, another problem, which becomes apparent when we compare the equations for adjacent cells, e.g. i and $i+1$:

$$\begin{aligned} \frac{dc_i}{dt} &= k_i(c_{i-1} - c_i) - k_{i+1}(c_i - c_{i+1}) \\ \frac{dc_{i+1}}{dt} &= k_{i+1}(c_i - c_{i+1}) - k_{i+2}(c_{i+1} - c_{i+2}) \end{aligned} \tag{3}$$

The term $k_{i+1}(c_i - c_{i+1})$ is calculated twice, first to determine the amount of the diffusing substance exported from cell i to cell $i+1$, and a second time to determine the amount of the substance received by cell $i+1$ from cell i. Superficially, performing the same calculation twice may seem merely redundant: computationally inefficient, but without any effect on the final result. However, the problem created by repeating this calculation is deeper. Suppose that the diffusion coefficients are random variables, which is well justified if the number of diffusing molecules is small. Calculating an expression with random variables twice will likely produce different results, and the amount of substance exported by cell i will be different from the amount received by cell $i+1$, violating the law of mass conservation.

We can solve this problem by computing fluxes between any pair of adjacent cells only once and using the result twice, to update concentration in each cell. While implementing this solution, we need to properly recognize the topology of the modeled structure, which is a sequence of cells *separated by walls*. This topology offers placeholders for all variables inherent in diffusion: concentrations c are associated with cells, and fluxes J with walls (Fig. 3). The system of equations (4) results:

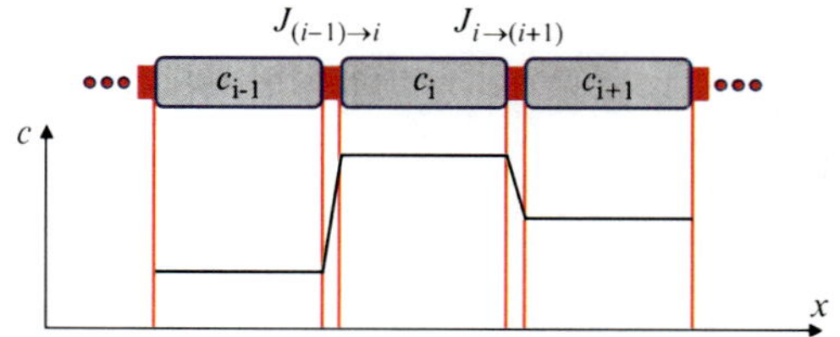

Fig. 3 Modeling a filament as a cell complex

$$\left.\begin{aligned} J_{(i-1)\to i} &= K_{(i-1)\to i}(c_{i-1} - c_i) \\ \frac{dc_i}{dt} &= J_{(i-1)\to i} - J_{i\to(i+1)} \end{aligned}\right\} \quad i = 2, 3, \ldots, n-1 \tag{4}$$

Although the flux through each wall is computed the same way as in (2) and (3), it is now computed only once. Consequently, mass is conserved even in the presence of random fluctuations of flux. Furthermore, if the system of equations (4) is evaluated numerically, for example using the forward Euler method with time step Δt,

$$\left.\begin{aligned} J^{t+1}_{(i-1)\to i} &= K^t_{(i-1)\to i}\left(c^t_{i-1} - c^t_i\right) \\ c_{t+1}i &= c^t_i + \left(J^t_{(i-1)\to i} - J^t_{i\to(i+1)}\right)\Delta t \end{aligned}\right\} \quad \begin{aligned} i &= 2, 3, \ldots, n-1 \\ t &= 0, 1, 2, \ldots \end{aligned} \tag{5}$$

mass will be conserved exactly in spite of the errors in estimating fluxes over finite time intervals that are inherent in numerical methods.

A sequence of cells separated by walls is an example of a one-dimensional *cell complex* [2]. Formally, such a complex is an interwoven sequence of objects of two types: line segments (cells) and points at which these segments meet (walls). Thinking in terms of cell complexes facilitates proper definition of discrete models, especially in higher dimensions. For example, in three-dimensional tissues built from polyhedral cells we distinguish three-dimensional cells, two-dimensional polygonal faces that are shared by pairs of cells, one-dimensional edges that bind these faces, and zero-dimensional vertices in which the edges meet. Each of these objects may provide a placeholder for different variables, organizing simulations of multicellular organisms in a systematic manner [5].

3 L-systems

Do cell complexes provide a good framework for describing processes such as diffusion or genetic regulation in multicellular systems? They are certainly a step in the right direction, but many problems remain. The key issue is how to identify the components of a cell complex, the variables related to each component, and the equations that relate these variables. One method is to use indices and specify neighborhood relations between the components with index arithmetic. For instance, if n components are arranged into a sequence indexed from 1 to n, the

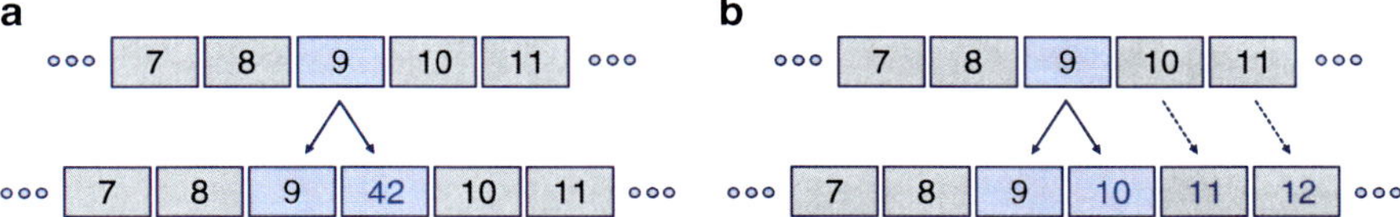

Fig. 4 Inadequacies of cell identification with indices in a growing filament. (**a**) Assignment of an arbitrary number to a newly created cell breaks index arithmetic. (**b**) Renumbering dissociates indices from the identity of cells

neighbors of component $i \in \{2, \ldots, n-1\}$ will have indices $i-1$ and $i+1$. Equations (4) and (5) are examples of this notation. It is so standard in mathematical practice that we tend to use it without much thought. Unfortunately, it does not work well for developing systems.

To see the problem, let us consider a filament with consecutively numbered cells [6]. A cell, say number 9, divides and becomes two cells. What indices should they have? One possibility would be to pass number 9 to one of the child cells and assign some arbitrary number that has not yet been used, say 42, to the other cell (Fig. 4a). Each cell will then have a unique identifier, but we can no longer rely on index arithmetic to find who is the neighbour of whom: it no longer suffices to add one to find the neighbour to the right, or subtract one to find the neighbour to the left.

Another possibility is to preserve index arithmetic (Fig. 4b). We can accomplish this, for example, by assigning the second new cell number 10, and renumbering all of the remaining cells to the right. The old cells 9-10-11 now become cells 10-11-12. In this case, we can perform index arithmetic on the new filament, but the identity of cells is no longer maintained. For example, cell 10 has become cell 11, and may become cell 12, 13 or higher in the future.

Analyzing these problems, we conclude that their source is not merely one or another indexing scheme, but the very attempt to use indices to identify cells in a growing organism. Paraphrasing Hermann Weyl, who said "The introduction of numbers as coordinates [...] is an act of violence" [7, p. 90], we can say " ... and so is the introduction of indices."

An alternative idea is to exploit the topological structure of the filament and introduce operators that will return the informational content of the neighbours. The possibility of accessing such context in a local manner, without globally indexing all components of the modelled structure, is one of key ideas behind L-systems, the formalism for describing and simulating development introduced in 1968 by Aristid Lindenmayer [1]. Using the notation for (context-sensitive) L-systems presented in [8], we can write (5) as:

$$\begin{aligned} C(c_L) < W(J) > C(c_R) &\rightarrow W(K \cdot (c_L - c_R)) \\ W(J_L) < C(c) > W(J_R) &\rightarrow C(c + (J_L - J_R)\Delta t) \end{aligned} \tag{6}$$

L-system expressions are called *rewriting rules* or *productions*, as in the theory of algorithms and formal languages. The first production above states that wall W is associated with a single variable, noted J. This variable represents flux though the wall. Its value is updated (arrow $\rightarrow$) considering concentrations c_L and c_R of the diffusing substance in cells C on the left and right side of wall W. These cells are indicated by the operators $<$ and $>$, respectively. The variable identifiers c_L and c_R are local to this production and can be arbitrary, but must be distinct. Given the wall and its context (i.e., the adjacent cells), the updated flux is calculated using the expression $K \cdot (c_L - c_R)$.

The second production works in a similar way. In order to update the concentration c of the substance in cell C, the walls W that delimit this cell on the left and right sides are considered. Each wall is characterized by a flux: J_L and J_R, respectively. Within time Δt associated with a single simulation step, concentration c changes by $(J_L - J_R)$ Δt. The L-system productions (6) thus express the same idea as Equation (5), but without involving indices or any other global enumeration of cells and walls. Instead, they use the operators that look for the context, or neighbourhood, of cells and walls.

Summarizing the ideas presented so far, we have introduced three notions of key importance to the modeling of multicellular structures. First, we formalized these structures as cell complexes. This notion provides a vehicle for assigning variables to proper elements of the structure, in our example concentrations to cells and fluxes to walls. Second, we used locally defined context, rather than globally defined indices (or any other global enumeration) to access information about the neighbours of any element in the complex (Figs. 4a and 4b). The third point is a little more subtle, but equally true: the arrow is an operator that relates what was before to what will come next, and thus indicates the neighbourhood in time. Thus, in contrast to Equation (5), L-system (6) needs no indices for time as well. An additional benefit of L-systems is that they naturally extend to another type of productions, which capture cell division. For example, the following production:

$$C(c) : \mathit{condition} \rightarrow C(c)\, W(0) C(c) \tag{7}$$

says that, if some *condition* is met, cell C will divide into two child cells with the same concentration c as the parent cell, separated by a wall.

4 Heterocyst Differentiation in *Anabaena*

To illustrate the presented concepts in a biological context, we will apply them to model morphogenesis in a growing filament. The chosen organisms, representing genus *Anabaena*, integrate some of the most fundamental processes linking patterning and growth. Consequently, they have been repetitively used to illustrate both the basic mechanisms of morphogenesis and diverse aspects of modeling with L-systems [9–15]. Here we focus on the integration of L-systems and cell complexes.

Anabaena is a genus of cyanobacteria, organisms that have been on Earth for over 3 billion years and are responsible for the introduction of oxygen into the atmosphere [16]. It creates multicellular filaments consisting of two basic types of cells. Vegetative cells are capable of photosynthesis and produce sugars. Heterocysts are capable of fixing nitrogen from the atmosphere, and produce nitrogenous compounds that the bacterium needs. Photosynthesis and nitrogen fixation are biologically difficult to reconcile, because the enzyme crucial to the fixation is inhibited by oxygen. Consequently, some cyanobacteria separate photosynthesis and nitrogen fixation in time: they photosynthesize during the day and fix nitrogen at night. Others, including *Anabaena*, separate these tasks in space [16].

On the average, consecutive heterocysts in an *Anabaena* filament are separated by about ten vegetative cells [17]. Heterocysts cannot divide, but vegetative cells do divide and grow, causing the filament to elongate. As the existing heterocysts are moved apart by the growing vegetative segments, new heterocysts differentiate in-between. The process that controls this differentiation has been extensively studied. A small protein called PatS is produced by heterocysts and diffuses through the vegetative segments of the filament, inhibiting the differentiation of new heterocysts [18]. As the existing heterocysts move apart, the concentration of PatS in the vegetative segments gradually decreases, eventually falling below a threshold near the center of the segment. This decline is detected by the genetic regulatory circuit that triggers the differentiation of a new heterocyst.

The maintenance of approximately constant spacing between heterocysts in a growing filament can be captured and explained using computational models that may represent different tradeoffs between biological accuracy and simplicity. Here we present a very simple L-system illustrating the use of cell complexes.

#define	H	0	*// Heterocyst cell type*
#define	V	1	*// Vegetative cell type*
#define	K	(**ran**(2.0))	*// Diffusion coefficient*
#define	v	0.5	*// Turnover rate*
#define	R	1.1	*// Cell growth factor*
#define	Θ	0.1	*// Threshold for heterocyst differentiation*
#define	s_{MAX}	0.8	*// Cell size at division*
#define	Δt	0.01	*// Time step*

Axiom: $\boldsymbol{C}(H, 1, 1)\ \boldsymbol{W}(0)\ \boldsymbol{C}(V, 1, 1)\ \boldsymbol{W}(0)\ \boldsymbol{C}(H, 1, 1)$

p_1: $\boldsymbol{C}(a_L, c_L, s_L) < \boldsymbol{W}(J) > C(a_R, c_R, s_R) \rightarrow \boldsymbol{W}(K \cdot (c_L - c_R))$

p_2: $\boldsymbol{C}(a, c, s) : a = H \rightarrow \boldsymbol{C}(H, 1, 1)$

p_3: $\mathbf{W}(J_L) < \boldsymbol{C}(a, c, s) > \mathbf{W}(J_R)$:

$$\{c \leftarrow c + ((J_L - J_R) - vc)\Delta t; s \leftarrow sR^{\Delta t};\}$$
$$c < \Theta \rightarrow \boldsymbol{C}(H, 1, 1)$$

p_4: $\mathbf{W}(J_L) < \boldsymbol{C}(a, c, s) > \mathbf{W}(J_R)$:

$$\{c \leftarrow c + ((J_L - J_R) - vc)\Delta t;\ s \leftarrow sR^{\Delta t};\}$$
$$s > s_{MAX} \rightarrow \boldsymbol{C}(V, c, s/2)\ \boldsymbol{W}(0)\ \boldsymbol{C}(V, c, s/2)$$

p_5: $\mathbf{W}(J_L) < \boldsymbol{C}(a, c, s) > \mathbf{W}(J_R)$:

$$\{c \leftarrow c + ((J_L - J_R) - vc)\Delta t;\ s \leftarrow sR^{\Delta t};\}$$
$$\rightarrow \boldsymbol{C}(V, c, s)$$

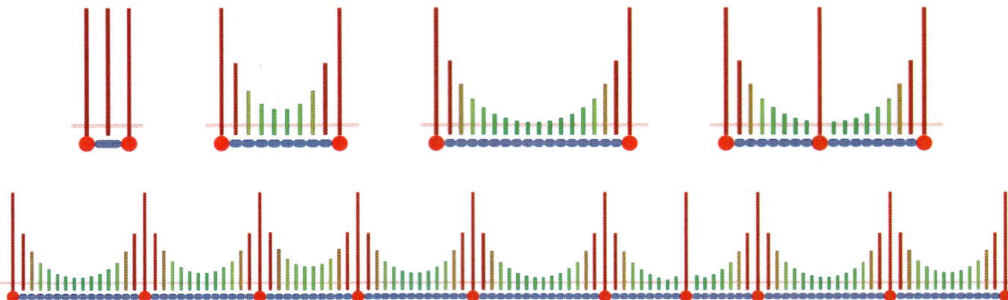

Fig. 5 Snapshots of a simulation of heterocyst differentiation in a growing *Anabaena* filament. Non-differentiated vegetative cells are shown in blue and heterocysts are shown in red. Vertical bars indicate concentration of a diffusing inhibitor produced by the heterocysts. The red horizontal line indicates the threshold of heterocyst differentiation

The axiom specifies that the initial structure consists of three cells C separated by walls W. The cells are characterized by three parameters: type a (H for heterocyst, V for a vegetative cell), inhibitor concentration c, and cell length s. Productions are ordered, and the first applicable production is used for each cell or wall. Production p_1 determines flux J of the inhibitor across a wall, as in L-system (6). Production p_2, applicable to heterocysts, sets both the inhibitor concentration and heterocyst length to 1. Productions p_3 to p_5 apply to vegetative cells. They describe changes in inhibitor concentration due to its diffusion and turnover, and changes in cell length due to growth. In addition, production p_3 specifies that the vegetative cell in which the inhibitor concentration falls below threshold Θ will differentiate into a heterocyst. Likewise, production p_4 states that a vegetative cell which exceeds maximum length s_{MAX} will divide.

Fig. 5 shows selected steps of a simulation using this L-system. As time progresses, the vegetative cell and its descendants divide, pushing the heterocysts apart. Concentrations of the inhibitor in the vegetative cells decreases with their distance from the heterocysts, as the diffusive supply of the inhibitor diminishes. When the concentration of the inhibitor falls below a threshold, the corresponding cell differentiates into a heterocyst. The average distance between heterocysts is thus maintained in spite of the filament's growth.

5 Leaf Development

The previous example was focused on the arrangement of cells of different types within a growing filament. This filament was visualized by placing cells along a line, as its overall form was not important. In the next example we use a more complex geometric interpretation of an L-system operating on a cell complex. The resulting computational model plausibly explains the form of a lobed leaf.

Many aspects of plant development are regulated by the plant hormone auxin [19], which is actively exported from cells by a family of proteins called PINFORMED—in short, PIN. Within this family, the PIN1 protein [20] plays the

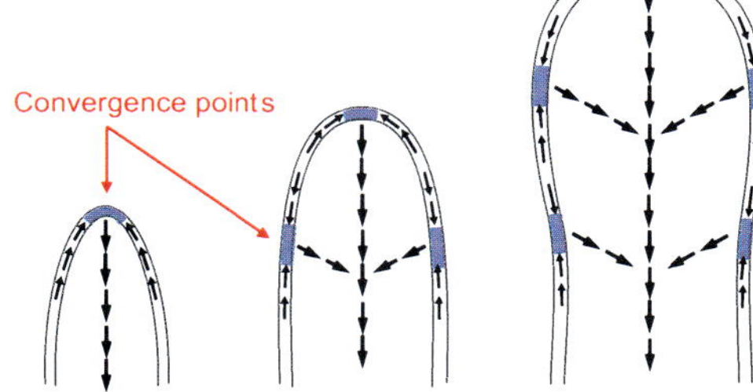

Fig. 6 A conceptual model of auxin flow in a growing leaf. Polarized auxin transport at the leaf margin (black arrows) leads to the emergence of auxin convergence points (blue rectangles). Canalized auxin flow from these points defines the pattern of main veins

dominant role in leaf development. Confocal microscopy images of diverse plants show that PIN1 proteins at the leaf margin export auxin towards discrete locations, called convergence points [21, 22]. From there, auxin propagates into the leaf blade, forming streams, or canals, which define the paths of future veins [23]. The first convergence point in a young leaf primordium is at the leaf tip. As the leaf grows, the distance between this convergence point and the leaf base increases, and new convergence points are gradually formed in the available space [23] (Fig. 6).

The above process can be compared to the differentiation of heterocysts in the growing filament of *Anabaena*. However, while the positioning of new heterocysts can be intuitively explained by the depletion of the diffusively transported inhibitor (PatS) between heterocysts moving apart, the molecular mechanism defining the spacing of convergence points on the leaf margin is not yet fully understood. The key assumption is that the concentration of PINs in the membrane of cells at the leaf margin depends on the concentration of auxin in the adjacent cells. The higher this concentration, the more PINs will be allocated to the abutting cell membrane [24, 25]. This feedback between PINs and auxin is illustrated in Fig. 7. Here cells are represented schematically as black contours, with auxin concentrations shown as filled blue squares. The size of these squares is proportional to the concentration of auxin within the cells. Auxin fluxes are shown as black arrows between the cells: the wider the arrow, the larger the flux. PIN concentrations are visualized as red rectangles running parallel to cell edges; the wider the line, the larger the PIN concentration at the corresponding cell membrane. The feedback loop of interactions is indicated by the green arrows. The top arrow shows that PINs in the membrane of cell i abutting cell j pump auxin towards cell j. The bottom arrow shows that the concentration of auxin in cell j affects the allocation of PINs in the membrane of cell i, and thus controls further flow of auxin into cell j.

To show that the postulated interactions between auxin and PINs can produce a pattern of approximately equidistant convergence points in a file of cells, we construct a simple computational model governed by three equations (for related models and their analysis see [23–27]). The first equation describes the flux $J_{i\to j}$ of auxin from cell i to the adjacent cell j as the sum of active and diffusive transport:

$$J_{i\to j} = Tc_i\,[PIN]_{i\to j} - Tc_j\,[PIN]_{i\leftarrow j} + K(c_i - c_j). \tag{8}$$

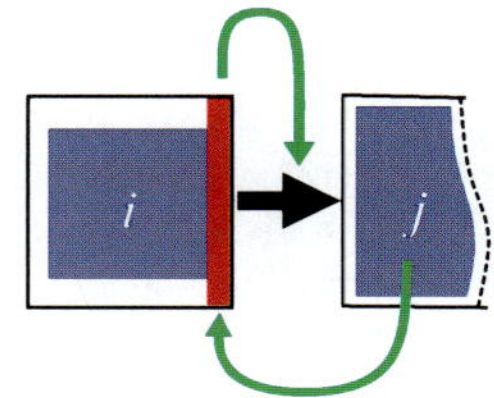

Fig. 7 Up-the-gradient model of PIN polarization and auxin transport. Green arrows indicate the feedback of interactions: PINs (red rectangle) promote auxin efflux from cell i (black arrow), while high auxin concentration in cell j (blue area) polarizes PINs in cell i toward j

The active transport from cell i to cell j is assumed to be proportional to the auxin concentration c_i in cell i, multiplied by the concentration $[PIN]_{i\to j}$ of PINs in the membrane of cell i abutting cell j. The coefficient of proportionality is T. An analogous term describes active transport of auxin from cell j to cell i. The last term represents diffusive transport with the diffusion coefficient K, as in Equation (4, top). The second equation describes the allocation of PINs to the membrane of cell i abutting cell j. It has the form

$$[PIN]_{i\to j} \sim [PIN]_i \cdot f(c_j), \tag{9}$$

where $[PIN]_i$ is the overall concentration of PINs in cell i, and f is some increasing function of auxin concentration c_j in cell j.

The third equation adds to the law of mass conservation (Equation (4), bottom) terms representing local auxin production with a constant absolute rate σ and local turnover with relative rate μ:

$$\frac{dc_i}{dt} = J_{x\to i} - J_{i\to x} + \sigma - \mu c_i. \tag{10}$$

To model a sequence of cells and walls obeying the above equations, we express them as an L-system operating on a cell complex:

#define	T	1.2	*// Polar transport coefficient*
#define	K	0.02	*// Diffusion coefficient*
#define	σ	0.1	*// Auxin production*
#define	μ	0.005	*// Auxin turnover*
#define	Δt	0.05	*// Time step*

Axiom: $\boldsymbol{C}(0,0,0)\ \boldsymbol{W}(0)\ \boldsymbol{C}(0,0,0) \cdots \boldsymbol{C}(0,0,0)\ \boldsymbol{W}(0)\ \boldsymbol{C}(0,0,0)$

$$p_1\text{:}\quad \boldsymbol{C}(\overleftarrow{p}_L, c_L, \overrightarrow{p}_L) < \boldsymbol{W}(J) > \boldsymbol{C}(\overleftarrow{p}_R, c_R, \overrightarrow{p}_R) \\ \rightarrow \boldsymbol{W}(T(c_L\overrightarrow{p}_L - c_R\overleftarrow{p}_R) + K(c_L - c_R))$$

$$p_2\text{:}\quad \boldsymbol{C}(\overleftarrow{p}_L, c_L, \overrightarrow{p}_L)\boldsymbol{W}(J_L) < C(\overleftarrow{p}, c, \overrightarrow{p}) > \boldsymbol{W}(J_R)\boldsymbol{C}(\overleftarrow{p}_R, c_R, \overrightarrow{p}_R) : \\ \rightarrow \boldsymbol{C}(f(c_L), c + (J_L - J_R + \sigma - \mu c)\Delta t, f(c_R))$$

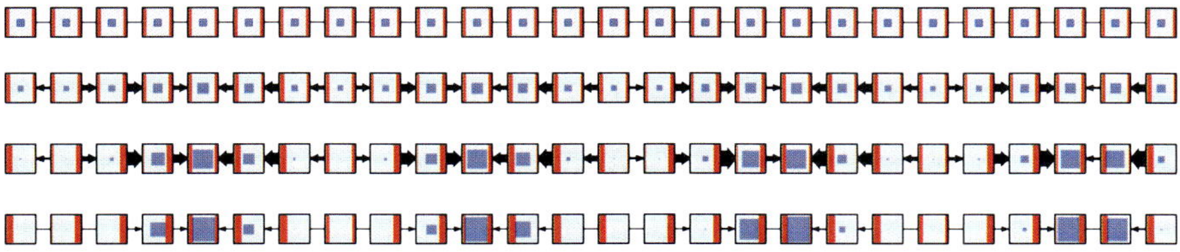

Fig. 8 The initial state and three stages of simulation of convergence point formation in a file of cells

Assuming uniform distribution of auxin throughout the file of cells, the initial state of the system can be visualized as the first row of Fig. 8. As expected, a stable pattern of discrete convergence points—maxima of auxin concentration with PINs oriented towards them—emerges as the simulation progresses. Three stages of simulation in a file of constant length are shown in Fig. 8.

We will now apply the above process to model the development of leaf form. Hay et al. [21] postulated that the convergence points on the leaf margin define the positions of accelerated leaf outgrowth. A limited but simple method for modeling such outgrowth is the boundary propagation method ([28], Chap. 1). It operates by moving the boundary of a shape in the normal direction in each simulation step (Fig. 9).

We model the leaf margin as a single file of cells, initially in a shape resembling a leaf primordium. The propagation rate of each cell is proportional to the concentration of auxin. In addition, we assume that cells reaching the threshold length divide as in the case of *Anabaena*. Fig. 10 shows an example of the resulting progression of the shapes of the growing margin and compares the final stage of the simulation with an ivy leaf.

The molecular details of ivy leaf development are not yet known. Nevertheless, a closely related model has been constructed and supported by experiments for *Arabidopsis* leaves [23] and it is likely that it extends to other plants, such as ivy. In summary, both the model of heterocyst differentiation in *Anabaena* and the model of leaf development illustrate the principles of computational model construction using L-systems and cell complexes. In spite of their simplicity, these models provide insights into pattern formation in nature.

6 Conclusions

Mathematics of multicellular development. Modeling multicellular systems in development requires a spatially discrete formalism, which sets it apart from the continuous treatment of time and space in classical mathematical physics. Cell complexes provide a convenient abstraction for representing topological relations between components of discrete structures. Variables describing a system can be associated in a natural manner with components of different dimensions within a

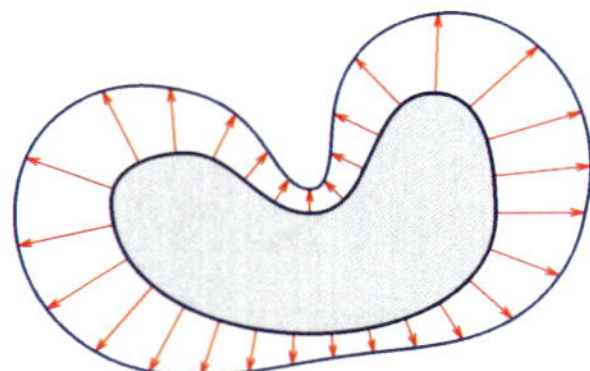

Fig. 9 Shape formation through boundary propagation. The outer shape results from the propagation of the inner shape in the normal directions, with a variable velocity depicted by red arrows

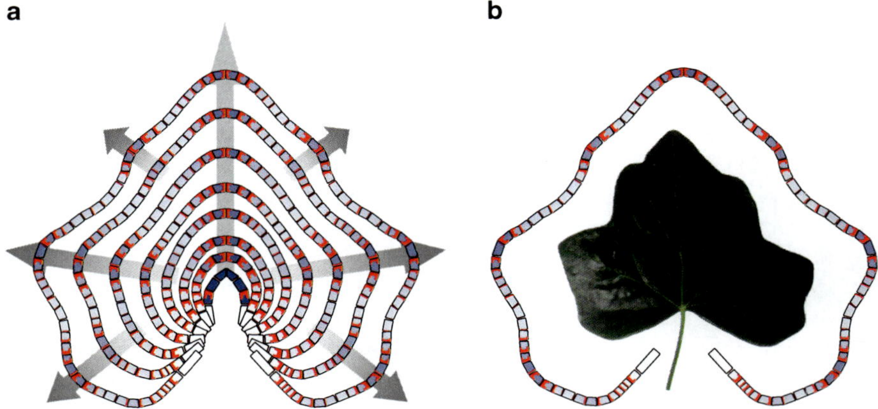

Fig. 10 A model of leaf development using the boundary propagation method. (**a**) Superposition of selected simulation steps. Propagation speed is controlled by auxin concentration and is the highest at the convergence points. (**b**) Comparison of the final shape generated by the model with an ivy leaf

cell complex, allowing for convenient storage of these variables in their topologically meaningful location [29]. Respecting the principle of locality, equations relating these variables may only refer to the variables in the neighboring components of the structure. The notion of cell complexes makes it possible to access these variables using an index-free notation. This is particularly important when dealing with systems in which structure changes dynamically, for example as a result of cell division. In the one-dimensional case, such changes can be conveniently expressed using the notion of L-systems. This provides an explanation of why L-systems work so well in modeling applications.

Molecular processes and pattern formation. We have illustrated modeling with cell complexes using examples of pattern formation in growing sequences of cells. Maxima of the concentration of a morphogen arise from an interplay between its local production and passive, diffusive transport, as in the model of *Anabaena*, or by reshuffling an existing or diffusely produced substance through active transport, as in the model of leaf margin. One question is whether these molecularly different mechanisms represent fundamentally different paradigms of pattern formation, or different implementations of a common principle. From the biological perspective it would also be interesting to know why such different mechanisms have evolved to create similar patterns.

Open problems. The confluence of L-systems and cell complexes provides a convenient framework for modeling one-dimensional developing structures. Modeling of higher-dimensional structures with dynamic cell complexes is substantially more difficult, and is a subject of ongoing research [5]. Difficulties extend to the visualization of three-dimensional models, where representations of the model's surface only provide partial information about the entire structure, and volumetric representations are often visually confusing.

Another problem open for further research concerns numerical methods for modeling structures with dynamic topology. Traditional formalisms for specifying and solving large systems of equations are based on matrix notation, which is not well suited for modeling multicellular organisms. First, matrices have fixed dimensions, so each time a cell divides, matrices describing the system globally have to be redefined. Second, the matrices are very sparse, since each cell can only be affected by a small number of neighboring cells due to the locality of interactions. General-purpose algorithms for solving systems of sparse equations use automatic techniques to identify which variables are connected through equations. However, constructing a sparse matrix and then identifying these connections represents unnecessary work, because a precise description of the connections between variables is already present in the topology of the complex. Thus, instead of expressing a structure using a matrix, and applying general methods for dealing with sparse matrices, it is better to operate directly on cell complexes [4, 29, 30]. The appropriate numerical methods have been devised in some contexts [12, 31–33], but a more complete toolbox of numerical methods designed for dynamic cell complexes is needed.

Acknowledgments This note is an edited transcript of a presentation by PP at the workshop *Pattern Formation in Morphogenesis* (IHES, Bures-sur-Yvette, January 11–14, 2010). The authors thank Linda Pritchard for preparing the original transcript, and Lynn Mercer for comments on the manuscript. Insightful discussions and continuing collaboration with Eric Mjolsness on the mathematical foundations of biological modeling, Carla Davidson on the development of *Anabaena*, and Miltos Tsiantis and Adam Runions on the modeling of leaves, are gratefully acknowledged. The reported research has been supported by the Natural Sciences and Engineering Research Council of Canada and the Human Frontier Science Program.

References

1. Lindenmayer A (1968) Mathematical models for cellular interaction in development, parts I and II. J Theor Biol 18:280–315
2. De Floriani L, Hui A (2007) Shape representation based on simplicial and cell complexes. In: Schmalsteig D, Bittner J (eds) Eurographics 2007 state of the art reports. The Eurographics Association, Prague, pp 63–87
3. Edelstein-Keshet L (1988) Mathematical models in biology. Random House, New York

4. Chard J, Shapiro V (2000) A multivector data structure for differential forms and equations. Math Comput Simulat 54:33–64, 15
5. Lane B, Harrison CJ, Prusinkiewicz P (2010) Modeling the development of multicellular structures using 3D cell complexes. In: DeJong T, Da Silva D (eds) Proceedings of the 6th international workshop on functional-structural plant models. University of California, Davis
6. Prusinkiewicz P (2009) Developmental computing. In: Calude C, Costa JF, Dershiwitz N, Freire E, Rozenberg G (eds) Unconventional computation. 8th international conference, UC 2009. Lecture notes in computer science, vol 5715. Springer, Berlin, pp 16–23
7. Weyl H (1949) Philosophy of mathematics and natural science. Princeton University Press, Princeton
8. Prusinkiewicz P (1986) Graphical applications of L-systems. In: Proceedings of Graphics Interface '86—Vision Interface '86. Canadian Information Processing Society, Toronto, pp 247–253
9. Baker R, Herman GT (1972) Simulation of organisms using a developmental model, parts I and II. Int J Bio Med Comput 3:201–215, and 251–267
10. Coen E, Rolland-Lagan A-G, Matthews M, Bangham A, Prusinkiewicz P (2004) The genetics of geometry. Proc Natl Acad Sci U S A 101:4728–4735
11. de Koster CG, Lindenmayer A (1987) Discrete and continuous models for heterocyst differentiation in growing filaments of blue-green bacteria. Acta Biotheor 36:249–273
12. Federl P, Prusinkiewicz P (2004) Solving differential equations in developmental models of multicellular structures expressed using L-systems. In: Bubak M (ed) Computational science—ICCS 2004 part II, vol 3037, Lecture notes in computer science. Springer, Berlin, pp 65–72
13. Hammel M, Prusinkiewicz P (1996) Visualization of developmental processes by extrusion in space-time. In: Proceedings of Graphics Interface '96, Canadian Information Processing Society. Toronto, Canada, pp 246–258
14. Lindenmayer A (1974) Adding continuous components to L-systems. In: Rozenberg G, Salomaa A (eds) L-systems, vol 15, Lecture notes in computer science. Springer, Berlin, pp 53–68
15. Prusinkiewicz P, Lindenmayer A (1990) The algorithmic beauty of plants. Springer, New York, with Hanan JS, Fracchia FD, Fowler DR, de Boer MJM and Mercer L
16. Haselkorn R (1978) Heterocysts. Annu Rev Plant Physiol 29:319–344
17. Haselkorn R (1998) How cyanobacteria count to 10. Science 282:891–892
18. Yoon H-S, Golden JW (1998) Heterocyst pattern formation controlled by a diffusible peptide. Science 282:935–938
19. Leyser O (2011) Auxin, self-organization, and the colonial nature of plants. Curr Biol 21: R331–R337
20. Gälweiler L, Guan C, Müller A, Wisman E, Mendgen K, Yephremov A, Palme K (1998) Regulation of polar auxin transport by AtPIN1 in *Arabidopsis* vascular tissue. Science 282:2226–2230
21. Hay A, Barkoulas M, Tsiantis M (2006) ASYMMETRIC LEAVES1 and auxin activities converge to repress *BREVIPEDICELLUS* expression and promote leaf development in *Arabidopsis*. Development 133:3955–3961, 16
22. Scarpella E, Marcos D, Friml J, Berleth T (2006) Control of leaf vascular patterning by polar auxin transport. Genes Dev 20:1015–1027
23. Bilsborough GD, Runions A, Barkoulas M, Jenkins HW, Hasson A, Galinha C, Laufs P, Hay A, Prusinkiewicz P, Tsiantis M (2011) Model for the regulation of *Arabidopsis thaliana* leaf margin development. Proc Natl Acad Sci U S A 108:3424–3429
24. Jönsson H, Heisler MG, Shapiro BE, Meyerowitz EM, Mjolsness E (2006) An auxin-driven polarized transport model for phyllotaxis. Proc Natl Acad Sci U S A 103:1633–1638
25. Smith RS, Guyomarc'h S, Mandel T, Reinhardt D, Kuhlemeier C, Prusinkiewicz P (2006) A plausible model of phyllotaxis. Proc Natl Acad Sci U S A 103:1301–1306

26. Draelants D, Broeckhove J, Beemster GTS, Vanroose W. Pattern formation in a cell based auxin transport model with numerical bifurcation analysis. Preprint available at arXiv:1202.5161v1 [math.DS]
27. Sahlin P, Söderberg B, Jönsson H (2009) Regulated transport as a mechanism for pattern generation: capabilities for phyllotaxis and beyond. J Theor Biol 258:60–70
28. Sethian J (1999) Level set methods and fast marching methods: evolving interfaces in computational geometry, fluid mechanics, computer vision, and materials science. Cambridge University Press, Cambridge
29. Desbrun M, Kanso E, Tong Y (2006) Discrete differential forms for computational modeling. In: Grinspun E, Schröder P, Desbrun M (eds) ACM SIGGRAPH 2006 course notes on discrete differential geometry. ACM, New York
30. Palmer R, Shapiro V (1993) Chain models of physical behavior for engineering analysis and design. Res Eng Des 5:161–184
31. Parter S (1961) The use of linear graphs in Gauss elimination. SIAM Rev 3:119–130
32. Prusinkiewicz P, Allen M, Escobar-Gutierrez A, DeJong T (2007) Numerical methods for transport-resistance source-sink allocation models. In: Vos J (ed) Functional-structural modeling in crop production. Springer, Dordrecht, pp 123–137
33. Sowell E, Haves P (1999) Numerical performance of the SPARK graph-theoretic simulation program. In: Proceedings of IBPSA building simulation '99, Kyoto, Japan

Multistability and Hysteresis-Based Mechanism of Pattern Formation in Biology

Alexandra Köthe and Anna Marciniak-Czochra

Abstract Multistability plays an important role in cell signalling. Coupled with the diffusion process, it may give rise to spatial patterns in chemical and biological systems. Such processes lead to nonlinear dynamical models with multiple steady states, which differ from the usual reaction–diffusion systems. To investigate mechanism of pattern formation based on these concepts we propose a model consisting of a reaction–diffusion equation coupled with an ordinary differential equation. The test organism for mathematical modelling is a fresh-water polyp *Hydra*. The model considered here is a minimal version of the receptor-based model with multistability proposed by Marciniak-Czochra (Math Biosci 199:97–119, 2006). It describes the dynamics of nonlinear intracellular signalling coupled to cell-to-cell communication via diffusive molecules and shows how bistability and hysteresis in the kinetic system may result in spatial patterning. In particular, it is shown that multistability without the hysteresis effect is not enough for creation of stable patterns. Biologically, this model explains results of experiments such as grafting which are not easily explicable by the pure reaction–diffusion (Turing type) patterning.

1 Developmental Pattern Formation on the Example of *Hydra*

Understanding the evolution of spatial patterns and the mechanisms which create them are among the crucial issues of developmental biology. During the development of multicellular organisms, embryonic cells choose differentiation programs based on positional information. This information is delivered by extracellular signalling molecules which are highly conserved and regulate the growth and differentiation of cells in all metazoans including humans [1]. In other areas of

A. Köthe • A. Marciniak-Czochra (✉)
Interdisciplinary Center of Scientific Computing (IWR) and BIOQUANT, University of Heidelberg, INF 267, Heidelberg 69120, Germany
e-mail: alexandra.koethe@bioquant.uni-heidelberg.de; anna.marciniak@iwr.uni-heidelberg.de

V. Capasso et al. (eds.), *Pattern Formation in Morphogenesis*, Springer Proceedings in Mathematics 15, DOI 10.1007/978-3-642-20164-6_13,

biology, such as neurophysiology or ecology, mathematical modelling has led to many discoveries and insights through a process of synthesis and integration of experimental data (see e.g. [2] and references therein). Also in developmental biology many different morphologies have been the subject of mathematical modelling, e.g., [3, 4]. Some biological systems have attained the status of "paradigm" in theoretical work.

One of the most frequently discussed organisms in theoretical papers on biological patterns formation is the fresh-water polyp *Hydra*. It is a simple organism which can be treated as a model for axis formation and regeneration in higher organisms [1, 5, 6]. *Hydra* has a tubular body about 5 mm long with a whorl of tentacles surrounding the mouth at the upper end and a disk-shaped organ for adhesion at the lower end. The longitudinal pattern is subdivided into a head, a gastric region, a budding zone (where new animals are generated by a process of natural cloning), a stalk and a foot. *Hydra* retains a population of stem cells which are activated when needed. Morphogenetic mechanisms active in adult polyps are responsible for the regenerative ability and the establishment of a new body axis. Research on *Hydra* might reveal how to selectively reactivate the genes and proteins to regenerate human tissues.

The developmental processes governing formation of the Hydra body plan and its regeneration are well understood at the tissue level [7]. Experiments performed on *Hydra* suggest that the function of cells is determined by their location. This can be shown by a simple cutting experiment (see Fig. 1): after a transverse cut, cells of the gastric region of the animal located at the upper end of the lower fragment differentiate and form the missing head, while the cells located at the lower end of the upper fragment regenerate the foot. Moreover, overlapping cut levels show that the same cells can form either the gastric region, or the head, or the foot, according to their position along the body axis. Experiments of this kind suggest that the cells respond to local positional cues that are dynamically regulated. The hypothesis is that cells differentiate according to positional information (compare Fig. 2). The question is how this information is supplied to the cells and which mechanisms are regulating formation of spatially heterogenous structures in the positional information, and consequently patterns of cell differentiation.

Although much is known at the tissue level, the molecular basis for self-organisation of the body plan of *Hydra* is not well understood. Since cell differentiation and proliferation is controlled by signalling pathways, it is necessary to bridge the gap between observations at the tissue level and at the cellular and subcellular level.

The advent of new techniques in molecular biology has made it possible to advance the understanding of the development of multicellular organisms. Large scale expression screening helps to identify new factors involved in embryonic development. Recently, expression analysis during regeneration and budding indicated a pivotal role of the Wnt (wingless gene) pathway in the *Hydra* head organizer [5, 6]. Also the evidence of Dkk (Dickkopf) signalling in *Hydra* regeneration was provided [1, 5, 6, 8]. Experimentally observed patterns of Wnt and Dkk gene expression give rise to many new questions. New aspects of Wnt signalling, such as bistability in Wnt dynamics and switches in the Wnt and Dkk functionality

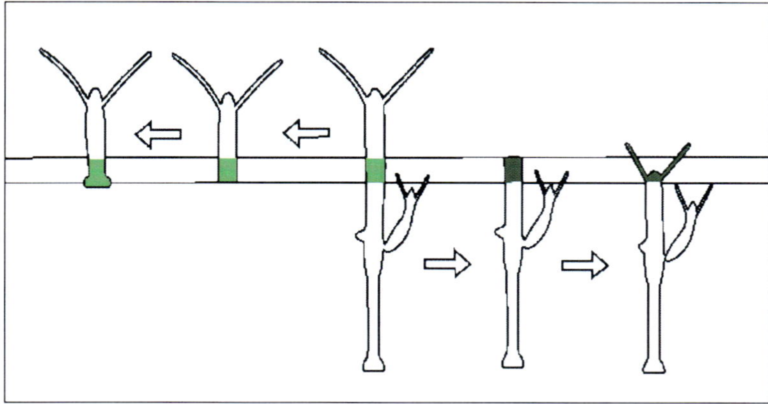

Fig. 1 Cutting experiment. *Hydra* regenerates after a transverse cut of cells of the gastric region (from both upper and lower half of the body column). The experiment shows regeneration abilities of the animal

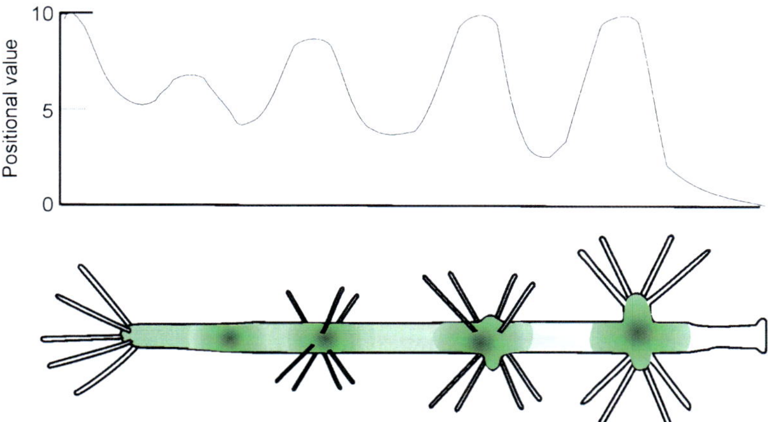

Fig. 2 The illustration of the idea of "positional value", which is supplied to the cells and interpreted by them. The hypothesis is that the formation of the head is determined by the high "positional value" (which is above some threshold). The figure shows the "positional value" for a supernumerary head structure

depending on the cellular context provide examples supporting the view that multistability plays an important role in signaling [9–11]. Mathematical modelling can help to integrate these concepts and observations into a new model of pattern formation controlled by the intracellular dynamics of Wnt signalling. The approach we propose in this paper is a step in this direction.

The exact molecular mechanism of pattern formation in *Hydra* is unknown and the proposed models are hypothetical. Experiments performed on *Hydra* provide an opportunity to test the abilities and limits of the models. We investigate what kind of mechanisms regulating the dynamics of cell surface receptors and diffusing biochemical molecules can explain the observed results of experiments. One of attempts is to understand what minimal processes are sufficient to produce patterns. In this work we focus on a minimal version of the receptor-based model with hysteresis proposed in [12] to describe *Hydra* head formation. We apply analysis and numerical simulations to shed the light on the properties of hysteresis-based mechanism of pattern formation, which is very different than Turing-type mechanism.

2 Mathematical Models of Pattern Formation

Classical mathematical models of pattern formation in cell assemblies have been constructed and developed using reaction–diffusion equations. They have provided explanations of pattern formation for animal coat markings, bacterial and cellular growth patterns, angiogenesis (blood vessels), tumour growth and tissue development.

The existing qualitative theory of reaction–diffusion systems is mainly focused on stability conditions for homogeneous steady states and concerns the destabilisation of the stationary spatially homogeneous state and the growth of spatially heterogeneous patterns [2, 13]. The mechanism responsible for such behaviour of model solutions is diffusion-driven instability (Turing-type instability). It has been applied to a variety of pattern formation problems in biology (see [2, 14] and references therein). The theory of systems functioning far from equilibrium is not very well developed and limited to particular cases, e.g., systems with monotone kinetics [15]. Moreover, the majority of theoretical studies focus on the analysis of the non-degenerated reaction–diffusion systems, i.e., with a strictly positive diffusion coefficient in each equation. However, in many biological applications, it is relevant to consider receptor-based systems involving ordinary differential equations.

2.1 Novel Pattern Formation Mechanism Based on the Multistability in Signalling Pathways

It is becoming increasingly clear that multistability plays an important role in cell signalling. Biological examples of bistable systems have included the hysteretic lac repressor system, cell-cycle transitions [16], and cell division cycles of fission yeast [17]. The first experimental demonstration of hysteresis in the eukaryotic cell engine was reported in [18]. These processes were modelled using ordinary differential equations. However, it is known that multistability in signalling coupled with the diffusion process may lead to spatial patterns in chemical and biological systems, such as Liesegang rings formed by precipitating colloids [7] and bacterial growth patterns [19]. Such processes lead to nonlinear dynamical models with multiple steady states, which differ from the usual reaction–diffusion systems. Also, processes containing switching between different pathways or states lead to new types of mathematical models, which consist of nonlinear partial differential equations of diffusion, transport and reactions, coupled with dynamical systems controlling the transitions.

2.2 Receptor-Based Models

In general, equations of such models can be represented by the following initial-boundary value problem

$$
\begin{aligned}
U_t &= D\Delta U + F(U,V), & \\
V_t &= G(U,V) & \text{in} \quad \Omega, \\
\partial_n U &= 0 & \text{on} \ \partial\Omega, \\
U(x,0) &= U_{init}(x), & \\
V(x,0) &= V_{init}(x), &
\end{aligned}
$$

where U is a vector of variables describing the dynamics of diffusing extracellular molecules and enzymes, which provide cell-to-cell communication, while V is a vector of variables localised on cells, describing cell surface receptors and intracellular signalling molecules, transcription factors, mRNA, etc. D is a diagonal matrix with positive coefficients on the diagonal, the symbol ∂_n denotes the normal derivative (no-flux condition), and Ω is a bounded region.

The receptor-based models for pattern formation follow the idea is that positional value of the cell can be determined by the density of bound receptors, which do not diffuse [29]. Regulatory and signalling molecules (ligands) act by binding and activating receptor molecules which are located in the cell membrane (or, with lipophilic ligands, in the cytoplasm). In the model, we assume that free and bound receptors exist only on the surfaces, while ligands are transported by diffusion within the extracellular space. A ligand reversibly binds to a free receptor, which results in a bound receptor that can be removed from the cell surface due to degradation or internalisation, or dissociate back to free receptors and ligands. Both ligands and free receptors are produced *de novo* and undergo natural decay. Binding of a soluble ligand to a cell surface receptor requires interaction of molecules diffusing in the three-dimensional space with some molecules attached to a two-dimensional surface. Since the size of a cell is very small compared to the dimension of the tissue, it is convenient to treat the systems including cells as multiscale systems (see Fig. 3). To study dynamics of the whole array of cells, it is convenient to derive a model which describes all the processes on the macro-scale level represented by a two- or three-dimensional sheet of cells. A rigorous

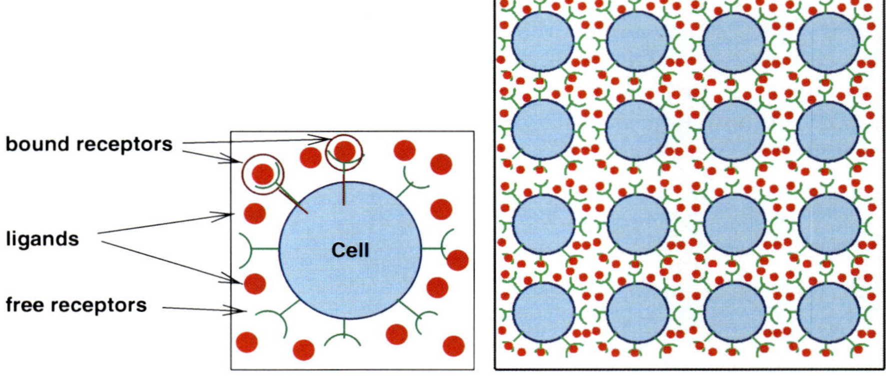

Fig. 3 Receptor-based model. The array of the cells (on the *right-hand side*) consists of periodic repetition of cells with the surrounding intercellular space. The model takes into account dynamics of free and bound receptors on cell membranes and diffusing signalling molecules. The effects of intracellular dynamics are modeled via nonlinear functions describing production of new signalling molecules and free receptors

derivation, using methods of asymptotic analysis (homogenisation), of the macroscopic reaction–diffusion models describing the interplay between the nonhomogeneous cellular dynamics and the signaling molecules diffusing in the intercellular space has been recently published in [20] and [21]. It is shown that receptor-ligand binding processes can be modelled by reaction–diffusion equations coupled with ordinary differential equations in the case when all membrane processes are homogeneous within the membrane, which seems to be the case in most of processes. If homogeneity of the processes on the membrane does not hold, equations with additional integral terms are obtained, see [20].

The receptor-based models give rise to interesting phenomena such as threshold behaviour and hysteresis when the steady state equation $G(U,V)=0$ has multiple solutions $V_i = H_i(U)$. These concepts proved to be basis for the explanation of the morphogenesis of *Hydra* [12], dorso-ventral patterning in **Drosophila** [22] as well as for modelling of the formation of growth patterns in populations of microorganisms [19].

3 Models with Multistability and Hysteresis for *Hydra*

Transplantation and tissue manipulation experiments provided data for models of patterning in Hydra. In turn, theoretical models had a strong influence on experimental design, starting with the positional information ideas of Wolpert [23], the activator-inhibitor model of Gierer and Meinhardt [3, 24] and, finally, receptor-based models of Marciniak-Czochra [12, 14].

As shown in [14] and also recently highlighted in [25] receptor-based models may also exhibit Turing-type instability and lead to stable spatially heterogenous patterns. Since the shape of regular Turing pattern depends on the size of domain and diffusion and rather than on initial conditions, one of the difficulties of the Turing-type models is their inability of reproducing the experiments resulting in multiple head formation in Hydra [14].

Transplantation experiments indicate co-existence of different stable patterns. The patterns may have multiple peaks, which depend on the local cues induced by the grafted tissue. Such experiments suggest a mechanism of pattern formation based on multistability in intracellular signalling. The model proposed in [12] can explain these experiments.

The model is based on the hypothesis that the density of bound cell-surface receptors regulates the expression of genes responsible for cell differentiation, i.e., provides positional information to the cells. However, importantly, it also includes a hysteresis-based relation in the quasi-stationary state in the subsystem describing receptor dynamics, $G(U,V) = 0$. It allows for formation of gradient-like patterns corresponding to the normal development as well as emergence of patterns with multiple maxima describing transplantation experiments (see Fig. 4).

The model suggests how the nonlinearities of the intracellular signalling may result in spatial patterning. Numerical simulations show the existence of stationary and oscillatory patterns, resulting from the existence of multiple steady states and switches in the production rates of diffusing molecules.

3.1 *The Prototype Model of Hysteresis-Based Pattern Formation*

In the remainder of this paper we investigate the role of multiple steady states and switches in the dynamics of the ODE-subsystem on the example of a basic model of a reaction–diffusion equation coupled to an ordinary differential equation.

We focus in the following on a basic reaction–diffusion model exhibiting the multistability and hysteresis-driven mechanism of pattern formation,

$$u_t = u_{xx} + f(u,v), \tag{1}$$

$$v_t = g(u,v), \tag{2}$$

for $x \in [0,L]$, with zero-flux boundary condition for u and the kinetic functions given by

$$f(u,v) = \alpha v - \beta u + \gamma,$$
$$g(u,v) = \mu u - p(v).$$

To model the hysteresis effect we choose $p(v)$ as a non-monotone polynomial of degree 3 with only one real zero at $v = 0$. These kinetics are heuristic and aim to

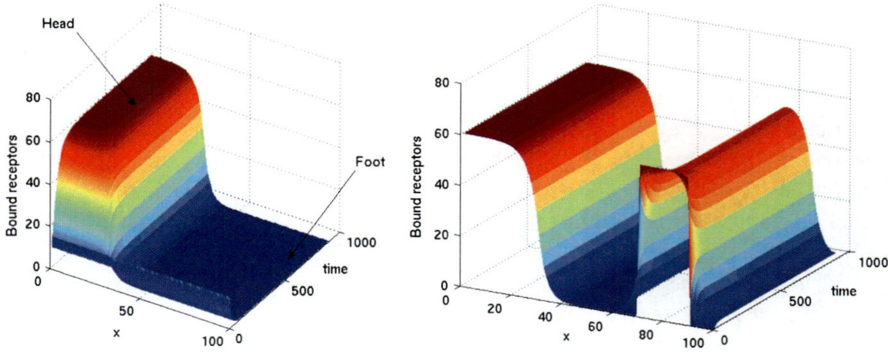

Fig. 4 Simulations of the receptor-based model with hysteresis: formation of a gradient-like pattern corresponding to a normal development and head formation in *Hydra* (*left panel*) and formation of two heads pattern for the initial conditions corresponding the transplantation experiment (*right panel*)

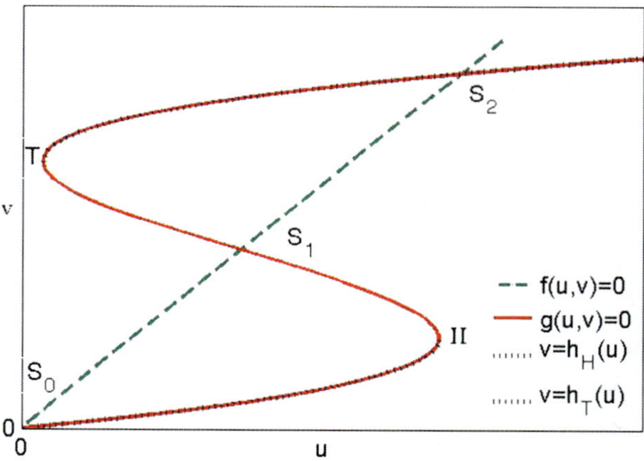

Fig. 5 A typical configuration of the kinetic functions

summarize the interactions of a complicated cascade of biochemical intracellular and extracellular reactions.

We assume that the local maximum $H = (u_H, v_H)$ and the local minimum $T = (u_T, v_T)$ of $v \longrightarrow p(v)$ have positive coordinates. Furthermore, we assume that there are three intersection points of $f = 0$ and $g = 0$ with nonnegative coordinates $S_0 = (u_0, v_0), S_1 = (u_1, v_1)$, which is between T and H, and $S_2 = (u_2, v_2)$. Without loss of generality, we can take $\gamma = 0$ and $\mu = 1$ using the transformation $u \longrightarrow u - u_0$ and $v \longrightarrow \frac{1}{\mu}(v - v_0)$. Consequently, we obtain $S_0 = (0, 0)$.

As $v \longrightarrow p(v)$ is non-monotone, it can be inverted only locally. We call $u \longrightarrow h_H(u)$ the branch of the inverse which is defined on $(-\infty, u_H]$ and similarly $u \longrightarrow h_T(u)$ is the branch, defined on $[u_T, \infty)$ (see Fig. 5).

The system has three spatially homogeneous steady states S_0, S_1, S_2. Based on linear stability analysis we conclude that S_0 and S_2 are stable, while S_1 is a saddle. Furthermore, the system cannot exhibit diffusion-driven instability (Turing-type instability).

The system of the two equations (1) and (2) can be interpreted as a basic model exhibiting the dynamics of the receptor-based model in [12] with u describing a density of diffusing ligands (signalling molecules) and v standing for the rate of their production. In fact, it is a minimal reaction–diffusion model which may describe formation of stable patterns. In the next section we show that replacing the nonlinearities so that $g(u, v)$ has no singular points, i.e., $\partial_v g$ is always strictly positive, leads to instability of the nonhomogeneous stationary solutions.

3.2 Instability of Spatial Patterns in Models Without Hysteresis

In this section we show that the system with multistability but reversible quasi-steady state the ODE subsystem cannot exhibit stable spatially heterogeneous patterns. Hysteresis is necessary to obtain stable patterns.

We consider system (1) and (2) with

$$f(u, v) = \alpha v - \beta u,$$
$$g(u, v) = u - p(v).$$

Again we assume, that there are three intersection points $S_0 = (0,0), S_1 = (u_1, v_1)$ and $S_2 = (u_2, v_2)$ with nonnegative coordinates of $f = 0$ and $g = 0$. But in contrast to the hysteresis case, we assume $p(v)$ to be a monotone polynomial of degree 3 (see Fig. 6a).

Theorem 1 *There exists no stable spatial pattern (i.e., spatially non-homogeneous stationary solutions) of system* (1) – (2) *with a monotone* $p(v)$.

Proof Since $u = p(v)$ is monotone, we can invert it globally. We call the inverse $v = h(u)$. Taking the steady state solution of (2), $v = h(u)$, and inserting it into (1) with u_t = 0 yields the following two point boundary value problem

$$0 = u_{xx} + f(u, h(u)) = u_{xx} + \alpha h(u) - \beta u, \tag{3}$$

with zero-flux boundary conditions. Along the trajectories it holds,

$$\frac{u_x^2(x)}{2} + Q(u(x)) = constant, \tag{4}$$

where

$$Q(u) := \int_0^u (\alpha h(\tilde{u}) - \beta \tilde{u}) d\tilde{u} = F(h(u)) \quad \text{with}$$
$$F(v) = \int_0^v (\alpha \tilde{v} - \beta p(\tilde{v})) p'(\tilde{v}) d\tilde{v}. \tag{5}$$

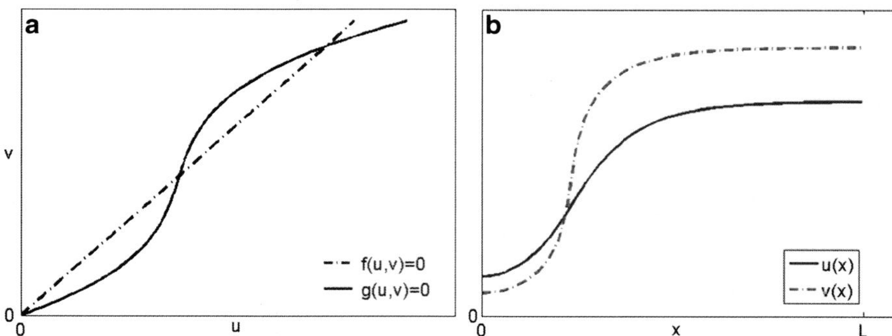

Fig. 6 The model with bistability but without hysteresis effect. *Panel A*: kinetic functions in steady state in case of monotone function $p(v)$. *Panel B*: stationary solutions for u and v, which is always unstable as shown in Theorem 1

As $p(v)$ is monotone, we obtain $\frac{\mathrm{d}}{\mathrm{d}v}F(v) = \alpha v - \beta p(v))p'(v)$, which is equal to zero in 0, v_1 and v_2. Consequently, $Q(u)$ has a minimum on the interval $(0, u_2)$, and it is possible to find values such that $0 < u_0 < u_{end} < u_2$ with $Q(u_0) = Q(u_{end})$. Furthermore we can calculate the so-called time-map, which gives the length of the interval one needs to travel from u_0 to u_{end}

$$L = \frac{1}{\sqrt{2}} \int_{u_0}^{u_{end}} \frac{du}{\sqrt{Q(u_0) - Q(u)}}. \tag{6}$$

When $u_0 \to 0$ (resp. $u_{end} \to u_2$) then L tends to infinity. Consequently, there exists exactly one solution, U, on each interval length L sufficiently large.

We now show that the solution, $(U(x), V(x) = h(U(x)))$, is unstable. Therefore we consider the linearisation $\mathcal{L}$ of the operator

$$\begin{pmatrix} u \\ v \end{pmatrix} \mapsto \begin{pmatrix} u_{xx} \\ 0 \end{pmatrix} + \begin{pmatrix} f(u,v) \\ g(u,v) \end{pmatrix}$$

around the steady state solution $(U(x), V(x))$.

$$\mathcal{L}\begin{pmatrix} \eta \\ \psi \end{pmatrix} = \begin{pmatrix} \eta_{xx} \\ 0 \end{pmatrix} + \begin{pmatrix} -\beta & \alpha \\ 1 & -p'(V(x)) \end{pmatrix} \cdot \begin{pmatrix} \eta \\ \psi \end{pmatrix}$$

The solution of the eigenvalue problem $\mathcal{L}\begin{pmatrix} \eta \\ \psi \end{pmatrix} = \lambda \begin{pmatrix} \eta \\ \psi \end{pmatrix}$ has to fulfill the equation

$$\lambda \eta = \eta_{xx} - \beta \eta + \frac{\alpha}{\lambda + p'(V)} \eta$$

with zero-flux boundary conditions. To show, that this problem has a positive eigenvalue, we define the operator $A(\lambda) : \eta \to \eta_{xx} - \beta\eta + \frac{\alpha}{\lambda+p'(V)}\eta$ and $\eta_x(0) = \eta_x(L) = 0$ and observe that $\frac{\alpha}{\lambda+p'(V)} - \beta$ is bounded independently of λ. Denoting $\mu(\lambda)$, respectively $\nu(\lambda)$, the spectrum of $A(\lambda)$ with zero-flux, respectively Dirichlet boundary conditions and using the Sturm comparison principle, we obtain that

$$\mu(\lambda) > \nu(\lambda).$$

Since $U_x(x)$ is an eigenfunction for the eigenvalue 0 of the Dirichlet problem,

$$0 = (U_x)_{xx} + \alpha h'(U)U_x - \beta U_x = (U_x)_{xx} + \left(\alpha \frac{1}{p'(V)} - \beta\right)U_x = A(0)U_x,$$

we conclude that $\nu(0) \geq 0$ and, therefore, $\mu(0)>0$. Furthermore, $\mu(\lambda)$ depends continuously on λ and it can be calculated by

$$\mu(\lambda) = \sup_{\eta\in W^{1,2}(0,L)||\eta||_{L_2}=1} \langle A\ (\lambda)\eta, \eta\rangle. \tag{7}$$

Finally,

$$\begin{aligned}\langle A\ (\lambda)\eta, \eta\rangle &= \int_0^L \eta_{xx}\eta dx + \int_0^L \left(\frac{\alpha}{\lambda + p'(V)} - \beta\right)\eta^2 dx \\ &\leq - \int_0^L \eta_x\eta_x dx + c\int_0^L \eta^2 dx \\ &\leq c \end{aligned} \tag{8}$$

The boundedness of $\mu(\lambda)$ yields the existence of $\bar{\lambda} > 0$ fulfilling $\mu(\bar{\lambda}) = \bar{\lambda}$. Consequently, there exists a function $\bar{\eta} \neq 0$ with

$$A\ (\bar{\lambda})\bar{\eta} = \mu(\bar{\lambda})\bar{\eta} = \bar{\lambda}\bar{\eta} \quad \text{with} \quad \bar{\eta}_x\ (0) = \bar{\eta}_x\ (L) = 0.$$

This theorem shows, that bistability in the reduced model is not sufficient for pattern formation. One necessarily needs hysteresis to get stable patterns.

4 Hysteresis-Driven Pattern Formation

In this section we investigate spatially heterogeneous stationary solutions of system (1) – (2) with hysteresis effect. The equation $g(u, v) = 0$ has three local solutions, but only the outer branches $v = h_T(u)$ and $v = h_H(u)$ can give rise to stable stationary solutions. We define $f_H(u) = f(u, h_H(u))$ for $u \leq u_H$ and $f_T(u) = f(u, h_T(u))$ for

$u_T \leq u$. Using phase plane analysis, we investigate the solutions of the two point boundary value problem, $u_{xx} + f_i(u) = 0$ for $i = H, T$ with zero-flux boundary conditions.

Proposition 1 *Neither* $u_{xx} + f_H(u) = 0$, *nor* $u_{xx} + f_T(u) = 0$ *has a nonconstant solution fulfilling zero-flux boundary conditions.*

Proof We rewrite the equations as systems of first order ODEs

$$\begin{aligned} u_x &= w \\ w_x &= -f_i(u) \quad \text{for} \quad i = H, T. \end{aligned}$$

By definition $f_H(0) = 0$ and $-f_H(u) > 0$ for $u \in (0, u_H)$. That means that the flux at the point $(u_0, 0)$ always points upwards and to the right. Therefore, a solution starting at $(u_0, 0)$ for $0 < u_0$ will never reach the w-axis again. Similarly, $f_T(u_2) = 0$ and $-f_T(u) < 0$ for $u \in (u_T, u_2)$. Consequently, all orbits ending at $(u_{end}, 0)$ with $u_{end} < u_2$ have started at some point with positive w-component.

We may conclude that there exist no stable patterns, which are continuous in v. Next, we construct transition layer solutions by gluing the phase planes together at an arbitrary chosen $\bar{u} \in [u_H, u_T]$ (see Fig. 7), by solving the problem

$$u_{xx} + f_{\bar{u}}(u) = 0 \qquad \text{with} \quad u_x(0) = u_x(L) = 0, \tag{9}$$

with

$$f_{\bar{u}}(u) = \begin{cases} f_H(u) & \text{for} \quad u \leq \bar{u} \\ f_T(u) & \text{for} \quad u > \bar{u}. \end{cases}$$

Here, we do not expect u to be a C^2-function on the whole interval [0, L], but there is some value $\bar{x}$ with u $\bar{x} = \bar{u}$, so that we have

$$u \in C^2([0, \bar{x})) \cap C^2((\bar{x}, L]) \cap C^0([0, L])$$

and

$$v \in C^0([0, \bar{x})) \cap C^0((\bar{x}, L])$$

as v is given by

$$v\ (x) = \begin{cases} h_H(u(x)) & \text{if} \quad u \leq \bar{u} \\ h_T(u(x)) & \text{if} \quad u > \bar{u}. \end{cases}$$

Crucial for the analysis of (9) is the behaviour of the potential

$$Q_{\bar{u}}(u) = \int_0^u f_{\bar{u}}(\tilde{u}) d\tilde{u},$$

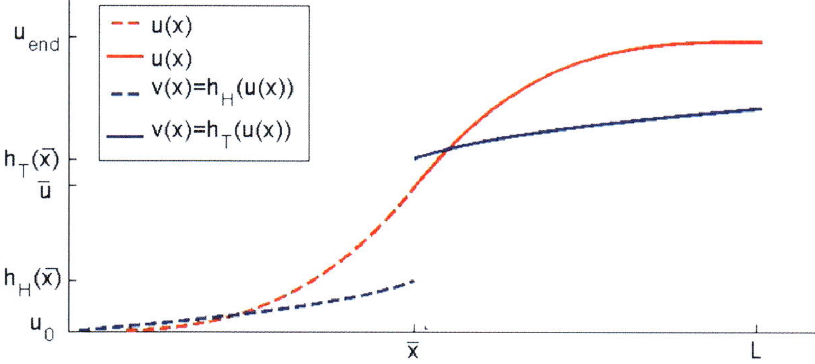

Fig. 7 An example of the solution u connecting u_0 with u_{end} and the corresponding $v(u)$

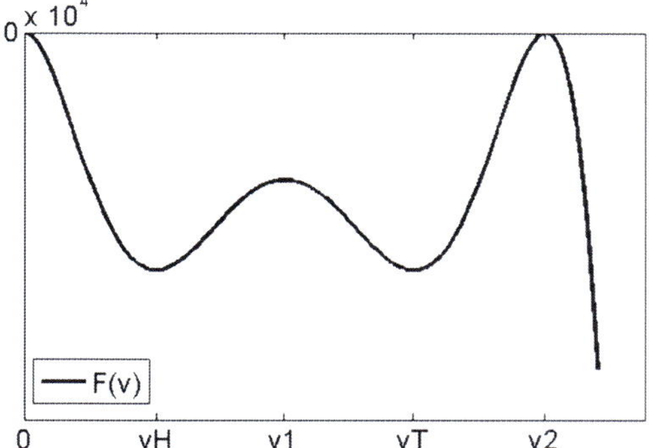

Fig. 8 $F(v) = \int_0^v f(p(\tilde{v}), \tilde{v})p'(\tilde{v})d\tilde{v}$ of model (1) and (2) with $\alpha = 11.8$, $\beta = 1$, $\gamma = 0$, $\mu = 1$ and $p(v) = 0.1v^3 - 6 - 3v^2 + 100v$. Corresponding potentials $Q_u(u)$ obtained for different $\bar{u}$ are plotted in Fig. 9

which fulfills $Q_{\bar{u}}(u) = f_{\bar{u}}(u)$ and $Q_{\bar{u}}(0) = 0$. Along the trajectories, it holds

$$\frac{u_x^2(x)}{2} + Q_{\bar{u}}(u(x)) = constant \quad \text{for all } x. \tag{10}$$

To investigate the behaviour of $Q_{\bar{u}}$, we define (see Fig. 8)

$$F(v) = \int_0^v f(p(\tilde{v}), \tilde{v})p'(\tilde{v})d\tilde{v}$$

and, after a change of variables $\tilde{u} \mapsto p(\tilde{u})$ in the definition of $Q_{\bar{u}}(u)$, we obtain

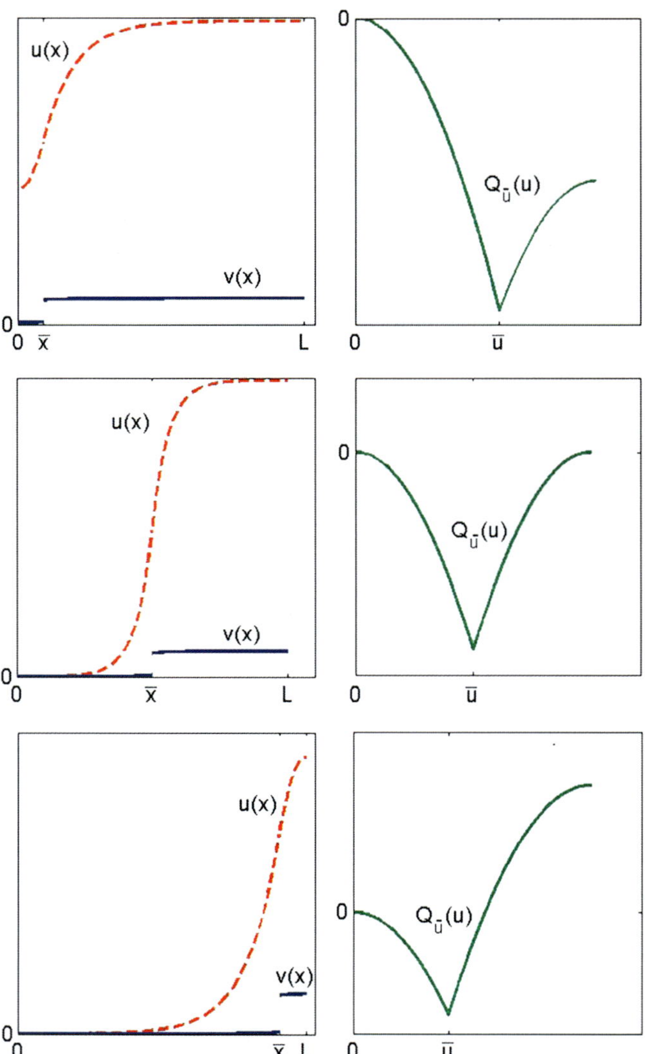

Fig. 9 *Left panel*: stationary solutions for u (transition layers) and v (discontinuous solution) of model (1)–(2) with $\alpha = 11.8$, $\beta = 1$, $\gamma = 0$, $\mu = 1$ and $p(v) = 0.1v^3 - 6.3v^2 + 100v$. *Right panel*: corresponding potentials $Q_{\bar{u}}(u)$. We observe that depending on the choice of $\bar{u}$, shape of $Q_{\bar{u}}(u)$ differs and, consequently, we obtain different shapes of the stationary solution. In case of $Q_{\bar{u}}(u_2)<0$, the pattern achieves near-maximum values on a long x-interval (*upper panel*); in the symmetric case $Q_{\bar{u}}(u_2) = 0$ both high and small values are expressed on similar subdomains (*middle panel*); in case of $Q_{\bar{u}}(u_2)>0$, the pattern achieves near-minimum values on a long x-interval (*lower panel*)

$$Q_{\bar{u}}(u) = \begin{cases} F(h_H(u)) & \text{if} \quad u \le \bar{u} \\ F(h_H(\bar{u})) - F(h_T(\bar{u})) + F(h_T(u)) & \text{if} \quad u > \bar{u}. \end{cases}$$

In particular, (10) yields the following relationships for the solution of (9):

- $Q_{\bar{u}}(u_0) = Q_{\bar{u}}(u_{end})$, which is equivalent to $F(h_H(u_0)) - F(h_H(\bar{u})) = F(h_T(u_{end})) - F(h_T(\bar{u}))$,
- $Q_{\bar{u}}(u)$ has a local minimum at $\bar{u}$, therefore, $\bar{u} < u$ is necessary.
- $Q_{\bar{u}}$ has a local maximum at u_2, so the possible range for u_0 and u_{end} depends on $Q_{\bar{u}}(u_2)$.

Additionally, we shall verify that a solution constructed above reaches u_{end} at $x = L$. Therefore, we consider the so-called "timemaps" and define

- $T_1(u_0)$ as the "time" x such that the orbit starting at $(u_0, 0)$ needs to reach for the first time the $u = \bar{u}$ axis and
- $T_2(u_{end})$ as the "time" such that the orbit starting at the $u = \bar{u}$ axis needs to reach $(u_{end}, 0)$ for the first time.

Using (10) and the boundary condition at $x = 0$, we obtain, for all $x \in [0, \bar{x}]$,

$$\frac{u_x(x)^2}{2} + Q_{\bar{u}}(u) = Q_{\bar{u}}(u_0)$$

and

$$x = \frac{1}{\sqrt{2}} \int_{u_0}^{u(x)} \frac{du}{\sqrt{Q_{\bar{u}}(u_0) - Q_{\bar{u}}(u)}}.$$

In particular, it holds

$$T_1(u_0) = \frac{1}{\sqrt{2}} \int_{u_0}^{\bar{u}} \frac{du}{\sqrt{Q_{\bar{u}}(u_0) - Q_{\bar{u}}(u)}} = \frac{1}{\sqrt{2}} \int_{u_0}^{\bar{u}} \frac{du}{\sqrt{F(h_H(u_0)) - F(h_H(u))}}.$$

Similarly, using the boundary condition at $x = L$, we obtain

$$T_2(u_{end}) = \frac{1}{\sqrt{2}} \int_{\bar{u}}^{u_{end}} \frac{du}{\sqrt{Q_{\bar{u}}(u_{end}) - Q_{\bar{u}}(u)}} = \frac{1}{\sqrt{2}} \int_{\bar{u}}^{u_{end}} \frac{du}{\sqrt{F(h_T(u_{end})) - F(h_T(u))}}.$$

It leads to the following condition, $T_1(u_0) + T_2(u_{end}(u_0)) = L$, where the dependence $u_{end}(u_0)$ is given by $Q_{\bar{u}}(u_0) = Q_{\bar{u}}(u_{end})$. Furthermore we obtain a condition for the value $\bar{x}$, which has to respect $\bar{x} = T_1(u_0)$.

Furthermore, the shape of the stationary solution depends on the symmetry properties of the potential $Q_{\bar{u}}$. In Fig. 9 we present three representative cases of

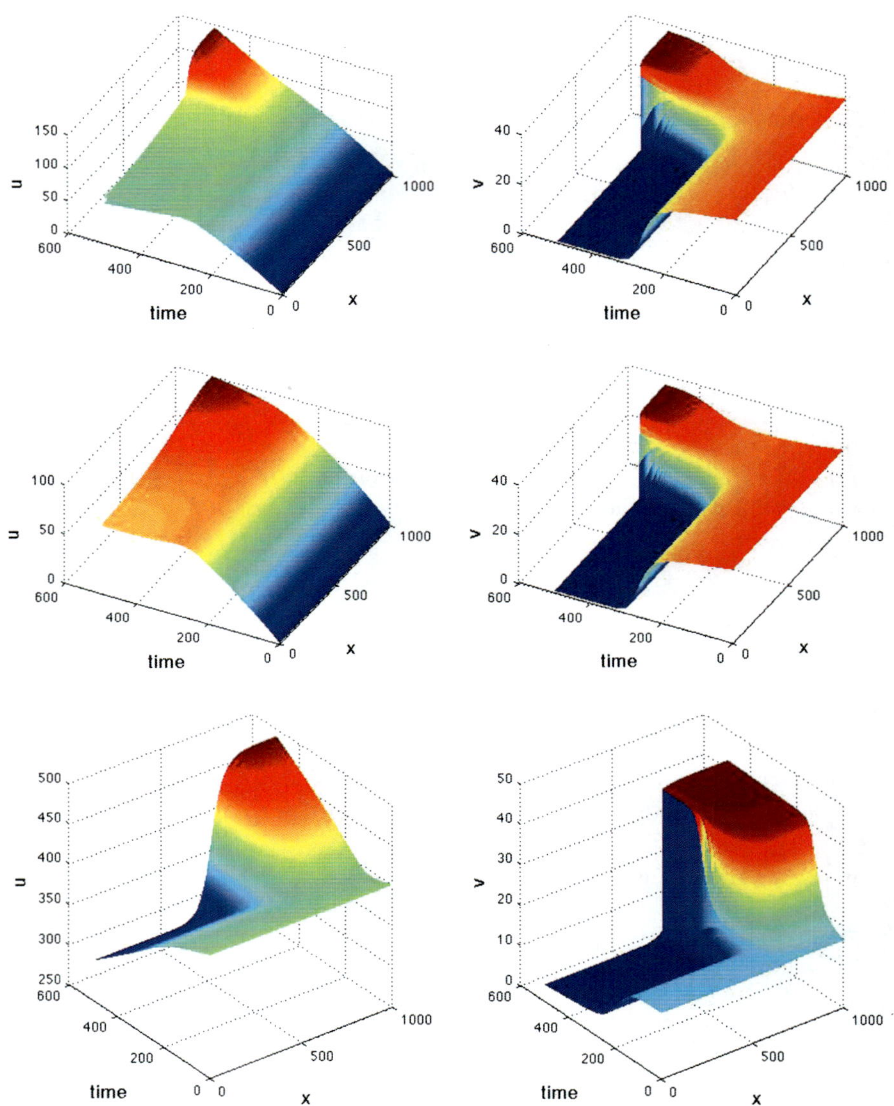

Fig. 10 Time-dependent simulation of model (1)–(2) showing emergence of a gradient-like pattern from a nearly homogeneous state. The simulations were performed for parameters given in Fig. 9 in the *first* and *second upper panel* and $\alpha = 17$ in the *third panel*. *First two panels* show differences between dynamics of the same system with different diffusion coefficients, which is $d = 1/(L^2) = 0.01$ and $d = 1/(L^2) = 1$ respectively. The simulations are performed for constant initial data $u_0 = 0.1$ and a gradient-line initial data v_0 with a very small amplitude (between $v_0 = 30.3$ and $v_0 = 30.4$ in the *two upper panels* and between 16 and 16.5 in the *lower panel*). The shape of emerging pattern depends strongly on the initial perturbation

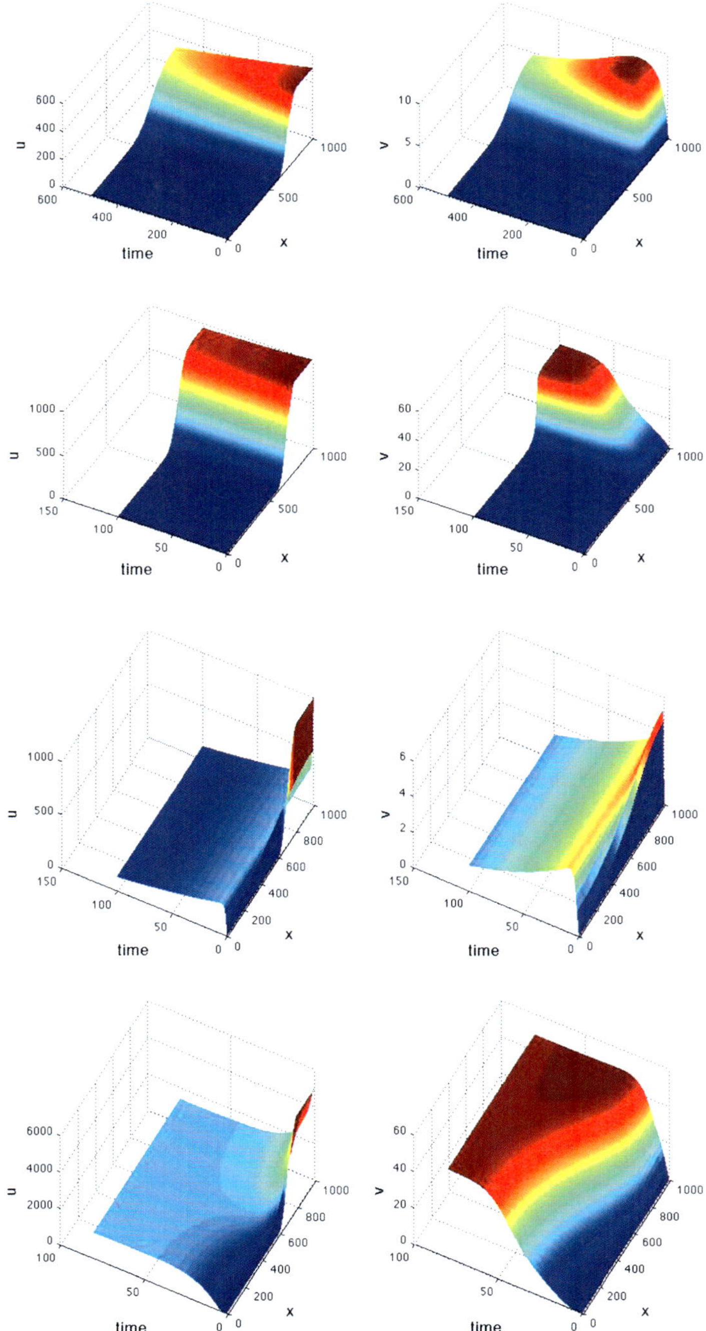

Fig. 11 Time-dependent simulations of model (1)–(2) with parameters given in Fig. 9 and diffusion coefficients $d = 1/(L^2) = 0.01$ in the *first* and *second upper panel* and $d = 1/(L^2) = 10$ in the *two lower panels*, respectively. The simulations were performed for the constant initial data

the stationary solutions obtained for different values of $\bar{u}$ and the same kinetics and corresponding potentials.

Let us notice that here we obtain the value of L (or equivalently diffusion size) for a chosen $\bar{u}$ and u_0. However, in the time-dependent problems the size of the domain (or diffusion) is given and values $\bar{u}$ and u_0 are established dynamically depending on the initial conditions. Numerical simulations of the time-dependent problem show that, for fixed parameters of the model, pattern selection depends strongly on the initial conditions, see Figs. 10 and 11. Consequently, the values of $\bar{u}$ and u_0 are determined by these data. On the other hand, for a given initial data, the solutions may tend to different stationary solutions (spatially homogeneous or heterogeneous) for different sizes of diffusion, see Fig. 12.

5 Conclusions

Novel predictions of the receptor-based models concerning the regeneration process, which assumed the existence of strong localised signal lead to new hypotheses, which may be tested experimentally. New aspects of Wnt signalling, such as bistability in Wnt dynamics guided us in our mathematical model of *Hydra* development. Interestingly, the same effects were recently found in other model organisms [22, 26]. These observations are consistent with the predictions of the proposed models [12]. A separate interesting question for mathematical modelling concerns the mechanism of the formation of a serial and overlapping expression pattern of different Wnt proteins, which transcripts have been isolated from a starlet sea anemone *Nematostella vectensis* [26, 27].

Some of the factors and mechanisms playing a role in developmental processes are also important in carcinogenesis. Since Wnt regulates proliferation in adult tissues, perturbations in Wnt signalling promote human degenerative diseases and cancer. An understanding of the spatio-temporal Wnt signalling in development also will have impacts in understanding the mechanisms of growth of colon and lung cancer, e.g., [28].

On the mathematical side, in this paper we showed how the nonlinearities of signalling pathways may result in spatial patterning. We studied a hysteresis-driven mechanism resulting from the existence of multiple steady states in the quasi-stationary subsystem. In comparison to diffusion-driven instability, hysteresis-driven pattern formation can take place in a model consisting of only one reaction–diffusion equation coupled with one ODE. It leads to the formation of transition layer patterns.

Fig. 11 (continued) $v_0 = 0.01$ and u_0 equal to 500, 1,000, 1,000 and 5,000 for $x \in [0.8, 1]$ connected in a smoothly to $u_0 = 0.1$ for $x \in [0, 0.6]$. We observe that in case of a large diffusion coefficient the solutions tend, depending on the initial data, to the lower or higher constant steady state. However, in case of small diffusion and large enough initial data, the solution may tend to a gradient-like pattern

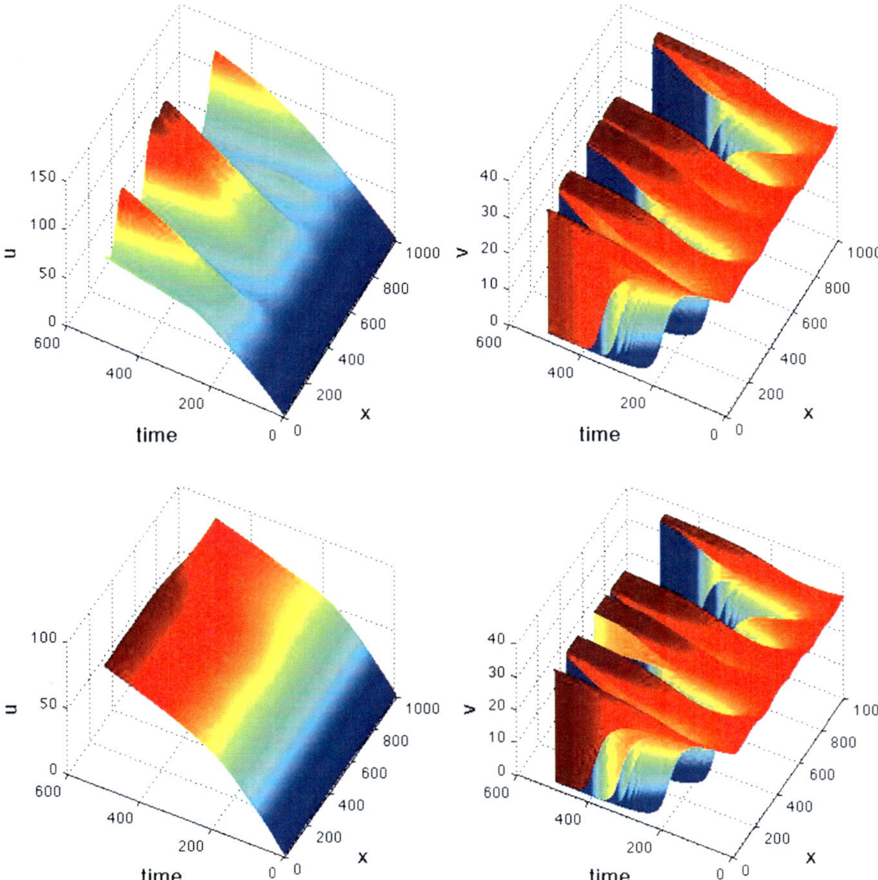

Fig. 12 Time-dependent simulation of model (1)–(2) showing emergence of a pattern containing multiple peaks. The simulations were performed for parameters given in Fig. 9 and different diffusion coefficients: $d = 1/(L^2) = 0.01$ on the *left hand-side* and $d = 1/(L^2) = 1$ on the *right hand-side*. The initial conditions correspond to a small heterogenous perturbation of a constant value $v_0 = 30$, while $u_0 = 1$ is set to be constant. The observed structures may correspond to formation of multiple organisation centres in cell reaggregates

We proposed a minimal receptor-based model with such stationary solutions, and showed that bistability is not enough for formation of stable patterns. The hysteresis effect in stationary kinetics system is essential for the observed patterning process. For the receptor-based model with hysteresis, we showed that no regular patterns are possible and found conditions for existence of transition layer stationary solutions, which link the shape of the pattern (minimal and maximal values and the transition point) with the size of domain (or equivalently diffusion coefficient). Analytical investigations were supported by numerical simulation of stationary and time-dependent problems.

Analysis of such basic models is a first step towards comprehensive classification of the mechanisms of formation of dynamical and stationary patterns, depending on the type of nonlinearity in the kinetics and relations between the rates of diffusion and reactions (different scaling). Many questions regarding pattern selection and stability remain open.

Acknowledgement A. M.-C. and A. K. were supported by European Research Council Starting Grant 210680 "Biostruct" and Emmy Noether Programme of German Research Council (DFG).

References

1. Holstein TW, Hobmayer E, Technau U (2003) Cnidarians: an evolutionary conserved model system for regeneration? Dev Dyn 226:257–267
2. Murray JD (2003) Mathematical biology, 2nd edn. Springer, New York
3. Gierer A, Meinhardt H (1972) A theory of biological pattern formation. Kybernetik 12:30–39
4. Baker RE, Schnell S, Maini PK (2009) Waves and patterning in developmental biology: vertebrate segmentation and feather bud formation as case studies. Int J Dev Biol 53:783–794
5. Guder C, Pinho S, Nacak TG, Schmidt HA, Hobmayer B, Niehrs C, Holstein TW (2006) An ancient Wnt-Dickkopf antagonism in Hydra. Development 133:901–911
6. Hobmayer B, Rentzsch F, Kuhn K, Happel CM, Laue CC, Snyder P, Rothbacher U, Holstein TW (2000) Wnt signaling and axis formation in the diploblastic metazoan *Hydra*. Nature 407:186–189
7. Müller S.C, Venzl G (1984) Lecture notes in biomathematics. In: Jäger W, Murray J (eds) Modeling of patterns in space and time. Springer-Verlag, Berlin, pp 254
8. Augustin R, Franke A, Khalturin K, Kiko R, Siebert S, Hemmrich G, Bosch TC (2006) Dickkopf related genes are components of the positional value gradient in Hydra. Dev Biol 296:62–70
9. Kestler HA, Kühl M (2011) Generating a Wnt switch: it's all about the right dosage. J Cell Biol 193:431–433
10. Kreuger J, Perez L, Giraldez AJ, Cohen SM (2004) Opposing activities of Dally-like glypican at high and low levels of Wingless morphogen activity. Dev Cell 7:503–512
11. Mao B, Niehrs C (2003) Kremen2 modulates Dickkopf2 activity during Wnt/LRP6 signaling. Gene 302:179–183
12. Marciniak-Czochra A (2006) Receptor-based models with hysteresis for pattern formation in hydra. Math Biosci 199:97–119
13. Smoller J (1994) Shock waves and reaction–diffusion equations, vol 258, 2nd edn, Grundlehren der Mathematischen Wissenschaften. Springer, New York
14. Marciniak-Czochra A (2003) Receptor-based models with diffusion-driven instability for pattern formation in hydra. J Biol Syst 11:293–324
15. Heinze S, Schweizer B (2005) Creeping fronts in degenerate reaction–diffusion systems. Nonlinearity 18:2455–2476
16. Sha W, Moore J, Chen K, Lassaletta AD, Yi CS, Tyson JJ, Sible JC (2003) Hysteresis drives cell-cycle transitions in *Xenopus laevis* egg extracts. Proc Natl Acad Sci U S A 100:975–980
17. Novak B, Pataki Z, Ciliberto A, Tyson JJ (2001) Mathematical model of the cell division cycle of fission yeast. Chaos 11:277–286
18. Cross FR, Archambault V, Miller M, Klovstad M (2002) Testing a mathematical model of the yeast cell cycle. Mol Biol Cell 13:52–70

19. Hoppensteadt F, Jäger W, Pöppe C (1983) A hysteresis model for bacterial growth patterns. In: Levin S (ed) Modelling of patterns in space and time, Lecture notes in biomathematics. Springer, Heidelberg
20. Marciniak-Czochra A, Ptashnyk M (2008) Derivation of a macroscopic receptor-based model using homogenisation techniques. SIAM J Math Anal 40:215–237
21. Marciniak-Czochra A (2012) Strong two-scale convergence and corrector result for the receptor-based model of the intercellular communication. IMA J Appl Math, DOI: 10.1093/imamat/HXS052
22. Umulis DM, Serpe M, O'Connor MB, Othmer HG (2006) Robust, bistable patterning of the dorsal surface of the Drosophila embryo. Proc Natl Acad Sci U S A 103:11613–11618
23. Wolpert L (1969) Positional information and the spatial pattern of cellular differentiation. J Theor Biol 25:1–47
24. Meinhardt H (1993) A model for pattern formation of hypostome, tentacles and foot in hydra: how to form structures close to each other, how to form them at a distance. Dev Biol 157:321–333
25. Klika V, Baker RE, Headon D, Gaffney EA (2012) The influence of receptor-mediated interactions on reaction–diffusion mechanisms of cellular self-organisation. Bull Math Biol. doi:10.1007/s11538-011-9699-4
26. Kusserow A, Pang K, Sturm C, Hrouda M, Lentfer J, Schmidt HA, Technau U, von Haeseler A, Hobmayer B, Martindale MQ, Holstein TW (2005) Unexpected complexity of the Wnt gene family in a sea anemone. Nature 433:156–160
27. Lee PN, Pang K, Matus DQ, Martindale MQ (2006) A WNT of things to come: evolution of Wnt signaling and polarity in cnidarians. Semin Cell Dev Biol 17:157–167
28. Mazieres J, He B, You L, Xu Z, Jablons DM (2005) Wnt signaling in lung cancer. Cancer Lett 222:1–10
29. W.A.Müller (1993). Pattern control in hydra: basic experiments and concepts. In Experimental and Theoretical Advances in Biological Pattern Formation, New York, Plenum Press

How to Knock Out Feedback Circuits in Gene Networks?

H. Gruber, A. Richard, and C. Soulé

The development of living organisms involves the successive activation of several gene networks. It is often known if activation or inhibition of a given gene occurs in these networks, but their global dynamic remains poorly understood. It is thus important to understand the function of motives contained in such a network, and for this one has to be able to perform knock out experiments.

The role of feedback circuits for the dynamic of gene networks is by now well established, both experimentally and theoretically. We shall discuss the possibility of knocking out these circuits.

To a gene network N is associated an *interaction graph* G, defined as follows: the vertices of G are the genes of N; there is a positive edge

$$A \rightarrow B$$

in G when (the product of) the gene A activates the expression of B; and there is a negative edge

$$A \dashv B$$

when the gene A inhibits the expression of B. Given an oriented circuit C in G, we say that C is positive (resp. negative) when it contains an even (resp. an odd) number of negative edges.

H. Gruber
Knowledgepark AG, Leonrodstr. 68, Munich D-80636, Germany
e-mail: info@hermann-gruber.com

A. Richard
I3S, 2000 routes des Lucioles, Les Algorithmes – Euclide B, Sophia Antipolis 06903, France
e-mail: richard@unice.fr

C. Soulé (✉)
IHÉS, 35 route de Chartres, Bures-sur-Yvette 91440, France
e-mail: soule@ihes.fr

V. Capasso et al. (eds.), *Pattern Formation in Morphogenesis*, Springer Proceedings in Mathematics 15, DOI 10.1007/978-3-642-20164-6_14,

R. Thomas conjectured that, if N has several stationary states (i.e. if N leads to differentiation), the associated graph G must contain a positive circuit; and, if N presents sustained oscillations, the graph G must contain a negative circuit of length at least two. Both Thomas rules were proved mathematically, for discrete and differential models of N (see [1, 2] and references therein).

In view of these results, it may be of interest to knock out some of the genes in N, in order to obtain a new network N', the graph of which contains only few circuits. For instance, if I is a set of genes such that $G\backslash I$ has no circuits, then knocking out all the genes in I will lead to a new network N' such that its interaction graph G' does not contain any circuit (since it is a subgraph of $G\backslash I$). The dynamic of N' has then to be very simple: it has to evolve towards a unique stationary state (this result is due to Robert [3] for discrete dynamical systems, and it holds for differential models with decay, as those treated in [4]). If this does not happen, one has to conclude that N does not provide a complete description of the system under study: new genes and/or new interactions have to be searched for.

Since the network N can contain many circuits (up to hundreds), it is of interest to find a good algorithm for knocking out all circuits in a graph G by deleting a minimal amount of vertices. This is a purely combinatorial problem, which is known to be **NP**-complete [5].

Recently, one of us (H.G.) found a fairly simple algorithm which, given G, gets rid of all circuits in G by deleting "few" vertices. This algorithm is the following. Given a vertex v in G, let $N(v)$ be the set of vertices W in G, $w \neq v$, such that vw is an edge, and let $d(G, v)$ be the number of vertices in $N(v)$. One defines as follows a finite sequence of subgraphs G_n in G. Given G_n, we let $v_n \in G_n$ be a vertex such that $d(G_n, v_n)$ is minimal. Then G_{n+1} is the graph spanned by $G_n - (N(v_n) \cup \{v_n\})$. The definition of G_n stops when there is no vertex left. Now, consider the subgraph H in G spanned by the vertices $v_1, v_2, \ldots, v_n, \ldots$ It can be shown [6] that H does not contain any oriented circuit, and that its set of vertices has order at least

$$\sum_{v \text{ vertex in } G} \frac{1}{1 + d(G, v)}.$$

It would be interesting to apply this algorithm to actual gene networks. When edges in G are endowed with a sign, one would also like to have an efficient algorithm for depriving G of its positive (resp. its negative) circuits, by knocking out a small set of vertices. We plan to address these questions elsewhere.

References

1. Kaufman M, Soulé C, Thomas R (2007) A new necessary condition on interaction graphs for multistationarity. J Theor Biol 248:675–685
2. Richard A (2010) Negative circuits and sustained oscillations in asynchronous automata networks. Adv Appl Math 44:378–392

3. Robert F (1986) Discrete iterations: a metric study, vol 6, Series in computational mathematics. Springer, Berlin/Heidelberg/New York
4. Soulé C (2006) Mathematical approaches to differentiation and gene regulation. C R Paris Biol 329:13–20
5. Karp RM (1972) Reducibility among combinatorial problems. In: Miller RE, Thatcher JW (eds) Complexity of computer computations. Plenum, New York, pp 85–103
6. Gruber H (2011) Bounding the feedback vertex number of digraphs in terms of vertex degrees. Discrete Appl Math 159(8):872–875

Formation of Evolutionary Patterns in Cancer Dynamics

Marcello Delitala and Tommaso Lorenzi

1 Introduction

The idea that cancer progression in *multicellular systems* is a *multistage process* [1] that develops through successive *somatic DNA alterations* is widely accepted in the scientific community. Apart from some unresolved controversies [2], the dynamics of the phenomenon is analogous to a Darwinian micro-evolution, where the single cells represent the competing individuals. At each successive progression stage, cells carrying new mutations can become increasingly able to escape both homeostatic regulation mechanisms and the immune system supervision. Thus, mutated cells gain a selective advantage that increases their fitness and their chance to proliferate, to detriment of the other cells. Hence, cells and their descendants can accumulate successive mutations that will cause the evolution from normal phenotypes to highly malignant ones (i.e. the conversion of normal cells into cancer ones). Genetic alterations that do not confer a selective advantage (i.e. neutral mutations) or even alterations that confer a selective disadvantage can also take place and they may play an active part in cancer evolution [3]. However, at this stage, we do not take into account these phenomena and we analyze only the role played by successive mutations that confer increasingly selective advantages to the mutated cells.

Somatic DNA alterations that take place at the molecular scale can cause the generation, at the cellular scale, of *multiple sub-populations* of cancer cells at different progression stages, i.e. at different malignancy levels [4, 5]. The co-presence of these sub-populations provides the basis for *intra-tumor heterogeneity* and their coalescence may result in the formation of solid aggregates composed of distinct sectors at

M. Delitala (✉) • T. Lorenzi
Department of Mathematics, Politecnico di Torino, Corso Duca degli Abruzzi 24, 10129 Torino, Italy
e-mail: marcello.delitala@polito.it; tommaso.lorenzi@polito.it

V. Capasso et al. (eds.), *Pattern Formation in Morphogenesis*, Springer Proceedings in Mathematics 15, DOI 10.1007/978-3-642-20164-6_15,

the tissue scale [6]. The geometrical features (i.e. shape and position) of solid tumor aggregates are closely related to the progression level of the disease [7] and an analysis of the resulting patterns can provide useful information on the tumor stage and can eventually suggest the best therapy to be used. For this reason, the dynamics of solid tumor aggregates is an important example of *morphogenetic phenomenon*. However, also the occurrence of the above mentioned successive somatic DNA alterations can be considered as *a broad sense morphogenetic phenomenon* that leads to the emergence of some *evolutionary patterns*. The genetic mutations that cause the evolution of normal cells into cancer ones can actually be considered as parts of some mutation pathways, which are progressively defined during the cancer progression process. These pathways can vary to a great extent from tissue to tissue, as well as from individual to individual [8], and they are highlighted by the formation of the aforementioned sub-populations of cells at different progression stages. Therefore, the creation of these sub-populations can be seen as the emergence of the evolutionary patterns that are defined by the mutation pathways that take place at the molecular scale, and so we can speak of the *morphogenesis of multiple sub-populations of cancer cells*. As a consequence, morphogenesis of tumor aggregates, that is observed at the tissue scale, can be considered as the result of evolutive morphogenetic phenomena which occur at the underlying scales and the whole carcinogenesis process can be interpreted as a *multiscale morphogenetic process*.

Over the last 50 years, many different approaches have been proposed to formalize cancer dynamics, at different representation scales, in mathematical terms. See for instance the review paper [9] and the references therein. This chapter is devoted to the presentation of a possible mathematical formalization of the process that leads to the formation of some sub-populations of cancer cells at different malignancy levels. The formalization relies on the mathematical structures that are provided by the *Kinetic Theory for Active Particles* (KTAP) [10], since they seem able to take into account heterogeneity phenomena, which are related to the wide range of particular functions expressed by malignant cells.

The contents of this chapter can be summarized as follows:

- Section 2 offers a mathematical framework for modeling the dynamics of some multicellular systems at the cellular scale. This framework is based on the mathematical structures of the Kinetic Theory for Active Particles.
- Section 3 proposes a particular model for the evolutive processes that lead to the morphogenesis of multiple sub-populations of cells during cancer progression. The model is developed through the framework introduced in the previous section and it refers to epithelial tissue.
- Section 4 shows the results of some simulations that have been developed with an exploratory aim. A critical analysis of the results is also provided.
- Finally, the conclusions are drawn in Section 5.

2 Mathematical Framework

The cells that belong to a complex living system, such as a given human tissue, can be considered as divided into distinct populations, which are composed of cells that express a peculiar collective and well defined biological function [11]. In general, we refer to n cell populations, which are labeled by an index $i = (1,\ldots,n)$. According to the Kinetic Theory for Active Particles [10], each cell in a population is considered as an *active particle*, whose state is generally identified, in addition to geometrical and mechanical variables (i.e. position and velocity), by a scalar variable u, that is called *activity*. This variable can be considered as included in a proper domain of the space of states that are associated to population i, and it is suitable to describe *the organized behavior of cells*, such as the expression of a particular biological function. If the distribution of cells is assumed to be uniform in space, i.e. all cells reside in equivalent positions, the only interactions that play a role are those that involve the activity variable. As a result, the *microscopic state* of a cell at a given instant of time can be identified just by the activity. Moreover, if a discrete representation of the states is consistent with the biological reality, the activity is assumed as a discrete variable and, as a consequence, its domain is represented by a discrete set $I_u = \{u_r\}_1^R$.

Individuals in state u_r of population i can be described, at time t, by a discrete distribution function:

$$f_i^r(t) = f_i(t, u_r) : [0, T] \times I_u \to \mathbf{R}^+,$$

where $[0, T]$ is the observation time interval and the independent variable t is normalized with respect to a suitable reference value. Since both *conservative and non-conservative phenomena* (i.e. phenomena that preserve or not, respectively, the total number of cells inside the system) are involved in the dynamics of complex multicellular systems, $f_i^r(t)$ should be considered as a distribution function in the sense that has been clarified in [10]. In particular, the summation over indexes r and i of $f_i^r(t)$ might be different from one at time $t > 0$ and, since we characterize the state of a cell by means of a discrete variable u_r, the value of $f_i^r(t)$ represents the number of cells that are found at time t in population i at progression stage u_r, normalized with respect to the total number of cells that are observed, at time $t = 0$, inside the system.

If the expression of f_i^r is known, macroscopic gross variables can be calculated as moments of this function. For example, the local size of a population $v_i = v_i(t)$ can be expressed as the zero order moment:

$$v_i(t) = \sum_{r=1}^{R} f_i^r(t), \tag{1}$$

and, as a consequence, the total size of all populations $N = N(t)$ can be defined as follows:

$$N(t) = \sum_{i=1}^{n} v_i. \tag{2}$$

The values of $v_i(t)$ and $N(t)$ are normalized with respect to the total number of cells that are found inside the system at the beginning of the observation time interval. As a consequence, $N(t = 0) = 1$.

We assume that cells are lead to change their state and/or their population because of two main classes of phenomena: (1) evolutive events that do not require interactions between cells (i.e. phenomena that are not mediated by interactions), such as renewal, proliferation and mutation; (2) phenomena that occur through interactions with other cells (i.e. phenomena that are mediated by interactions), such as cellular competition. Interactions between cells take place through complex signaling processes. However, within KTAP formalism [10] they are usually described as instantaneous interactions between cells, in order to reduce the modeling complexity.

The net inlet and outlet of the cells in state u_r of population i can be described by means of a suitable mathematical framework. If conservative and creation phenomena that are not mediated by interactions, as well as, destructive interactions are taken into account, while the effects of external actions are ignored, the following framework can result:

$$\frac{df_i^r}{dt}(t) = C_h^i[\mathbf{f}_{I_u}](t) + K_i[\mathbf{f}_{I_u}](t) - D_{ij}[\mathbf{f}_{I_u}](t), \tag{3}$$

where $\mathbf{f}_{I_u}$ is the vector of all distribution functions, i.e. f_i^r for $r = (1, \ldots, R)$ and $i = (1, \ldots, n)$, and:

- $C_h^i[\mathbf{f}_{I_u}]$ denotes the net gain of cells in the u_r state of population i as a consequence of *conservative phenomena that are not mediated by interactions between cells*;
- $K_i[\mathbf{f}_{I_u}](t)$ describes the creation of cells in the u_r state of population i as a result of *non-conservative phenomena that are not mediated by cellular interactions*;
- $D_{ij}[\mathbf{f}_{I_u}]$ models the destruction of cells with state u_r in population i as a consequence of *non-conservative phenomena that are mediated by cellular interactions*.

The expressions of the right-hand side terms in (3) rely on the formal structures proposed in [10]. However, since we are also dealing with phenomena that are not mediated by cellular interactions, these structures are tailored in the following way:

$$
\begin{aligned}
C_h^i[\mathbf{f}_{I_u}](t) &= \sum_{h=1}^{n}\sum_{p=1}^{R} \sigma_h(u_p) A_h^i(u_p \rightarrow u_r) f_h^p(t) - \sigma_i(u_r) f_i^r(t), \\
K_i[\mathbf{f}_{I_u}](t) &= \sigma_i^*(u_r)\kappa_i(u_r) f_i^r(t) \\
D_{ij}[\mathbf{f}_{I_u}](t) &= \sum_{j=1}^{n}\sum_{q=1}^{R} \eta_{ij}(u_r, u_q)\mu_{ij}(u_r, u_q) f_i^r(t) f_j^q(t),
\end{aligned} \tag{4}
$$

where:

- $A_h^i(u_p \rightarrow u_r)$ is a probability density function, which describes the probability, net of death phenomena, that a cell in state u_p of population h acquires the u_r state of population i after a conservative phenomenon that occurs at a rate $\sigma_h(u_p)$. The following property applies to A for all populations i and all states u_r:

$$
\sum_{h=1}^{n}\sum_{p=1}^{R} A_i^h(u_r \rightarrow u_p) = 1. \tag{5}
$$

- $\kappa_i(u_r)$ denotes the creation rate of cells in state u_r of population i as a result of phenomena that are not mediated by cellular interactions and that occur at a rate $\sigma_i^*(u_r)$.
- $\eta_{ij}(u_r, u_q)$ is the interaction rate, namely the rate at which a cell in state u_r of population i interacts with other cells in state u_q of population j.
- $\mu_{ij}(u_r, u_q)$ denotes the destruction rate of cells in state u_r of population i as a consequence of interactions with other cells in state u_q of population j.

A model of a specific biological system can be derived from the mathematical framework (3) by defining the expressions of the above functions, which depend on the biological system of interest. These phenomena should be selected, looking for a compromise between mathematical tractability and biological consistency. This task could be properly performed through the formal structures that might be offered by a *biomathematical theory* [12], whose definition represents a challenging goal that catalyzes the interest of mathematicians, physicists and biologists.

3 The Model

The mathematical model here proposed refers to the dynamics, at the cellular scale, of cancer pathologies that affect epithelial tissues, where cells can be assumed to be homogeneously distributed in space. This model is derived by the mathematical framework proposed in the previous section and describes the creation of multiple sub-populations of cancer cells, which can be interpreted as a morphogenetic phenomenon, as it has been previously clarified.

Normal and mutated cells inside a sample of epithelial tissue are seen as being grouped into three populations, according to their differentiation stage: *stem cell* population, *transit-amplifying cell (TAC)* population (i.e. those cells that descend from stem cells through cellular differentiation and that can be considered as semi-differentiated cells) and *highly-differentiated cell* population (i.e. those cells that result from the further differentiation of TACs and that die at the end of their life cycle). These populations are labeled by index $i = (1,\ 2,\ 3)$, respectively. The state of each cell is characterized by means of the discrete variable u_r, which is related to the progression stage of the cell (i.e. its degree of malignancy). Four possible progression stages are taken into account. In fact, we assume $r = 1$ if a cell in population i is a normal cell, and $r = 2, r = 3, r = 4$ or $r = 5$ if it is a mutated cell at increasing malignancy stages. As a result, the values of u_r range over the interval $I_u = \{u_r\}_1^5$. As previously noted, we only focus on malignant genetic alterations that provide mutated cells with a selective advantage. Therefore, in the following, we refer to mutated cells both as malignant cells and cancer cells.

According to the KTAP formalism, the state of population i can be characterized by means of the distribution function $f_i^r(t) = f_i(t, u_r)$, where the time variable t is normalized with respect to the life-cycle duration of cells in state u_1 of population $i = 1$ (i.e. the life-cycle duration of normal stem cells). The value of $f_i^r(t)$ is related to the number of cells that are found at time t in population i at progression stage r, normalized with respect to the total number of cells that are observed at time $t = 0$ inside the system.

The effects of the action expressed by the *immune system* against cancer cells are taken into account by the model here proposed. Although many exhaustive models have been defined to describe the dynamics of immune cells and their competition with tumor cells (see [13] and references therein), we do not introduce a detailed dynamics for immune cells and we assume that the distribution function related to the immune cells that attack the malignant cells in state u_r, with $r > 1$, of population i, at time t, is proportional to $f_i^r(t)$.

An evolution equation for the time-dynamics of f_i^r can be derived by the mathematical framework (3). In this way the effects of both conservative and non-conservative phenomena can be taken into account. The phenomena under consideration are briefly summarized hereafter together with the selected modeling strategies, which are introduced to reduce biological complexity (see also Fig. 1 for a brief figurative description). Under the following assumptions only the dynamics of stem and transit-amplifying cells, during carcinogenesis, can be described in a detailed way. In fact, since we assume that highly-differentiated cells do not play an active part in cancer progression, we look at their population just as a basin collecting differentiated TACs:

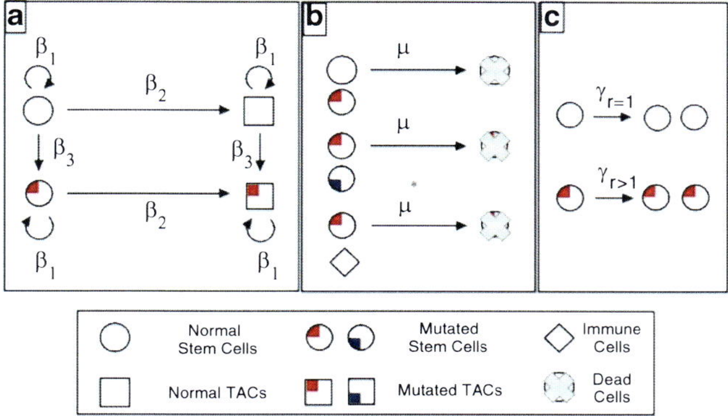

Fig. 1 *Biological phenomena and related parameters*. (**a**) Renewal (β_1), differentiation (β_2) and mutation (β_3). (**b**) Destruction events due to cancer progression (μ) and immune competition (μ^*). (**c**) Proliferation of normal cells ($\gamma_{r=1}$) and proliferation of mutated cells ($\gamma_{r>1}$)

Renewal, Differentiation and Malignant Mutations. Differentiation and renewal, as well as malignant mutations, can be considered, net of apoptotic events, as conservative phenomena that lead a cell in state u_p of population h to acquire the u_r state of population i with a certain probability that is described by parameters β_1, β_2 and β_3. Parameter β_1 models the *renewal rate* of the cells (i.e. the probability of a cell being replaced by a daughter cell in the same state of the same population), β_2 stands for the *differentiation rate* (i.e. the probability of a cell being replaced by a daughter cell in the same state of a different population, $i = h + 1$) and β_3 is the *mutation rate* (i.e. the probability of a cell being replaced by a daughter cell in a different state, $r = p + 1$, of the same population). These parameters are all positive. Since we assume that renewal, differentiation and mutation phenomena preserve the total number of cells, the above parameters are such that $\beta_1 + \beta_2 + \beta_3 = 1$. The net gain of cells in state u_r of population i due to renewal, differentiation and mutation is described by the $C_h^i[\mathbf{f}_{I_u}](t)$ term of (3), with $\sigma_h(u_p) = 1$ for all values of h and p and:

$$A_h^i(u_p \to u_r) = \begin{cases} \beta_1 & \text{if } h = i = (1,2) \text{ and } 1 \le p = r < 5 \\ \beta_1 + \beta_3 & \text{if } h = i = (1,2) \text{ and } p = r = 5 \\ \beta_2 & \text{if } h = i - 1 = (1,2) \text{ and } 1 \le p = r \le 5 \\ \beta_3 & \text{if } h = i = (1,2) \text{ and } 1 \le p = r - 1 < 5 \\ 1 & \text{if } h = i = 3 \text{ and } 1 \le p = r \le 5 \\ 0 & \text{otherwise.} \end{cases} \tag{6}$$

Proliferation of Mutated Cells. Mutated cells are able to bypass mitotic control mechanisms and to divide with a higher frequency than normal cells. On the other hand, as a consequence of the proliferation of tumor cells, normal cells are able to repopulate the system. We assume that these proliferation phenomena do not preserve the total number of cells in the system. Therefore, a positive parameter

γ_r, smaller than one, for every value of r, is introduced, that stands for the rate at which a cell is replaced by two cells belonging to the same state of the same population. The creation of cancer cells in state u_r of population i is described by the $K_i[\mathbf{f}_{I_u}](t)$ term of (3), where $\sigma_i^*(u_r) = 1$ for all values of r and i and

$$\kappa_i(u_r) = \begin{cases} \gamma_r & \text{if } i = (1,2) \text{ and } 1 \leq \mathrm{r} \leq 5 \\ 0 & \text{otehrwise.} \end{cases} \tag{7}$$

Destruction Events due to Cancer Progression and Immune System. Even though competition phenomena between normal and cancer cells, and cancer cells themselves, are mediated by the surrounding environment, for the sake of simplicity we model these phenomena as non-conservative interactions between a cell in state u_r and a cancer cell belonging to the same population in state u_q. In order to reduce the complexity of the model, we assume $q = r + 1$. These interactions can cause the destruction of the cell in state u_r with a rate μ, where μ is a positive parameter smaller than one. On the other hand, with reference to immune competition, we can suppose that interactions between a cancer cell in state u_r and a related immune cell can lead to the death of the cancer cell at a rate that is represented by a positive parameter μ^* (with μ^* smaller than one). The destruction of cancer cells in state u_r of population i is described by the $D_i[\mathbf{f}_{I_u}](t)$ term of (3), with $\eta_{ij}(u_r, u_q) = 1$ for all values of (r, i) and (q, j) and

$$\mu_{ij}(u_r, u_q) = \begin{cases} \mu & \text{if } i = j = (1,2) \text{ and } 1 \leq r = q - 1 < 5 \\ \mu^* & \text{if } i = j = (1,2) \text{ and } 1 < r = q \leq 5 \\ 0 & \text{otherwise.} \end{cases} \tag{8}$$

The evolution, in time, of the distribution functionf_i^r is described by the following set of non-linear ODEs, where$p = (2, 3, 4)$, which can be derived from (3) by means of the above listed assumptions:

$$\begin{cases} \frac{df_1^1}{dt}(t) = (\gamma_1 - \beta_2 - \beta_3)f_1^1(t) - \mu f_1^1(t) f_1^2(t) \\ \frac{df_1^p}{dt}(t) = (\gamma_p - \beta_2 - \beta_3)f_1^p(t) + \beta_3 f_1^{p-1}(t) - f_1^p(t)(\mu f_1^{p+1}(t) + \mu^* f_1^p(t)) \\ \frac{df_1^5}{dt}(t) = (\gamma_5 - \beta_2)f_1^5(t) + \beta_3 f_1^4(t) - \mu^* f_1^5(t) f_1^5(t) \\ \frac{df_2^1}{dt}(t) = \beta_2 f_1^1(t) + (\gamma_1 - \beta_2 - \beta_3) f_2^1(t) - \mu f_2^1(t) f_2^2(t) \\ \frac{df_2^p}{dt}(t) = (\gamma_p - \beta_2 - \beta_3)f_2^p(t) + \beta_2 f_1^p(t) + \beta_3 f_2^{p-1}(t) - f_2^p(t)(\mu f_2^{p+1}(t) + \mu^* f_2^p(t)) \\ \frac{df_2^5}{dt}(t) = (\gamma_5 - \beta_2)f_2^5(t) + \beta_2 f_1^5(t) + \beta_3 f_2^4(t) - \mu^* f_2^5(t) f_2^5(t) \\ \frac{df_3^1}{dt}(t) = \beta_2 f_2^1(t) - \zeta f_3^1(t) \\ \frac{df_3^p}{dt}(t) = \beta_2 f_2^p(t) - \zeta f_3^p(t) \\ \frac{df_3^5}{dt}(t) = \beta_2 f_2^5(t) - \zeta f_3^5(t). \end{cases} \tag{9}$$

Parameter ζ models the death rate of cells in population $i = 3$.

The expression of f_i^r is obtained by the solution of the initial value problem that is generated by linking (9) to the following initial conditions:

$$\begin{cases} f_i^r(t=0) = f_1^0 & \text{if } i = 1 \text{ and } r = 1 \\ f_i^r(t=0) = f_2^0 & \text{if } i = 2 \text{ and } r = 1 \\ f_i^r(t=0) = f_3^0 & \text{if } i = 3 \text{ and } r = 1 \\ f_i^r(t=0) = 0 & \text{otherwise,} \end{cases} \tag{10}$$

where f_1^0, f_2^0 and f_3^0 are positive constant values. These initial conditions mean that the system is assumed to be in healthy equilibrium at $t = 0$ and that carcinogenesis occurs at $t > 0$.

4 Simulations and Critical Analysis

Some numerical simulations of the mathematical problem that is defined by linking model (9) to initial conditions (10) are developed. Attention is paid to *emerging phenomena* that can appear within a complex living system, such as a multicellular system. The simulations are developed with the aim of:

- providing deeper insight into the role that differentiation, mutation and proliferation phenomena play on cancer dynamics;
- studying how these phenomena affect the morphogenesis of multiple subpopulations of cancer cells at different progression stages.

A sample of cells that satisfies some biological consistent initial conditions $f_1^0 = 0.05$, $f_2^0 = 0.45$ and $f_3^0 = 0.5$ (see for instance [6]) is chosen as reference system. We are interested in studying the role played by the model parameters in cancer dynamics. Therefore their values are defined with an explorative aim and they are made to vary case by case, according to the purposes of the related analysis. In the future, biological consistent values will be determined through suitable laboratory trials and appropriate statistical analysis.

Only figures referring to the dynamics of transit-amplifying cells are presented in the following. Since the number of TACs is much greater than that of stem cells, these figures seem more suitable to depict the dynamics of the whole system. The time dynamics of TACs is represented through the dynamics of the discrete distribution function $f_2^r(t)$ for $r = (1, \ldots, 5)$. As it has been stressed, the value of $f_2^r(t)$ is related to the number of transit-amplifying cells that are found at time t at progression stage r, referred to the total number of cells observed at time $t = 0$ (i.e. until healthy conditions occur) inside the system. The obtained results are depicted by Figs. 2 and 3, and they are here briefly discussed. The numerical results here presented can be analytically supported by means of standard methods of the Dynamical Systems Theory.

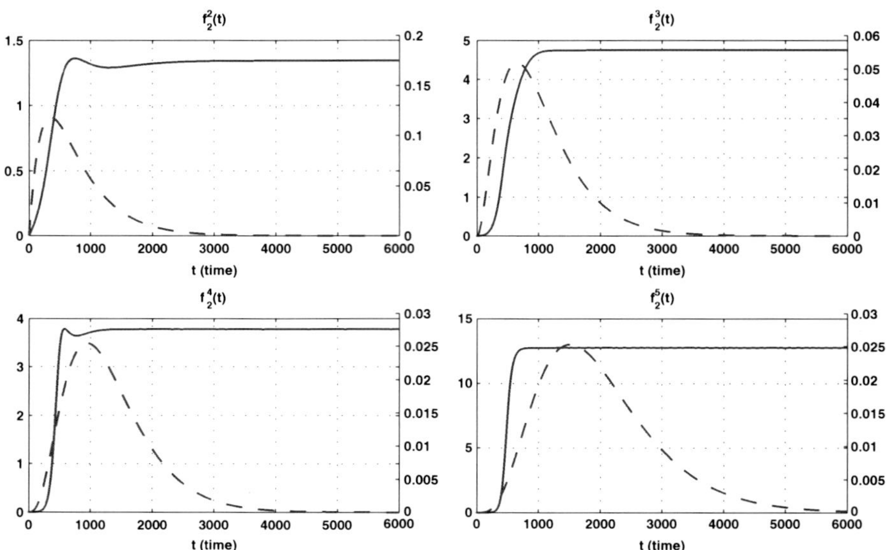

Fig. 2 *Differentiation-proliferation phenomena and cancer dynamics*. The dynamics of $f_2^r(t)$ for $r=(2,3,4,5)$, i.e. time dynamics of the distribution function of the TACs that are found inside the system at progression stage $r \geq 2$. The *dashed lines* are referred to the y-axis on the *right* and they are related to numerical results that are obtained with the proliferation rate $\gamma_r < \beta_2$ ($\gamma_r = 0.0001r$, for $r \geq 2$). On the other hand, the *solid lines* are referred to the y-axis on the *left* and they are related to simulations that are developed with $\gamma_r > \beta_2$ ($\gamma_r = 0.005r$, for $r \geq 2$). In both cases the values of the other parameters of the model are defined as follows: $\gamma_1 = 0.001$, $\beta_2 = 0.002$, $\beta_3 = 0.002$, $\mu = 0.001$, $\mu^* = 0.002$. The t-axis is normalized with respect to the life-cycle duration of normal stem cells, which is assumed equal to 1 day

Fig. 2 depicts the time-dynamics of f_2^r for $r = (1, \ldots, 5)$ when the proliferation rate γ_r is less than the differentiation rate β_2 (dashed lines, y-axis on the right) and when γ_r is greater than β_2 (solid lines, y-axis on the left), for every value of $r \geq 2$. If $\gamma_r \leq \beta_2$ for every value of r, that is, if genetic alterations provide mutated cells with a proliferation rate that is lower than the differentiation rate, f_2^r tends to zero over time and the number of cancer cells at progression stage r approaches the zero value. On the other hand, if $\gamma_r > \beta_2$ for $r \geq 2$, f_2^r reaches a finite value. Therefore, a finite number of cancer cells at different progression stages can be found inside the system. This implies tumor development and provides the basis for the creation of multiple sub-populations of cancer cells at different progression stages. Therefore, *if the genetic alterations do not provide mutated cells with a sufficiently high proliferation rate, the number of cancer cells at each progression stage tends to zero in time and the morphogenesis of sub-populations of cells at different malignancy levels does not occur*. Since the scenario described for $\gamma_r \leq \beta_2$ cannot be qualitatively altered by varying the value of the mutation rate β_3, we can conclude that, although cancer cannot initiate without mutations, tumor development requires an increase in the proliferation rate of mutated cells rather than in the mutation rate. This is in agreement with the considerations drawn in [5, 14].

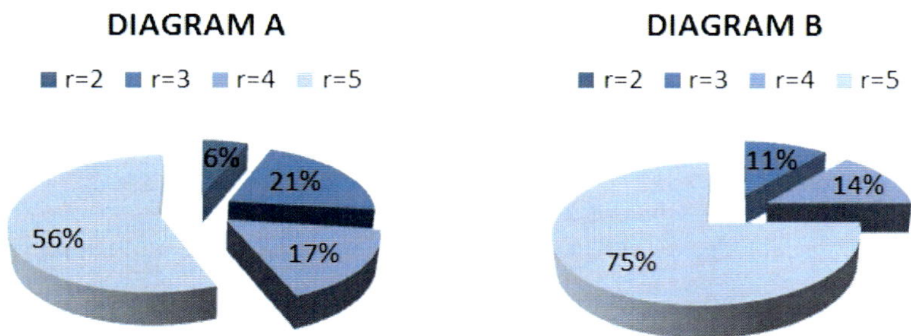

Fig. 3 *Mutation phenomena and cancer dynamics*. Percentage distribution of TACs over progression stages u_r, with $r = (2, 3, 4, 5)$, for two different values of the mutation rate, at time $t = 4000$ (time variable t is normalized with respect to the life-cycle duration of normal stem cells, which is assumed equal to 1 day). The blue-scale refers to progression stage u_r and ranges from *dark-blue* ($r = 1$) to *light blue* ($r = 4$). The two diagrams refer to the same value of γ_r (i.e. $\gamma_r = 0.005r$ for every value of $r \geq 2$), which is chosen so that cancer growth can occur. DIAGRAM A is obtained by setting the mutation rate β_3 equal to 0.002, while DIAGRAM B refers to a greater value of the mutation rate ($\beta_3 = 0.007$). In both cases the values of the other parameters of the model are defined as follows: $\gamma_1 = 0.001$, $\beta_2 = 0.002$, $\mu = 0.001$, $\mu^* = 0.002$

Moreover, for a given value of the proliferation rate γ_r, the lower is the value of the differentiation rate β_2, the greater is the chance of cancer development. *This result is consistent with the experimental observations for a reduction of cellular differentiation in cancer cells.*

Fig. 3 shows the percentage distribution of TACs, over the considered progression stages, for two different values of the mutation rate β_3, when the proliferation rate is high enough to allow cancer development (i.e. $\gamma_r > \beta_2$ for every value of $r \geq 2$). *If the mutation rate is higher, most of the cells are found at the same progression stage. Thus, we observe a concentration of the cells around a given progression stage, rather than the morphogenesis of some sub-populations of cancer cells at different malignancy levels*, and intra-tumor heterogeneity is lower (see DIAGRAM B). On the other hand, *if the mutation rate is lower, the cells are spread over different progression stages, that is, the morphogenesis of sub-populations of cancer cells is more effective* and intra-tumor heterogeneity is higher (see DIAGRAM A). As a result, we can conclude that the value of the mutation rate can have a great influence on the dynamics of the morphogenetic processes that lead to the formation of multiple sub-populations of cancer cells, as well as, on intra-tumor heterogeneity.

5 Conclusions

This chapter originated from the idea that carcinogenesis in multicellular systems can be considered as a multiscale morphogenetic process and, in particular, that the creation of multiple sub-populations of cells at different progression stages, which is observed during cancer progression, can be considered as a morphogenetic phenomenon.

A general mathematical framework for modeling cell dynamics in multicellular systems has been proposed by means of the Kinetic Theory for Active Particles. Then, a specific model for cancer evolution in epithelial cells has been defined through the proposed mathematical framework.

Simulations, developed with an exploratory aim, offer insights into the role played by mutation, proliferation and differentiation phenomena on cancer dynamics and on the morphogenesis of multiple sub-populations of cancer cells at different progression stages. In particular it seems that, rather than an increase in the mutation rate, the acquisition of an additional growing power and a reduction in the differentiation rate are the indispensable prerequisites for cancer development, as well as, for providing the basis for the creation of sub-populations of cancer cells. On the other hand, the evolution of this morphogenetic process seems to be strongly influenced by the value of the mutation rate. In particular, low values of the mutation rate promote the co-existence of cells at different progression stages inside the same multicellular system and make the formation of sub-populations of cancer cells more effective.

References

1. Hanahan D, Weinberg RA (2000) The hallmarks of cancer. Cell 100:57–70
2. Gatenby RA (2006) Commentary: carcinogenesis as Darwinian evolution? Do the math! Int J Epidemiol 35:1165–1167
3. Nowak MA, Komarova NL, Sengupta A, Jallepalli PV, Shih I, Vogelstein B, Lengauer C (2002) The role of chromosomal instability in tumor initiation. Proc Natl Acad Sci U S A 99:16226–16231
4. Nowell PC (2002) Tumor progression: a brief historical perspective. Semin Cancer Biol 12:261–266
5. Martins ML, Ferreira SC Jr, Vilela MJ (2007) Multiscale models for the growth of avascular tumors. Phys Life Rev 4:128–156
6. Weinberg RA (2007) The biology of cancer. Garland Science, New York
7. Kim Y, Stolarska MA, Othmer HG (2007) A hybrid model for tumour spheroid growth in vitro. I: theoretical development and early results. Math Model Method Appl Sci 17:1773–1798
8. Kinzler KW, Vogelstein B (1996) Lessons from hereditary colorectal cancer. Cell 87:159–170
9. Bellomo N, Li NK, Maini PK (2008) On the foundations of cancer modeling. Math Model Method Appl Sci 18:593–646
10. Bellomo N, Bianca C, Delitala M (2009) Complexity analysis and mathematical tools towards the modeling of living systems. Phys Life Rev 6:144–175
11. Hartwell HL, Hopfield JJ, Leibner S, Murray AW (1999) From molecular to modular cell biology. Nature 402:c47–c52
12. Bellouquid A, Delitala M (2006) Mathematical modeling of complex biological systems. A kinetic theory approach. Birkhäuser, Boston
13. Bellomo N, Delitala M (2008) From the mathematical kinetic, and stochastic game theory to modeling mutations, onset, progression and immune competition of cancer cells. Phys Life Rev 5:183–206
14. Tomlinson IPM, Novelli MR, Bodmer WF (1996) The mutation rate and cancer. Proc Natl Acad Sci U S A 93:14800–14803

How Cell Decides Between Life and Death: Mathematical Modeling of Epigenetic Landscapes of Cellular Fates

Andrei Zinovyev, Laurence Calzone, Simon Fourquet, and Emmanuel Barillot

Abstract We present a mathematical model of cell fate decision between survival, necrosis and apoptosis as a concrete implementation of the Waddington's metaphor of epigenetic landscape determining cellular behaviour. We describe the principles of the model construction and in silico experiments performed on it. The genetic network underlying cell fate decisions is reconstructed in the form of an influence diagram together with logical rules defining possible system state changes, while the epigenetic landscape is represented as a state transition graph generated from the discrete cell fate decision model. Stochastic cellular decision making is modeled as a random walk on the state transition graph, assuming equal probabilities of any possible system state update. The probability to reach a particular attractor (stable state) of the state transition graph, starting the random walk from a "physiological" cellular state, is interpreted as a probability of having a particular phenotype (outcome) in a biological experiment. As a result, one can predict the phenotypic probabilities and their changes as a result of specific perturbations. We show that such a "generic" cell model can recapitulate and explain experiments conducted on mice and cell lines and predict the outcome of not yet done experiments. In the discussion, we compare the design principles of the cell fate decision model with the principles of designing engineered devices and underline some important differences.

1 Introduction

In 1957 Conrad Hal Waddington suggested a mathematical metaphor of morphogenesis and introduced the notion of *epigenetic landscape* [1]. According to this metaphor, a cell develops through a series of irreversible bifurcations at some cell

A. Zinovyev (✉) • L. Calzone • S. Fourquet • E. Barillot
INSERM U900, Institut Curie, Mines Paris Tech, Paris, France
e-mail: andrei.zinovyev@curie.fr

V. Capasso et al. (eds.), *Pattern Formation in Morphogenesis*, Springer Proceedings in Mathematics 15, DOI 10.1007/978-3-642-20164-6_16,

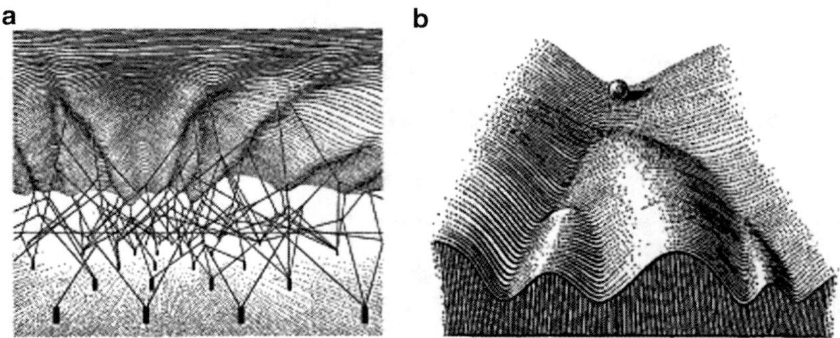

Fig. 1 The famous metaphor of canalization on the epigenetic landscape by C. H. Waddington [1]: (**a**) epigenetic landscape is formed and supported by dynamical properties of a complex system of interacting genes (gene network); (**b**) epigenetic landscape directs a cell in development through a number of bifurcation points where irreversible decisions on the cell fate take place

states in which crucial decisions on the cell fate are made. According to the Waddington's idea, the landscape of all possible paths leading to all cell fates (see Fig. 1) is formed by the properties of a complex network of interacting genes.

Recently, systems biology provided a number of concrete implementations of the Waddington's cell fate decision paradigm, through mathematical modeling of various biological systems (for example, [2–4]). A particular interest is to model cell fate decisions between life and death in mammalian cells. Understanding principles of this type of cellular decisions is crucial for understanding functioning of some tissues (such as gut epithelium) and for understanding tumor development, for which the tightly regulated mechanism of balancing between survival and death is violated towards survival.

Recent progress in studying the mechanisms of cell life/death decisions revealed its enormous complexity. Among many, one can mention three difficulties on the way to characterize, describe and create formal mathematical formulations of this mechanism.

First, the molecular mechanism allowing a cell to react to an external stress (such as damage of DNA, nutrient and oxygen deprivation, toxic environment) is assembled from highly redundant pathways which are able to compensate each other in one way or another. For example, there exist at least seven distinct and parallel survival pathways associated with action of AKT protein [5]. Disruption of one of these pathways in a potential cell death-inducing cancer therapy can be in principle compensated by the others. Thus, understanding and modeling the survival response in its full complexity is a daunting task.

Second, cellular death is an extremely complex phenotype as well, which can not be reduced to a simple disaggregation of cellular components driven by purely thermodynamical laws. Several distinct modes of cell death were identified in the last 10 years [6], such as necrosis, apoptosis and autophagy. Importantly, all these cell death modalities are controlled by cellular biochemical mechanisms, activated in response to diverse types of stress: roughly speaking, a cell is usually pre-programmed to die in a certain manner, sending appropriate signals to its surrounding and trying to reuse its remaining resources when possible. *Necrosis* is a type of cell death associated with a lack of important cellular resources such as ATP, which makes functioning of many biochemical pathways impossible. This is why it was usually thought of as an uncontrolled and purely thermodynamics-driven degradation of cellular structures. However, recent research showed that necrosis is accompanied by activation of certain pathways and can even be triggered by certain signals [6]. By contrast, *apoptosis* as a form of cellular suicide was, from the very beginning, known as a mode of cell death requiring investment of energy for disrupting mitochondria membranes and cleavage of intracellular structures. *Autophagy* remains a relatively poorly understood cell death mechanism which serves, paradoxically, as a survival mechanism for cell populations (during autophagy, some cells or cell parts are sacrificed to obtain deficit resources such as amino acids).

The third difficulty can be attributed not directly to the complexity of the biochemical mechanisms but rather to our capabilities of apprehending the design principles used by biological evolution. Inspired by an engineering approach, we have a tendency to investigate complex systems by dissecting them into relatively independent modules and associating well-characterized non-overlapping function to each molecular detail. In biology this view is not that easy to apply. Some cellular molecular machineries cannot be naturally dissected or associated with a well-defined function: a set of overlapping functions can be distributed among a group of molecular players. We will discuss this issue further in more details.

In this paper, we describe a mathematical model of cell fate decision between life, survival and two alternative modes of cell death: through apoptosis and necrosis, following [2]. Not having the ambition to deal with the whole complexity of cell fate decisions in vivo, we decided to concentrate on modeling the outcome of a classical and rather well-defined experiment of inducing cell death: adding to a cell culture a specific ligand (Tumor Necrosis Factor, TNF, or other members of its family such as FASL). The ligand can affect the so-called cell death receptors and trigger cell death through apoptosis or necrosis in some cell types and survival mechanisms in others [7]. The exact outcome of this experiment depends on many circumstances: cell type, dose of the ligand, duration of the cell treatment, certain mutations in cell genomes, etc. Trying to characterize the biochemical response of a cell to this relatively simple kind of perturbation allows to understand certain cell fate decision mechanisms.

2 Genetic Network of Cell Fate Decisions

Let us refer to a cell in a resting state as "naive". The naive state corresponds to the case when the cell is not under any life-threatening stress conditions and not receiving death receptor inducing signals. Such a cell remains alive but we shall not call this state "a survival", since survival in our description will be associated with active resistance to stressful conditions. In the experiment with adding TNF ligand to a cell culture, it is known that most cell types survive in a manner that depends on the activation of the protective transcription factor NFκB [8].

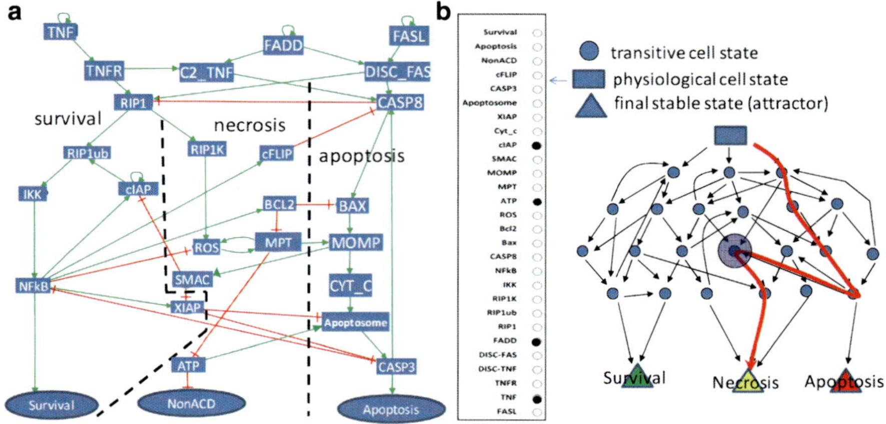

Fig. 2 Genetic network (**a**) and epigenetic landscape (**b**) of the cell fate decision machinery. (**a**) Biological diagram of molecular interactions involved in cell fate decisions derived from the biological literature. By *dashed lines* the diagram is roughly divided into three modules corresponding to three submechanisms of cell fate decisions. Notations: (1) Proteins: TNF, FADD, FASL, TNFR, CASP8, cIAP, cFLIP, BCL2, BAX, IKK, NFκB, CYT_C, SMAC, XIAP, CASP3; (2) States of proteins: RIP1ub (unibquinated form of RIP1), RIP1K (kinase function of RIP1); (3) Small molecules: ATP, ROS (reactive oxygen species); (4) Molecular complexes: apoptosome, C2_TNF, DISC_FAS; (5) Molecular processes: *MPT* mitochondria permebialization transition, *MOMP* mitochondrial outer membrane permeabilization; (6) Phenotypes: survival, apoptosis, NonACD (Non-apoptotic cell death). (**b**) State transition graph. Each node on this graph corresponds to a state, characterized by a sequence of "0" or "1" values. An example of one particular state is given on the *left* (a "physiological state"). Note that this is not a real state transition graph but only an illustration, since the graph contains 2^{28} nodes, which is impossible to visualize and even to fit into the computer's memory. An example of a systems trajectory is shown by *thick curve*. An example of a state corresponding to a "point of no return" is shown by the *big circle* (from this state the system will be determined to finish in the middle stable state). Compare this figure to Fig. 1

After a tremendous simplification (Fig. 2a), our knowledge about functioning of NFκB pathway in these conditions is the following: NFκB transcription factor is activated through the destruction of its inhibitor, IκB (which is not explicitly shown in the diagram), with the help of a specific kinase called IKK. In turn, IKK receives signals from the death receptors through another protein called RIP1, and, more specifically, the ubiquitinated form of the protein referred to as RIP1ub on the diagram. When NFκB is activated and transported into the nucleus, it can transcriptionally activate a plethora of genes, among which there are cFlip, BCL2, ROS scavenging enzymes coding genes, and members of the IAP (inhibitors of apoptosis) family cIAP1, cIAP2 (together referred to as cIAPs), and XIAP.

Our knowledge about necrosis and apoptosis can be also represented in an extremely simplistic fashion. In reality, these two mechanisms are tightly coupled and have significant overlap in participants, thus in the Fig. 2, the suggested separating line represents a naive simplification made for clarity reasons here.

In very few words, after a cell receives TNF signal, the apoptosis pathway is activated when specific death inducing complexes C2_TNF pass a signal to CASP8, by cleaving its initial unprocessed molecular form (uncleaved caspase 8). As a result, CASP8 can initiate disruption of the outer mitochondria membrane with the help of a special family of BAX proteins. As a result, many components of intramembrane space is released into the cytosol. Among them, cytochrome C (CYT_C) is the most important for inducing apoptosis, because it is necessary to form the so called Apoptosome and, finally, activate execution caspases such as CASP3 that will do the actual work for apoptosis, by cleaving and destroying many intracellular structures including membranes and DNA. Another important ingredient of the mitochondria intermembrane space is the second mitochondria-derived activator of caspase (SMAC), for its role in inhibiting survival.

Unlike apoptosis, the necrosis signaling passes through reactive oxygene species (ROS) which, when present in high concentrations, can severely damage mitochondria and cause mitochondrion permeability transition (*MPT*). This event, in turn, generates additional ROS in the cytosol, leading to the avalanche of *MPT*s in all cell's mitochondria. ROS level can be raised in a cell for many reasons, connected with, for example, toxic conditions. However, in response to activating death inducing receptors, RIP1 kinase can transduce specific signals leading to intracellular ROS production. Thus, necrosis can be initiated in a programmed fashion. Disrupting mitochondria in a cell leads to severe deficit of ATP, the main biochemical currency, with lethal consequences.

It is extremely important to underline that three pathways mentioned: apoptosis, necrosis and survival through NFκB activation, are connected by multiple crosstalk relations (Fig. 2) at all levels of the pathways (upstream and downstream). As a matter of fact, these crosstalks raise a question about the definition of each of the pathways, since, after full assembly of the cell fate decision machinery, it is difficult to decompose it back into independent modules without an a priori definition of pathway borders.

In [2] we summarized the current knowledge on the interactions between cell fate decision mechanisms in a simplistic wiring diagram where a node represents either a protein (TNF, FADD, FASL, TNFR, CASP8, cIAP, cFLIP, BCL2, BAX, IKK, NFκB, CYT_C, SMAC, XIAP, CASP3), a state of protein (RIP1ub, RIP1K), a small molecule (ROS, ATP), a molecular complex (Apoptosome, C2_TNF, DISC_FAS), a group of molecular substances (BAX and BCL2), a molecular process (Mitochondria permeabilization transition, *MPT*, Mitochondrial outer membrane permeabilization, *MOMP*) or a phenotype (Survival, Apoptosis, Non-apoptotic cell death, NonACD). Each directed and signed edge represents an influence of one molecular entity on another, either positive (arrowed edge) or negative (headed edge). These types of diagrams are ubiquitous in the biological literature.

Note on the diagram that the phenotype nodes are in fact simple interpretations of the following molecular conditions: (1) activated NFκB is read as survival state; (2) lack of ATP is read as nonapoptotic cell death state; (3) activated CASP3 is read as apoptotic cell death. Absence of any of such conditions is interpreted as a "naive" cell state, corresponding to the fourth cellular phenotype.

3 Biological Phenotype as an Attractor of a Dynamical System

The diagram drawn and recapitulating the biological knowledge can be used in some types of reasoning on the cell fates, however, it remains ambiguous in many aspects, preventing from application of mathematically strict ways of coming from an assumption to a conclusion. In other words, some formal language is needed to associate the biological diagram with a mathematical object, properties of which can be analyzed in a formal way.

One of the simplest way to introduce such a language, is to consider the so called "Boolean" or "discrete" approach [9]. In this approach, the "activity" of each node on the diagram is characterized by a number which can take some discrete values. In the simplest case it is "0" (inactive) or "1" (active) values, or, in the logical interpretation, "false" (inactive) and "true" (active). Then, to introduce the first formalization of the problem, one has to specify how a state of a given node depends on the state of the other nodes pointing at it. When a node has only one upstream neighbor, this is a fairly simple task, however, the number of possibilities grows exponentially with the number of affectors. There exists a compact way to represent the update rules, which consists in using standard logical formulas. The rules are inspired by the biological knowledge about the mechanisms of action of the affectors (their co-operativity, etc.) on the target node. More detailed description of the logical rules in the cell fate decision model is given in [2], here we provide them for completing the model definition (see Table 1).

Let us introduce a notion of a *state transition graph*, associated with the discrete dynamical system defined by the logical rules in Table 1. On this graph, each node represents a state of the system which in this case can be encoded by a sequence of 0s and 1s. The states are connected by a directed edge if a transition between two states is allowed, accordingly to any of the logical update rules in Table 1. In principle, the state transition graph can be defined independently and without the biological diagram, however, this would require a tremendous amount of empirical knowledge which is not available. Hence, the biological diagram with associated logical rules is used as a compact representation and a tool to generate the state transition graph. Detailed instructions on this procedure can be found in [9].

The set of all possible states provides a discrete phase space of the system. The state transition graph contains all possible ways of the systems dynamics (trajectories). In other words, it is the *multidimensional epigenetic landscape* of the cell fate decision network. Note that the state transition graph is assumed to be rather sparse compared to the fully connected graph where any two state transitions would be possible. Hence, on this landscape, one can determine bifurcating states, points of no return, etc. The properties of the landscape are determined by the properties of the complex biological gene network described in the previous section, further illustrating the old Waddington's idea.

Table 1 Logical update rules defining the discrete dynamical model shown on the Fig. 2

DISC_TNF′ = TNFR **AND** FADD	TNF′ = TNF
RIP1′ = (TNFR **OR** DISC_FAS) **AND** (**NOT** CASP8)	FADD′ = FADD
CASP8′ = (DISC_TNF **OR** DISC_FAS **OR** CASP3) **AND** (**NOT** cFlip)	FAS′ = FAS
RIPub′ = RIP1 **AND** cIAP	TNFR′ = TNF
cIAP′ = (NFkB **OR** cIAP) **AND** (**NOT** SMAC)	RIP1K′ = RIP1
BAX′ = CASP8 **AND** (**NOT** BCL2)	cFlip′ = NFkB
ROS′ = (RIP1K **OR** MPT) **AND** (**NOT** NFkB)	IKK′ = RIP1ub
MPT′ = ROS **AND** (**NOT** BCL2)	BCL2′ = NFkB
MOMP′ = BAX **OR** MPT	SMAC′ = MOMP
NFkB′ = IKK **AND** (**NOT** CASP3)	CYT_C′ = MOMP
XIAP′ = NFkB **AND** (**NOT** SMAC)	DISC_FAS′ = FAS **AND** FADD
Apoptosome′ = CYT_C **AND** ATP **AND** (**NOT** XIAP)	ATP′ = **NOT** MPT
CASP3′ = Apoptosome **AND** (**NOT** XIAP)	

Some states of the state transition graphs are distinguished from the others. First of all, there are so called "sink" nodes. A "sink" node is a node that any allowed transition from this node leads to itself. We will interpret such nodes as the final stable system states corresponding to the biological phenotypes. Mathematical analysis of the logical equations (Table 1), using GINsim software [10] shows that for our model there are 27 states. These states can be represented as a set vectors with 0-1 components in the phase space of the dynamical system. Their distribution can be visualized in 2D using, for example, principal component analysis (PCA), which is performed in Fig. 3. From this visualization, one can conclude that the stable states of the system can be grouped in four compact clusters, each corresponding to a particular cellular phenotype ("naive", survival, apoptosis, necrosis). Notice that the fact of existence of four well-separated clusters of the stable states is independent of our interpretation. It allows us to robustly match clusters to phenotypes. It happens that this matching is meaningful and the profiles of the stable states in terms of variable activities reflect our current knowledge of the functioning of the cell fate decision machinery.

Comparing the existing knowledge with the profiles of stable states allows to formulate some predictions. That way one can notice that RIP1 protein is always inactive in the cluster of stable states matched with the necrosis phenotype. It is known, however, that RIP1 plays an essential role in initiating necrosis. As a result, we predict that the role of RIP1 in necrosis, although being essential, is nevertheless transient. This type of predictions can be verified in an experiment.

4 Probability of a Phenotype

The state transition graph allows to address more complicated and interesting questions. One of these questions is the following: *Starting from a distinguished state of a cell, what is the probability to arrive to each of the stable states?*

This type of questions can be solved by converting the state transition graph into a Markovian process of random walk on a graph, and analyzed by the

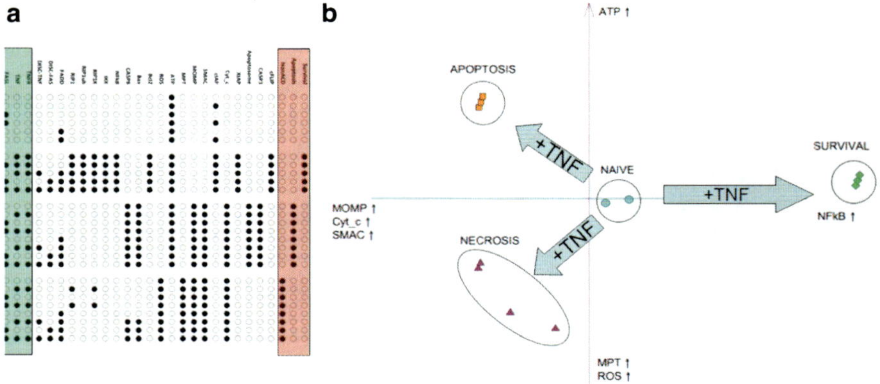

Fig. 3 Grouping stable states of the discrete system described in the Table 1 into four phenotypic clusters. A table of stable states of the discrete dynamical system of cell fate decision (*left*). First three and last three columns are inputs and outputs of the model, the rest are internal variables. Principal component analysis (PCA) of the table of stable states made on the values of the internal variables (*right*)

corresponding classical techniques [11]. However, there is one essential ingredient missing for performing such type of analysis: the state transition probabilities. Once again, defining these probabilities directly on the graph from some empirical observations is impossible at present time. Hence, these probabilities should be derived from the logical model with the use of some additional assumptions.

The simplest assumption is to consider all transitions firing from a given state as *equiprobable*, and it was used further to compute probabilities with which a cell will arrive at a certain phenotype. Biological interpretation of such an assumption is not simple. In a way, we consider a "generic" cell in which all possible system trajectories take place with equal probabilities (without dominance, i.e. any preferable route). One can argue that in any particular concrete cell, this would not be true anymore and that the generic cell is not representative of anything real observed in an experiment. Having in mind this difficulty, we avoid direct interpretation of absolute values of probabilities, concentrating rather on relative changes of them in response to some system modifications such as removing a node or fixing a node's activity. It happens that such a "generic" cell model is already capable of reproducing a number of known experimental facts.

At the same time, one can suggest various ways to overcome the "equiprobability" assumption. Typically, some nodes on the diagram can be considered as being "faster" or "slower" than others. This approach is implemented in [10]. Another way to make the probabilities of transitions outgoing from a state inequal, is to distinguish "slow" and "fast" influences on the biological diagram. For example, transcriptional influences can be considered as slow in comparison to influences explained by formation of molecular complexes. Investigation of the possibilities of this approach which has more clear biological interpretation is in process.

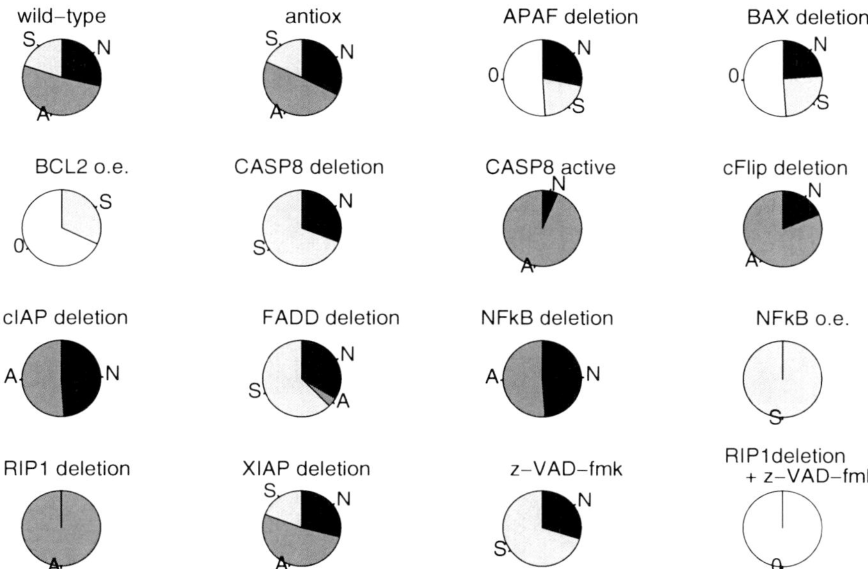

Fig. 4 Changes in the phenotype probabilities from the random walk on the state transition graph, starting from the initial physiological state. Various "mutant" modifications of the dynamical system are tested here. Here "A" denotes apoptosis, "N" denotes necrosis and "S" denotes survival, "0" denotes naive state. "O.e." stands for over expression of a protein, "antiox" means treatment by antioxidants, "z-VAD fmk" simulates the effect of caspase inhibitor z-VAD-fmk

When the state transition graph is parameterized by transition probabilities, one can, using the standard techniques (which, however, should be adapted here, given the size of the graph, see [2, 12]), compute the probability of hitting a given stable state, considering that a random walk starts from a given initial state. Then this probability is associated with a probability of observing a particular phenotype in an experiment. For doing this, it is convenient to define a unique initial state which is referred to as the "physiological state". In the model of Fig. 2 it is the state in which all elements are inactive except ATP, FADD and cIAP. This is a stable state, which looses its stability when TNF variable is changed from 0 to 1 and the dynamical system starts to evolve in time.

Using this approach, we performed a number of *in silico* experiments in which the probability of arriving to stable states was computed on a modified, or "mutant", model (as opposed to the initial "wild-type" model). Typical model modifications that were considered here are to remove a node from the diagram with all associated edges or to fix some nodes' activities to 1. For our cell fate decision model, the result is provided in Fig. 4. In [2] this table was systematically compared with the experimental data of the cell death phenotype modifications observed in various mutant experimental systems, including cell cultures and mice. The model was able to qualitatively recapitulate all of them and to suggest some new yet unexplored experimentally mutant phenotypes. The most interesting in this setting would be to consider synthetic interactions between individual mutants, when several nodes on the diagram are affected by a mutation simultaneously.

5 Conceptual Model of Cell Fate Decisions

To complete our study of cell fate decision, we reasoned on the simplest model of cell fate that can be deduced from the master model described above.

The purpose here was to simplify the network to obtain a formal representation of the logical core of the network. We have selected three components to represent the three cellular fates: NFκB for survival, *MPT* for necrosis and CASP3 for apoptosis. Based on reduction techniques and on the identification of all possible directed paths between these three components [13], a three-node diagram was deduced from the master model (see Fig. 5).

In this compact model, each original path (including regulatory circuit) is represented by an arc whose sign denotes the influence of the source node on its target. In some ambiguous cases (e.g. influence of *MPT* on CASP3 or of NFκB on *MPT*), the decision on the sign of the influence is based on the Boolean rules and not on the paths only. Indeed, two negative and one positive paths link NFκB to *MPT*. Therefore, the sign of the arc depends not only on the states of BCL2 and ROS, both feeding onto *MPT*, but also on the rule controlling *MPT* value. Since the absence of BCL2 and the presence of ROS (Boolean 'AND' gate) participate in the activation of *MPT*, if BCL2 is active, then *MPT* is set to 0, even when ROS is 1. By extension, if NFκB is 1, then *MPT* is 0, justifying the choice for a negative influence. In the case of mutations eliminating all the negative influences, however, a positive arrow must be considered.

The resulting molecular network is symmetrical: each node is self-activating and is negatively regulated by the other nodes. This is a conceptual picture representing the general architecture of the master model that can help address specific questions. Even for this relatively simple regulatory graph, there is a finite but quite high number of possible logical rules. For now, we use a simple generic rule involving the AND and NOT operators. For example, the logical rule for CASP3 is: NOT MPT AND NOT NFκB AND CASP3. This compact model has four stable states, each corresponding to one cell fate, along with the "naive" state. This is coherent with what was observed from the analysis of the complete model.

From this minimal complexity model, one can conclude that the initial model contains three mutually inhibiting and self-activating "modules" assuring stability and separability of the phenotypes. However, it is not easy to dissect and clearly separate the modules. In other words, the modules are not defined at the level of the structure of the biological diagram and gene interaction network (Figs. 2a and 1b) but rather on the epigenetic landscape itself (Figs. 2b and 1a). The structure of the genetic network does not necessarily contains well-separated structural components corresponding and determining these "dynamics" modules.

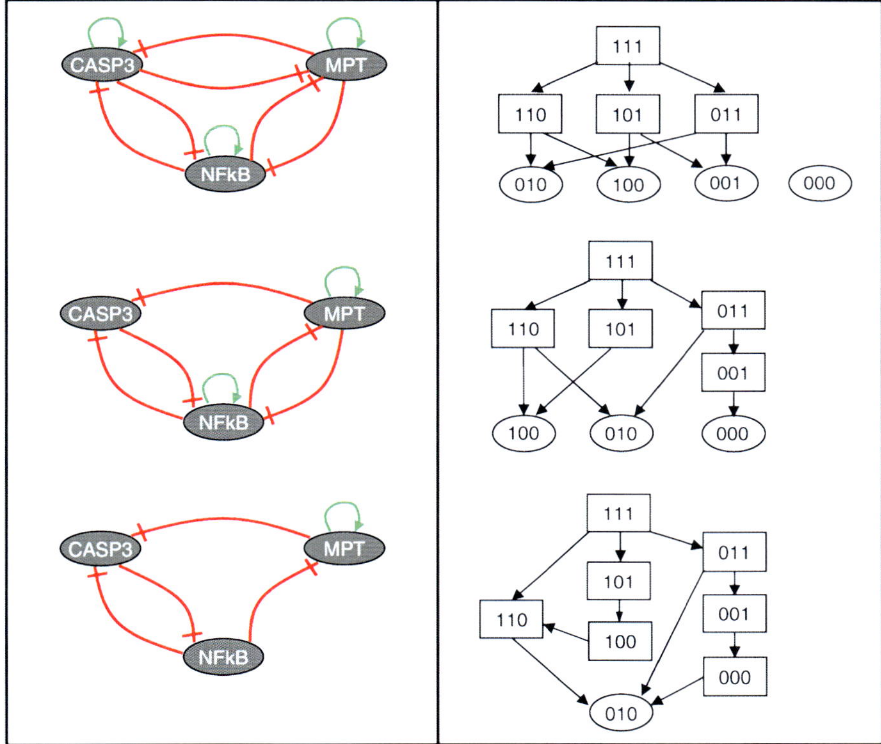

Fig. 5 A minimal model of three alternative cell fate decisions. *Upper row*: minimal conceptual model of cell fate decision machinery (*left*) and its state transition graph (*right*). *Middle row*: a modification of the minimal model, obtained by removing CASP8 from the biological diagram. *Bottom row*: a modification of the minimal model, obtained by removing both CASP8 and cIAP from the diagram (an example of synthetic interaction)

In the language of engineers, the motif shown on Fig. 5 can be called a "three-stable trigger". Bistable triggers were shown to be a typical motif in other molecular mechanisms, regulating, for example, the cell cycle [14].

6 Discussion

Mathematical models provide a way to test biological hypotheses *in silico*. They recapitulate consistent heterogeneous published results and assemble disseminated information into a coherent picture using an appropriate mathematical formalism (discrete, continuous, stochastic, hybrid, etc.), depending on the questions and the available data. Then, modeling consists of constantly challenging the obtained model with available published data or experimental results (mutants or drug

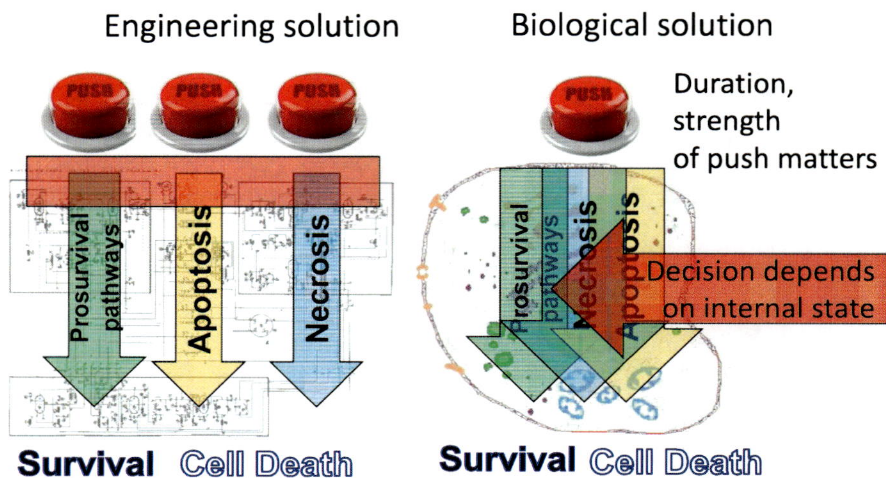

Fig. 6 Comparing engineering and biological approaches to deciding between three alternative cell fates. Engineering solutions are characterized by clear separation of modules, maximum predictability of the response. Features of the biological solution features are overlapping modules, a response to a perturbation which is probabilistic and dependent on the internal system state

treatments, in our case). After several refinement rounds, a model becomes particularly useful when it can provide counter-intuitive insights or suggest novel promising experiments.

Here, we have conceived a mathematical model of cell fate decision, based on a logical formalization of well-characterized molecular interactions. Former mathematical models only considered two cellular fates, apoptosis and cell survival. In contrast, we include a non-apoptotic modality of cell death, mainly necrosis, involving RIP1, ROS and mitochondria functions.

In recent literature there exists a number of reports in which analogies and differences between engineering and biological implementations of specific functions are discussed (for example, see [15]). Let us have a look at the process of cell fate decision between survival, necrosis and apoptosis from this prospective.

Let us imagine ourselves in an engineer's shoes who is asked to construct a device capable of performing three mutually exclusive functions. The most probable design for this problem appears in Fig. 6, left. It contains three modules performing the functions, each module is activated by a separate button. To improve the response, some analysis of the state of the buttons would be needed: for example, to determine what happens if when one button is already pressed, another is pressed and so on. The device will be usable and practical if the response to pressing one of the buttons will be maximally predictable.

By contrast, the design that is used by nature to solve these kinds of problems is quite different (Fig. 6, right). First, there is no evident separation of the signal processing into independent modules: as we have seen, the modules are connected by extensive crosstalk at all levels of the fate decision. In this case the definition of the modules becomes difficult, not mentioning the decoupling of their dynamics. Second, nature might use only one button (receptor) but the decision will depend on the strength of pressing it, duration of pressing and the internal state of the system at the moment when the button is pressed. As a result, the response will be stochastic and characterized by a non-zero probability of having several resulting phenotypes. If many buttons are used then their functions can hugely overlap. The effect of pressing one button might depend on the states of other buttons, etc. Using such a device in every-day life would be difficult.

It is not evident which design would be advantageous in naturally evolving systems. To understand the drawbacks and advantages, one has to formulate the optimality criteria. For example, one can ask which design is more resistant (able to perform its function after a perturbation) to a random withdrawal of certain number of elements. Systematic comparison is required to answer this question, and the answer is expected not to be simple. Another possible criterion to consider: which design is more probable to appear in the process of random tinkering? For example, one can assume that a typical mechanism of evolution consists in recurrent application of two operations: (1) duplicating already existing and working elements with further partial specialization and partial conservation of a common function; and (2) *ad hoc* re-using already existing details for a new and completely different purpose. In this scenario, huge overlap in the hypothetical modules is unavoidable and expected. For example, why does cytochrome C play a double role both in respiratory chain of mitochondria and in the apoptosis signaling after mitochondria membrane permeabilization? Is it an advantage in some way to have the crosstalk between cellular metabolism and apoptosis signaling, or is it a mere random consequence of the ad hoc use of cytochrome? How does this decision, when fixed, constrain further evolution of the machinery, etc.?

All these questions make the analogy between engineered devices and biological systems, although useful in some simple considerations, but fragile and sometimes misleading concept. Much more should be understood in order to define the correct optimality criteria used by evolution and, hence, to separate the role of chance from the role of design and selection in biological evolution as compared to engineered devices.

Acknowledgements We would like to acknowledge support by the APO-SYS EU FP7 project. All the authors are members of the team Systems Biology of Cancer, Equipe labellisee par la Ligue Nationale Contre le Cancer. The study was also funded by the Projet Incitatif Collaboratif Bioinformatics and Biostatistics of Cancer at Institut Curie.

References

1. Waddington CH (1957) The strategy of the genes. George Allen & Unwin, London
2. Calzone L, Tournier L, Fourquet S, Thieffry D, Zhivotovsky B, Barillot E, Zinovyev A (2010) Mathematical modelling of cell-fate decision in response to death receptor engagement. PLoS Comput Biol 6(3):e1000702

3. Kirouac DC, Madlambayan GJ, Yu M, Sykes EA, Ito C, Zandstra PW (2009) Cell–cell interaction networks regulate blood stem and progenitor cell fate. Mol Syst Biol 5:293
4. Manu SS, Spirov AV et al (2009) Canalization of gene expression and domain shifts in the Drosophila blastoderm by dynamical attractors. PLoS Comput Biol 5(3):e1000303
5. McCormick F (2004) Cancer: survival pathways meet their end. Nature 428(6980):267–269
6. Kroemer G et al (2008) Classification of cell death: recommendations of the Nomenclature Committee on Cell Death 2009. Cell Death Differ 16(1):3–11
7. Van Herreweghe F, Festjens N, Declercq W, Vandenabeele P (2010) Tumor necrosis factor-mediated cell death: to break or to burst, that's the question. Cell Mol Life Sci 67(10): 1567–1579
8. Karin M (2006) Nuclear factor-kappaB in cancer development and progression. Nature 441 (7092):431–436
9. Chaouiya C, de Jong H, Thieffry D (2006) Dynamical modeling of biological regulatory networks. Biosystems 84(2):77–80
10. Naldi A, Berenguier D, Faure A, Lopez F, Thieffry D, Chaouiya C (2009) Logical modelling of regulatory networks with GINsim 2.3. Biosystems 97(2):134–139
11. Feller W (1968) An introduction to probability theory and its applications, chapter 3, vol 1. Wiley, New York
12. Tournier L, Chaves M (2009) Uncovering operational interactions in genetic networks using asynchronous boolean dynamics. J Theor Biol 260(2):196–209
13. Zinovyev A, Viara E, Calzone L, Barillot E (2008) BiNoM: a Cytoscape plugin for using and analyzing biological networks. Bioinformatics 24(6):876–877
14. Santos SD, Ferrell JE Jr (2008) On the cell cycle and its switches. Nature 454(7202):288–289
15. Lazebnik Y (2002) Can a biologist fix a radio? – or, what I learned while studying apoptosis. Cancer Cell 2(3):179–182

Part III
Ideas, Hypothesis, Suggestions

From Hydra to Vertebrates: Models for the Transition from Radial- to Bilateral-Symmetric Body Plans

Hans Meinhardt

Abstract The development of a higher organism needs many coupled pattern-forming reactions. Crucial are interactions in which a local self-enhancing reaction is coupled to an antagonistic reaction of longer range. Using the pattern of head, tentacle and foot formation in the small freshwater polyp hydra as a model system it is shown (1) how polar pattern can emerge; (2) how a polar pattern can be maintained during substantial growth; (3) how structures next to each other can be generated and (4) how two organizing regions can be forced to appear at a maximum distance from each other at the two terminal poles. The understanding of the organization along the single axis of the radial-symmetric hydra was a key to understand the evolution of bilateral-symmetric body plans. Many observations can be explained under the assumption that the body of hydra-like ancestors evolved into the brain of higher organisms, that generation of a midline was a subtle patterning process for which evolution has found different solutions, that the ancestral hydra-type organizer became the organizer for the AP axis in higher organisms, and that the trunk is a later evolutionary addition.

1 Introduction

During early development, different structures emerge at different positions although the genetic information is the same in all cells. In pre-molecular times, important information has been derived from perturbation experiments and observing the subsequent pattern regulations. In the recent decades it has become possible to isolate genes and signaling molecules involved in pattern formation and to determine their expression patterns. Both approaches have shown, for instance, that small nests of cells, the so-called organizing regions, play a crucial role in the

H. Meinhardt (✉)
Max-Planck-Institut für Entwicklungsbiologie, Spemannstr. 35, Tübingen 72076, Germany
e-mail: hans.meinhardt@tuebingen.mpg.de

V. Capasso et al. (eds.), *Pattern Formation in Morphogenesis*, Springer Proceedings in Mathematics 15, DOI 10.1007/978-3-642-20164-6_17,

spatial organization. However, neither from the observation that lost parts regenerate nor from the local activation of particular genes at particular positions one can directly derive how pattern formation is achieved. Our intuition is very limited for processes that are based strong positive and negative feedback loops. Formulating models in a mathematically precise way together with simulations allows one to check whether the assumed mechanism is able to account for both expression patterns and pattern regulations. Models are an appropriate tool to bridge the gap between observations on the one hand and the understanding of the underlying principles on the other. Many of our models were proposed before a molecular approach became feasible. They have received much support from results obtained using the more recent molecular-genetic approaches, thus showing that modeling is a powerful tool.

In the first part of the present article, the small freshwater polyp hydra will be used as model system to demonstrate basic properties of pattern formation. Understanding axis formation in this evolutionary early radial-symmetric animal was for me the basis for a better understanding of how in higher organisms the two main body axes, anteroposterior and dorsoventral, are formed with an orthogonal orientation relative to each other. The resulting model is discussed in the second part. Hydra is most famous for its almost unlimited capability for regeneration [1–4]. Even more dramatic, hydra tissue can be dissociated into individual cells and, after re-aggregation, clumps of cells form again viable organisms [5]. Obviously, pattern formation does not require any initiating asymmetry and can proceed in a self-organizing way. The region around the opening of the gastric column, the so called hypostome, is an organizing region [6]. Transplantation of a small piece of tissue from this region into the body column can induce in the host a new head including tentacles, showing that these cells are able to control the patterning in the surroundings. Meanwhile firm evidence has been obtained that organizer formation in hydra is under the control of the Wnt-pathway [7–9]. Although much is known about the *Wnt*-pathway and although our previous models make firm predictions about how the patterning works, there is still a substantial gap between the theory and the understanding of the patterning process from a molecular point of view.

The overall structure of the polyp is simple: a head with tentacles is located at one end of the tube-shaped body column and a 'foot', a sticky holdfast, at the other (Fig. 1a). When a hydra reaches a certain size, lateral buds form that later detach. Both the simplicity and the well-investigated regulatory features inspired us a long time ago to propose a general theory of biological pattern formation [3, 14–16]. The key proposal was that pattern formation can occur if and only if a local self-enhancing reaction is coupled with an antagonistic reaction that acts over a longer range. Basic features and applications of this model to hydra patterning will be discussed in the first part of this article. In fact, Alan Turing showed in his pioneering paper that interactions of two substances that spread with different diffusion rates can lead to patterns in space [17]. However, most reactions that satisfy this condition do not form patterns at all. The two basic requirements we found, local self-enhancement and long-range inhibition, are not mentioned in Turing's paper, although one can show that Turing's equations satisfy our condition. To understand the essential

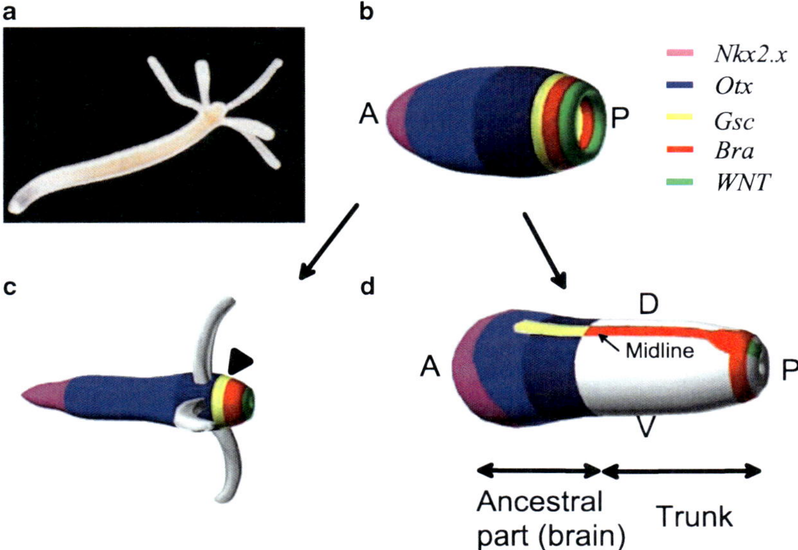

Fig. 1 The freshwater polyp hydra (**a**, **c**) and the proposed equivalence of the patterns along the anteroposterior (AP) axis of a hydra-like ancestor (**b**) and contemporary bilaterians (**d**): The hydra head is the most posterior structure. The small opening of the gastric column, the organizer in hydra, became in vertebrates a huge ring that acts as organizer for the AP axis. The marginal zone in amphibians is an example. The foot corresponds to the most anterior structures including the heart (*Nkx2*-related genes, *pink*). The central part of the gastric column (*Otx*, *blue*) evolved into the brain. A major step in evolution was the formation of a midline along the entire AP axis, the precondition to organize the dorsoventral axis. In vertebrates, midline formation is initiated by a second organizer – the Spemann-type organizer – and occurs by elongation into opposite directions, towards anterior (prechordal plate, *yellow*) and towards posterior (notochord, *red*). The formation of such a solitary midline organizer is a subtle patterning process (see Fig. 4). Another major evolutionary step was the formation of the trunk (*white*). In vertebrates, *Goosecoid* (*yellow*) is involved in brain patterning, *Brachyury* (*red*) in trunk and tail formation. Thus, the *Goosecoid*/*Brachyury* border (*arrowhead* in c) marks the presumptive site of trunk insertion (From [10])

requirements was a key to handle highly non-linear reactions and reactions, in which more than two components were involved.

More recently, hydra again came into the focus of interest, this time from an evolutionary point of view. *Coelenterates*, to which hydra belongs, are located at a very basal position of the evolutionary tree. A comparison of the expression patterns of homologous genes in hydra and higher organisms indicated that the body column of ancestral hydra-like organisms evolved into the brain of higher organisms (Fig. 1; [10]). Moreover, the hydra foot is under the control of the same master gene as the vertebrate heart (*Nkx2.5*), suggesting a common ancestry for these two so differently looking organs [18]. The hydra foot also acts as a pump, in this case, for circulating the gastric fluid. The understanding of the radial-symmetric patterning in hydra and the equivalence of the corresponding structures in higher organisms was for me a key to understanding the generation of the two body axes in higher organisms. In vertebrates, usually, only a *single* organizer is assumed to exist: the Spemann-type organizer. However, how can a single organizer provide positional information for *two* axes, and how is their orthogonal orientation achieved? This paradox resolves if, in addition to the well-known Spemann-type organizer, an ancestral hydra-type organizer is assumed to be involved. This second organizer was overlooked for a long time, since it does not have a patch-like extension, as usually presumed for an organizing region, but instead has the geometry of a huge ring. These two organizers together are able to provide the positional cues for the organization of the two major body axes in their orthogonal orientation [19]. Corresponding models for the transition from a radial- to a bilateral-symmetric body plan will be discussed at the end of this article.

2 Pattern Formation by Local Self-Enhancement and Long-Ranging Inhibition

A straightforward realization of local self-enhancement and long-ranging inhibition consists of an autocatalytic activator that produces a long-ranging inhibitor, which, in turn, antagonizes the self-activation [14–16]. If such an interaction is at work and the tissue has reached a minimum extension, a homogeneous distribution is unstable. The only pattern that can emerge at this stage is a graded distribution (Fig. 2a). This is a very important step: cells become exposed to different signaling concentrations in a systematic and reproducible way. Although the genetic information is the same in all cells, different genes can be activated at different positions. A region of high activator concentration can act as organizer. In even larger fields periodic and stripe-like distributions can emerge (Figs. 2a and 4b), patterns that are also frequently encountered during development. Meanwhile, several such activator-inhibitor pairs have been found in diverse systems. The interaction of *Nodal* and *Lefty2* is one of them. In vertebrates this system is involved in the mesoderm formation and in the left-right patterning, in sea urchins in the formation of the oral field [20, 21]. Another example is the signaling system that leads to the periodic spacing of heterocysts in Anabaena [22] (see [16] for modeling).

2.1 A Mathematical Description

Biological processes are assumed to be driven by the interaction of molecules. A theory of biological pattern formation has to describe the changes of concentrations in space and time as function of the local concentration of the relevant substances involved. The following set of equations describes the local change of an activator $a(x)$ and an inhibitor $h(x)$ per time unit [14], for simplicity written here for a one-dimensional array of cells:

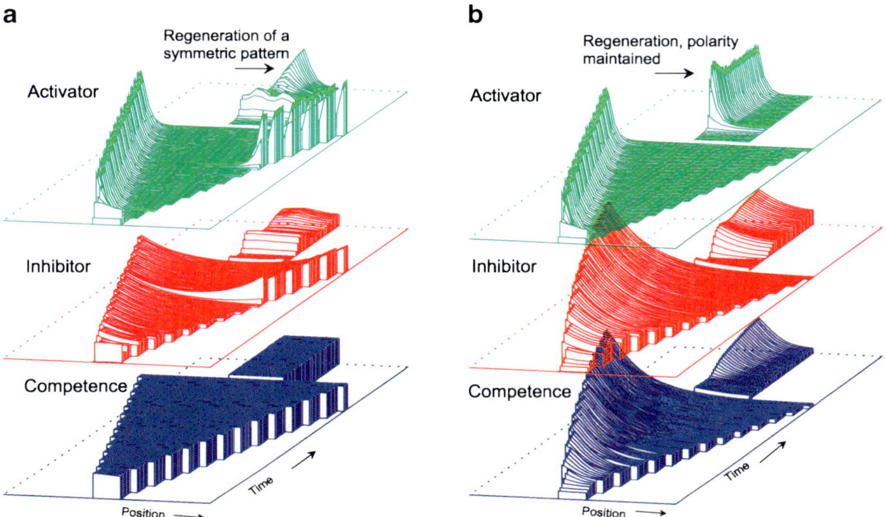

Fig. 2 Pattern formation by an activator-inhibitor system and the generation of a strong apical dominance by a feedback of the pattern onto the competence. (**a**) If a certain size is exceeded, a graded activator distribution emerges. Without feedback on the competence (flat blue profile at the *bottom*), further maxima appear during growth whenever the inhibitor level becomes low enough. Then, the overall pattern is no longer polar. The polarity in a regenerating piece can be random. (**b**) Due to the feedback (3), cells distant to the organizer loose more and more the competence, i.e., the ability to perform the pattern-forming reaction [ρ in (1), (2) and (3)]. Regions farther from the organizer have a much reduced chance of overcoming the inhibition that spreads from the organizing region. The polar structure is maintained in spite of substantial growth. During regeneration, due to its long time constant, the competence remains essentially unchanged, causing regeneration to occur reliably with the original polarity

$$\frac{\partial a}{\partial t} = \frac{\rho a^2}{h(1+\kappa a^2)} - \mu a + D_a \frac{\partial^2 a}{\partial x^2} + \rho_a \tag{1}$$

$$\frac{\partial h}{\partial t} = \rho a^2 - \nu h + D_h \frac{\partial^2 h}{\partial x^2} \tag{2}$$

Such equations are easy to read. Equation 1 states that the concentration change of the activator a per unit time ($\partial a/\partial t$) is proportional to a non-linear autocatalytic production term, a^2. The autocatalysis is slowed down by the action of the inhibitor, $(1/h)$. The second term, $-\mu a$, describes the degradation. The number of activator molecules that disappear per unit time is proportional to the number of activator molecules present (just as the number of people dying per year in a given city is, on average, proportional to the number of inhabitants).

The autocatalysis must be non-linear (a^2) since it must overcome the linear decay ($-\mu a$). This condition is satisfied if the active component is not the activator itself but a dimer of two activator molecules. The factor ρ, the *source density*, describes the general ability of the cells to perform the autocatalytic reaction. Its function is close to what is described as 'competence' in the biological literature. Slight asymmetries in the source density can have a strong influence on the *orientation* of the emergent pattern. The concentration change of a and h also depends on the exchange of molecules with neighboring cells. This exchange is assumed to occur by simple diffusion but other mechanisms are conceivable as well. The term κ leads to a saturation of the self-enhancing reaction and thus to an upper limit of activator production. This allows, for instance, the formation of stripes ([23], see Fig. 4b). The last term in (1a), ρ_a, is a small activator-independent activator production. This term ensures that the concentration of the activator never sinks to zero and enables the triggering of an activator maximum after removal of an established maximum. It is important to initiate the autocatalytic reaction at low activator concentrations as it is required during regeneration (Fig. 2). Equation 2 can be read in an analogous way. A necessary condition to enable spatial pattern formation is that the inhibitor spreads more rapidly than the activator, i.e., the condition $D_h \gg D_a$ must be satisfied. In addition, the inhibitor must have a more rapid turnover, i.e., $\nu > \mu$, otherwise the system will have the tendency to oscillate.

3 Pattern Regeneration

This mechanism is not only able to account for pattern generation but also for self-regulation after a perturbation. After removal of the activated region, the remnant inhibitor decays until a new maximum is formed via autocatalysis (Fig. 2). The polarity need not to be maintained. Polarity reversal has been observed in sea urchins after bisection and in fragments of stolons of a marine hydroid ([24, 25], see [15] for modeling). Polarity reversal can result from the inhibitor distribution remnant in the fragment. It is lowest at the pole opposite to the original organizing region, providing an elevated chance to win the competition. For reasons discussed in the next section, polarity reversal never occurs in hydra.

4 The Wavelength Problem

Usually embryos grow during development. In a standard reaction–diffusion system a graded concentration profile can be maintained only over a range of about a factor of two. With increasing field size, the tendency exists to change from a monotonic into a symmetric and eventually into a periodic distribution either by insertion of new maxima (Fig. 2a) or by splitting of existing maxima [29]. In other words, these patterning reactions have a certain wavelength. For a creature like hydra a loss of polarity due to growth would be inadequate since several heads would appear instead of one.

Some basic observations in hydra provide hints how this wavelength problem was solved by nature. Small fragments of less than 1/10 of the normal body size are

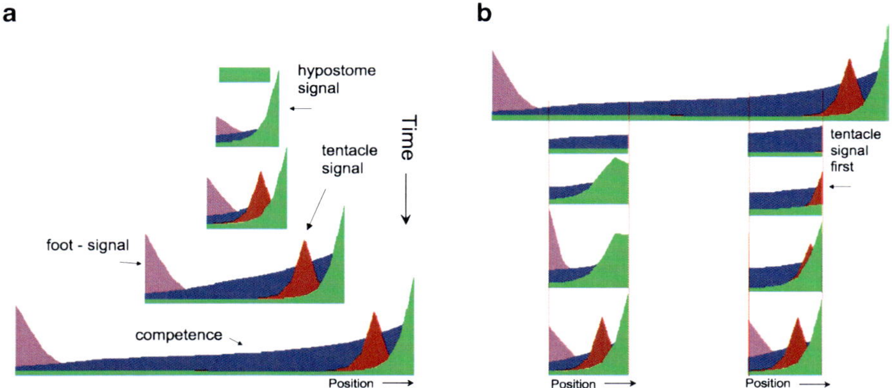

Fig. 3 A model for hydra: generation of complex patterns by linkage of several pattern-forming reaction [26]. Activator-inhibitor systems are assumed to generate the signals for hypostome, tentacle and foot formation. Primary head activation (right-most signal, *green*) and foot activation (left-most signal, *pink*) appear at opposite ends of the field due a coupling via the competence (*blue*). Tentacle activation (*red*) appears close to the hypostome since it requires a high competence but it is locally suppressed by the head signal. (**a**) Establishment and maintenance of the signals during growth. (**b**) Behavior during regeneration. The fragment at the *right* shows signal regeneration in a near-head fragment. Tentacle activation precedes head activation; the new tentacle signal appears first at the tip, the region of the highest competence. Later, the rising head signal shifts the tentacle signal to its final position. *Left*: in more basal fragments head activation occurs first. Tentacle activation takes place later after the competence has attained a threshold value and occurs at its final position, in agreement with the observations [27, 28]

still able to regenerate. In such a fragment the regeneration of the head occurs always at the side pointing towards the original head. Obviously, there is a systematic change in the otherwise uniformly-appearing body column that leads to an advantage in the competition to form the new head-forming signal. Tissue originally closer to the head has a head start and wins the competition. It is the *relative* position that matters.

In our first publication we already pointed out that symmetry breaking in biology is more the exception than the rule [14], although our model can account for it. In most situations there is some asymmetry. We called the ability to perform the self-enhancing reaction the 'source density'. It is closest to the term 'competence' as it is used in the literature. In the hydra literature, this feature is usually termed 'head activation gradient'. In the rest of this article, I will stick to the term 'competence' for describing the capability of performing the pattern-forming reaction.

The wavelength problem can be solved by a positive feedback of the head organizer onto the competence [26]. Under this condition, with increasing distance to an existing maximum, not only the inhibitor concentration but also the competence decreases (Fig. 2b). Consequently, the cells become less and less able to establish a new centre as long as the primary organizer is present. This dramatically enhances the dominance of the primary organizing region. The fading competence is an essential process in making development reproducible; otherwise, new organizing regions would emerge in an uncontrolled way.

The graded competence has another important consequence. To generate a graded activator profile in a homogeneous tissue, the range of the activator must be in the order of the extension of the tissue. Therefore, the organizing region occupies a substantial fraction of the field. This limits the range over which size regulation is possible. In contrast, if a graded competence exists, the range of the activator can be small; the sharp peaks will appear nevertheless at a terminal position (compare Fig. 2a, b). This is most important if small fragments from a large animal should regenerate in a polar fashion, as is the case in hydra or planarians.

The feedback of the activator a on the competence ρ can have the following form:

$$\frac{\partial \rho}{\partial t} = \gamma a - \mu_\rho \rho + D_\rho \frac{\partial^2 \rho}{\partial x^2} + \rho_\rho \qquad (3)$$

This type of feedback has been used for the simulations in Fig. 2a, b.

According to this model, the organizer has two apparently contradictory functions. On the one hand, it suppresses other organizing regions in the surrounding tissue; on the other hand it enhances the competence of the tissue to enable the formation new organizing regions if regeneration is required. Why don't the two features cancel each other? Both effects are expected to have about the same range, but completely different time constants. The inhibitor must have a high turnover to allow a rapid regeneration after head removal and keep the patterning system from entering into an oscillating mode. In agreement, new peaks of the *Wnt* pathways reappear after about 1 h [8]. In contrast, as determined by classical transplantation experiments, a change in the polarity takes several days [30]. That means that during head regeneration the competence remains nearly unchanged.

As shown in Fig. 3, the other main structures, foot and tentacles can also be correctly placed via the graded competence. Hydra is under control of two organizing regions located at opposite ends of the tissue, the head and the foot. This is common in many morphogenetic fields. A simple cross-inhibition would be inappropriate since in small (young) animals head and foot would mutually repress each other. For instance, foot regeneration in a small head-containing fragment would be impossible due to this cross-inhibition of the foot system by the head system. This problem disappears if the coupling of the head and foot system is achieved via the competence: the head activation appears at the position of the highest competence and can elevate the competence still further. If the foot system has the opposite behavior, i.e., if it appears at the lowest level of the competence, it emerges at the maximum distance from the head (Fig. 3). In principle, foot formation is possible everywhere. However, the head system generates a strong preference for foot formation at the antipodal position.

In contrast, hypostome and tentacles appear next to each other. A controlled neighborhood of structures is enforced if one structure activates the other at long range but excludes it locally [31]. The periodic nature of the tentacle pattern indicates that tentacle formation is controlled by a separate patterning process and not by a certain level of a head-derived signal. Many observations can be explained under the assumption that a high level of the competence, generated at longer range under the influence of the head organizer, is the precondition for tentacle initiation. Locally, however, the head organizer excludes tentacle formation. Thus, tentacles can only appear near the hypostome (Fig. 3). This model found support by recent molecular observations. With the drug Alsterpaullone it is possible to stabilize *β-catenin*. This causes that nuclear accumulation of *β-catenin* occurs throughout the organism, and not only in the head region. This treatment leads to tentacle formation all over and, as determined by transplantation experiments, to a higher competence level [7]. In terms of the model, with this treatment the near-head zone looses its privileged position and tentacle formation becomes possible everywhere.

Long-range activation and local suppression of a particular structures is a common theme in different developmental situations. For instance, the main players in *Drosophila* segmentation follow the same principle: *engrailed* and *wingless* exclude each other locally but the two genes activate each other at longer range by the secreted *hedgehog* molecules and *Wnt*-containing vesicles. In this case the activation is mutual, which leads to adjacent stripes of *wingless* and *engrailed* activations.

5 A Second Off-Axis Organizer Leads to Bilaterality

The development of a bilateral body plan requires the establishment of a second axis orthogonal to the (presumably ancestral) AP axis. To specify the dorsoventral (DV) axis, all higher organisms use the same molecular system that includes *Chordin* and *BMP's* [32]. Both molecules are already present in hydra, *Chordin* in the head region [52] and *BMP5/8* in the foot region [34], i.e., antipodal to the *Chordin* expression. This suggests that the systems that organize the embryonic AP and DV axes in higher organisms were already present in the radial-symmetric ancestor and parallel to each other along the only existing axis. Thus, the formation of the second axis in bilaterians was presumably not based on the invention of a new patterning systems but on a mutual reorientation of two existing patterning systems [10, 13].

The posterior region at which ectoderm and endoderm meets is usually called the blastopore. In hydra, the blastopore is the small hypostomal region. The corresponding structure in vertebrates, e.g., the marginal zone in *Xenopus* (see Fig. 5) or the germ ring in the zebra fish, is a large ring. As in hydra, in vertebrates the blastopore is a source region of components of the *Wnt* pathway that organize the more anteriorly located tissue, the fore- and midbrain [38, 39]. This does not require a Spemann-type dorsal organizer [40]. The widening of the small patch-like organizing blastopore to a large ring has an important consequence. The formation of the Spemann-type organizer on this large ring assures an off-axis position that provides the first step towards the establishment of a new axis perpendicular to the primary AP axis.

6 The Generation of a Solitary Midline: A Crucial Step Towards the Generation of Near-Cartesian Coordinate Systems

An off-axis spot-like organizer such as the Spemann-organizer is, on its own, insufficient for the DV-organization of an long-extended organisms. The organizing region for the DV axis must extend over the entire AP axis with only a small DV extension, i.e., it must have the geometry of a narrow stripe (Fig. 4a). The notochord and the floor plate, which act as organizer for the DV (or mediolateral axis), indeed have this geometry. The formation of a single, long and narrow source region requires specific non-trivial mechanisms (Fig. 4). If lateral inhibition is involved, why does the midline not disintegrates into individual patches? If in a pattern-forming reaction the self-enhancement saturates at moderate activator levels, stripe-like instead of spot-like distributions emerge [23]. Due to the saturation, the activated regions become larger although at a lower activator level. Since lateral inhibition is required for pattern formation, an activation is most stable if non-activated cells are nearby. These two requirements, large patches and non-activated cell(s) in the immediate neighborhood, seem to exclude each other. However, both requirements are satisfied in a stripe-like pattern. When triggered from random fluctuations, multiple stripes are formed that bifurcate, bend, and can have any orientation (Fig. 4b). In contrast, only a single straight stripe is required for the midline organizer.

Modeling has revealed that by a coupling of at least two pattern-forming systems a single solitary straight stripe can emerge: a patch-forming system acts on a stripe-forming system in such a way that only a single stripe can form. Different strategies evolved in different phyla (Fig. 4). In vertebrates, a dorsal organizer initiates and elongates the midline; the midline and the central nervous system forms at the same side as the organizer, i.e., *dorsal*. In the course of time, the midline is left behind the moving organizer just as a vapor trail is left behind a high-flying airplane. The detailed biological implementation is discussed in Figs. 5 and 6. In contrast, in insects (*Drosophila*) a dorsal organizer represses the midline. Thus, the midline and the central nervous system form *ventrally*. This mode of midline formation was predicted on the base of theoretical considerations [11] and since then has received strong support [12]. The dorsal elongation in vertebrates versus the compression to the ventral side in insects was proposed [10] to be the deep-seated reason for the well-known DV-VD reversal in the localization of the central nervous system [42].

In addition to patterning along the AP and DV axes, vertebrates have a left-right patterning [43–45]. A systematic asymmetry is generated by mechanisms that depend on molecular asymmetries in cila causing a "nodal flow". An asymmetry of the left-right type emerges even in the absence of this bias, although with a random orientation [46]. The local elongation of the midline suggested a mechanism for the left-right patterning in vertebrates [41]. In the model, if the organizer and the incipient midline system activates a third patterning system at a longer range and excludes it at a shorter range, the activation of the third system will

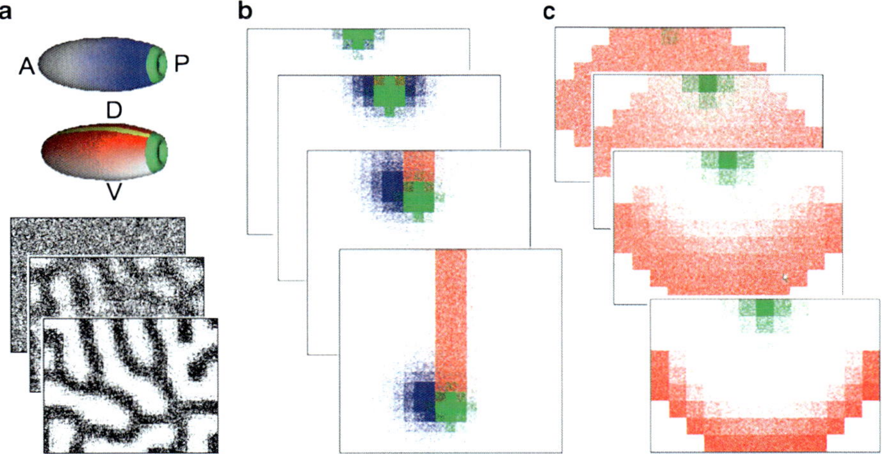

Fig. 4 The midline problem and its solutions. (**a**) For the transition from a radial-symmetric to a long extended bilateral-symmetric body plan, the formation of a midline organizing region was inevitable. The midline organizer (*yellow*) has to extend over the whole AP axis but must have only a small extension along the DV axis. (**b**) Pattern-forming systems in which the self-enhancement has an upper bound [$\kappa > 0$ in (1)] can generate stripe-like distributions. However, these stripes both bend and bifurcate, and the widths of stripe and interstripe regions are of the same order, properties incompatible with a solitary midline. (**c**, **d**) Different phyla found different mechanisms for generating a single straight midline organizer. In vertebrates (**c**), a moving spot-like organizer leaves behind a single stripe-shaped region. In this simulation, an activator-inhibitor system that generates a spot-shaped maximum (*green*) triggers a stripe-forming system (*red*), which, in turn, repels the organizer and causes thereby its shift. A third pattern-forming system (*blue*), repelled from the midline, allows left-right pattering [41]. (**d**) According to the model for insects, a patch-forming system has an *inhibitory* influence on a stripe-forming system. Thus, the midline forms at the largest possible distance from the dorsal organizer. The midline has from early stages onwards the full AP extension but sharpens along the DV axis. This prediction [11] is fully supported by the observation in *Tribolium* [12]. Planarians found still another way to form a midline and to keep the AP and DV axes perpendicular to each other [13]

appear on one side of the organizer and the midline (blue in Fig. 4c). Such a mechanism allows amplification of a minute bias towards the left and leads to symmetry breaking in the absence of such a bias. The predicted mechanism was found to be realized by the *Nodal/Lefty2* system [47]. It is interesting that such squeezing of a signal out of the midline is inappropriate if the midline is generated by a sharpening towards ventral over the entire AP axis, as it occurs in insects [16]. This could be the reason why a pronounced left-right patterning does not exist in insects.

It is conceivable that the invention of different modes of midline formation constituted the point of no return in the separation of the different phyla. In contrast to the urbilaterian concept, these models suggest that the second axis existed before the advent of bilateral-symmetric organisms. Different modes evolved for the

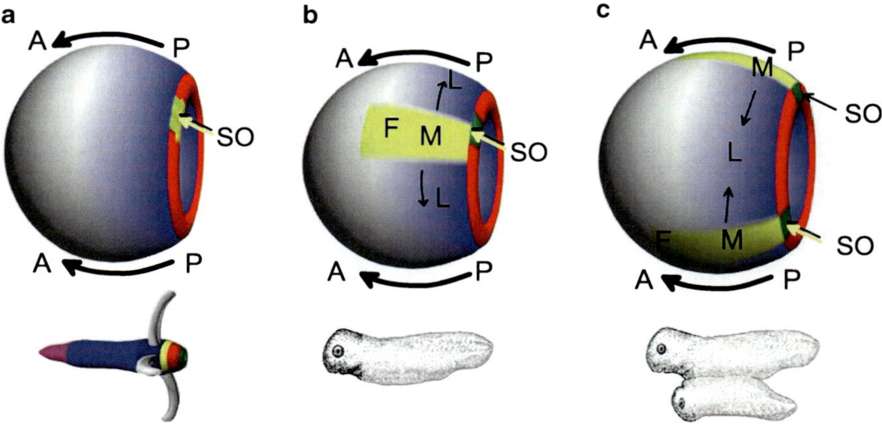

Fig. 5 The generation of a near-Cartesian coordinate system – the amphibian embryo as an example. (**a**) AP patterning of the early gastrula is assumed to be accomplished by a system that was already involved in the body patterning of ancestral organisms. The marginal zone (*red*) is assumed to be equivalent to the hydra organizer. These blastoporal rings, small in hydra and huge in amphibians, are the source region of *Wnt* signaling, which controls anteroposterior determination in a gradient-based manner (*blue* gradient). The Spemann organizer (SO, *yellow*), a small patch of cells with signaling properties, is located on this marginal zone, i.e., on the ancestral hydra-type organizer. (**b**) With the movement of the mesodermal cells of the marginal zone between the inner endoderm and outer ectoderm, the organizer-derived cells form the prechordal plate (*yellow*) in the future brain region. Its acts as the midline organizer for the mediolateral (L-M-L) patterning and induces neuronal development in the overlying ectoderm. The distance from this midline determines the mediolateral specification in the brain, for instance, the distance and symmetrical arrangement of the two eyes. Both signaling sources, the ring of the marginal zone and the prechordal plate, have a stripe-like extension with orthogonal orientation. Together, they provide a near-Cartesian positional information system for patterning the fore- (*F*) and midbrain (*M*). (**c**) Induction of a second organizer by transplantation of cells from the dorsal organizer to an antipodal position induces a new midline. Two embryos are formed that are fused at the ventral side. Note that the induction of a second Spemann-type organizer does not lead to a second AP axis, contrary to what is frequently stated in the literature. The patterning of the trunk occurs in a different way (see Fig. 6); (Figures from [29])

realignment of the two originally parallel axes and different mechanisms evolved to solve the midline problem [13].

7 Axis Formation in Vertebrates Occurs in Two Parts

As mentioned above, the brain of a higher organism can be regarded as an evolutionary ancestral part that evolved from the body pattern of ancestral organisms (Fig. 1). The organizing region for this part, the blastopore or the marginal zone in amphibians, is located at the most posterior position (Fig. 5).

Wnt molecules, spreading from this posterior organizer, specify fore- and midbrain. Modeling of gene activation under the influence of a morphogen has been proposed to proceed by a stepwise irreversible activation of genes that specify more and more posterior (or distal) structures [48]. This mode corresponds to the posterior transformation observed by Nieuwkoop [49] on the basis of his transplantation experiments. Thus, early steps of brain patterning can be regarded as proceeding via a concentration-dependent gene activation by a morphogen gradient but not all details are yet clear [38].

At the early gastrula stage, when the basic brain pattern is laid down, the trunk is not yet present. The formation of the trunk proceeds by a transformation of the ring-shaped region near the blastopore into a rod-shaped axial structure. In the cells near the blastopore, a sequential, time-dependent posterior transformation takes place, achieved by the sequential activation of Hox-genes [35, 50]. Activation of the new Hox-genes occurs in a horseshoe-like region: cells near the marginal zone leaving out the Spemann organizer. Cells of this zone move towards the organizer, leave the surrounding of the marginal zone and elongate the incipient midline, causing in this way the ring-to-rod transformation, a special case of the convergence-extension mechanism [51]. The function of the organizer at this stage is that the cells leave progressively the zone in which posterior transformation takes place (Fig. 6). With their departure, their AP specification is fixed. Note that this view leads to a model for dorsoventral patterning that differs from usual scheme as given in textbooks. The standard view is that the dorsoventral patterning occurs along the organizer–anti-organizer axis (blue double arrow in Fig. 6b; see, for instance, [36]). According to the model proposed here, the distance from the midline is decisive (red arrows in Fig. 6c), not the distance from the organizer.

A time-dependent sequential posteriorization was predicted in a model that also included the formation of the periodic trunk patterning [15]. In the course of time, two patterns have to be generated that are precisely in register, the periodic pattern of somites or segments on the one hand and the different specification of these periodic elements on the other. In short-germ insects the segments are formed near the posterior pole. This process can be modeled by an oscillation between anterior and posterior specifications at the posterior pole. Cells leaving the critical zone during outgrowth maintain their anterior or posterior compartmental specification (Fig. 6). Thus, the oscillation in time leads to a periodic pattern in space [15]. Somite formation in *Amphioxus* likewise takes place near the posterior pole [33]. In vertebrates, however, the somites are formed at some distance from the posterior pole [53]. To construct a model that retains the oscillation between anterior and posterior specifications previously derived for insect segmentation, I assumed that somites are subdivided into anterior and posterior half-somites and that their formation is based on an oscillation between anterior and posterior specifications. Corresponding activations at the posterior pole were assumed to move in a wave-like manner towards anterior and come to rest in the somite-forming zone. Each full cycle was assumed to add one pair or anterior and posterior half-somites (Fig. 6f). The predicted subdivision into half-somites was discovered 2 years later, after the appropriate staining techniques became available [54]. It took about 15 years for

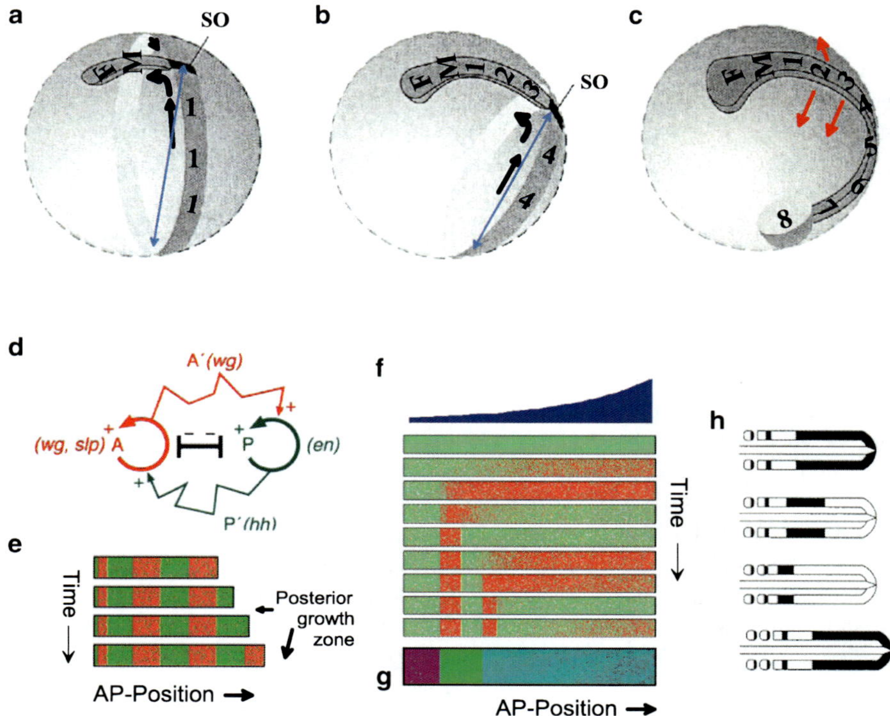

Fig. 6 Pattern formation in the trunk: While determination for the forebrain (F) and midbrain (M) occurs under the influence of a *Wnt*-gradient (Fig. 5), trunk determination (1, 2, 3, ...) occurs by a time-dependent posteriorization in the cells near the blastoporal ring [35]. These cells move towards the organizer and the incipient midline (*black arrows*). In the course of time the blastoporal ring perpendicular to the AP axis shrinks while the axial rod, oriented parallel to the AP axis, elongates [16, 19, 51]. The organizer is required for cells to leave the blastoporal region in which the stepwise posterior transformation takes place. Note that this model leads to a different view of the dorsoventral patterning than that of recently published models [36], according to which the dorsoventral patterning occurs along the organizer – anti-organizer axis (*blue double arrow*). According to the model proposed here, the distance from the midline is decisive (*red arrows*), not the distance from the organizer. Therefore, the dorsal-ventral pattering takes place after cells left the marginal zone. (**d**, **e**) The mechanism of mutual long-range activation of locally exclusive states can lead an oscillating patterns that gives rise to a stable periodic pattern during posterior outgrowth. (**f**) Somite formation was predicted to be achieved by oscillations at the posterior pole and a wave-like spread of these activities towards anterior [15]. (**g**) Each full cycle leads to the activation of a new gene. (**h**) This predicted mechanism was found with the observation of *c-hairy* activation in the chick [37]; (Figure in part from [19], [29] and [37])

this oscillation and wave-like spread towards anterior to be demonstrated experimentally, with the elucidation of the temporal-spatial progression of *c-hairy* activation in the chick [37] (Fig. 6h). The predicted AP subdivision of somites also turned out to be important. Nerves from the spinal cord grow only through anterior half-somites [53, 55].

An oscillation for somite formation was proposed even earlier in the "Clock-and-Wavefront" model [58]. Therein a very different mechanism for somite formation was postulated. A wave front was assumed to move slowly down the presomitic mesoderm, from anterior to posterior. Further, cells were postulated to oscillate in synchrony. Whenever this oscillation reached a maximum, a rapid morphological change was assumed to take place at the actual position of the wavefront, causing the separation of one somite from the next. Hence, in this model, the purpose of the oscillations was to separate one entire morphological somite from the next. The formation of the internal pattern of the somites was not addressed. Whether such a synchronous oscillation takes place in the whole organism or only in the pre-somitic mesoderm does not matter. Waves such as those actually observed, starting posteriorly and come to rest at the somite-forming zone, were not part of the model. The clock-and-wavefront model was not mathematically formulated, and thus computer simulations are not available. An important finding of Cooke was that reducing the number of available cells in early *Xenopus* embryos leads to *shorter* embryos, not to *taller* embryos. To account for that, Cooke and Zeeman assumed that in smaller embryos the oscillation frequency increases, so that the AP extension of the individual somites is reduced. However, somewhat later, Cooke realized that the oscillation frequency is size-independent [56].

In my original model [15], I assumed that the oscillation leads not only to the periodic structure but also affects the sequential activation of genes that specify more and more posterior structures (Fig. 6g). Just as the periodic movement of the pendulum of a grandfather's clock leads to sequential advancement of the pointer, it was assumed that the alternation between anterior and posterior specifications during the oscillation leads to a sequential activation of genes that specify more and more posterior structures. These genes are now known to be the HOX genes [50]. Meanwhile it became evident that not every full cycle leads to the activation of a new HOX gene, although some coupling between the oscillation and Hox gene activation has indeed been found [57].

8 Conclusion

Modeling reveals the minimum requirements that are necessary to accomplish essential steps in development. Using this tool, we found for several important developmental steps the crucial interactions before the molecular effectors were identified. During attempts to formulate the molecularly feasible interactions that account for particular steps, one can stumble over problems in the molecular realizations that would otherwise remain unnoticed. In this way I realized that the formation of a midline was asubtle step. Modeling revealed that during the evolution of higher organisms, nature found different solutions for this problem, assumed to be the branch point for the formation of different phyla. Such minimum models help one to deduce the underlying logic in spite of the overt complexity of the system and provide insights into why development is as complex as we observe.

References

1. Bode HR (2003) Head regeneration in Hydra. Dev Dyn 226:225–236
2. Bosch TCG (2007) Why polyps regenerate and we don't: towards a cellular and molecular framework for Hydra regeneration. Dev Biol 303:421–433
3. Gierer A (1977) Biological features and physical concepts of pattern formation exemplified by hydra. Curr Top Dev Biol 11:17–59
4. Holstein TW, Hobmayer E, Technau U (2003) Cnidarians: an evolutionarily conserved model system for regeneration? Dev Dyn 226:257–267
5. Gierer A, Berking S, Bode H, David CN, Flick K, Hansmann G, Schaller H, Trenkner E (1972) Regeneration of hydra from reaggregated cells. Nat New Biol 239:98–101
6. Browne EN (1909) The production of new hydrants in Hydra by insertion of small grafts. J Exp Zool 7:1–23
7. Broun M, Gee L, Reinhardt B, Bode HR (2005) Formation of the head organizer in hydra involves the canonical Wnt pathway. Development 132:2907–2916
8. Hobmayer B, Rentzsch F, Kuhn K, Happel CM, Cramer von Laue C, Snyder P, Rothbacher U, Holstein TW (2000) Wnt signalling molecules act in axis formation in the diploblastic metazoan Hydra. Nature 407:186–189
9. Lengfeld T, Watanabe H, Simakov O, Lindgens D, Gee L, Law L, Schmidt HA, Özbek S, Bode H, Holstein TW (2009) Multiple *Wnts* are involved in Hydra organizer formation and regeneration. Dev Biol 330:186–199
10. Meinhardt H (2002) The radial-symmetric hydra and the evolution of the bilateral body plan: an old body became a young brain. Bioessays 24:185–191
11. Meinhardt H (1989) Models for positional signalling with application to the dorsoventral patterning of insects and segregation into different cell types. Development (Supplement 1989):169–180
12. Chen G, Handel K, Roth S (2000) The maternal nf-kappa b/dorsal gradient of *Tribolium castaneum*: dynamics of early dorsoventral patterning in a short-germ beetle. Development 127:5145–5156
13. Meinhardt H (2004) Different strategies for midline formation in bilaterians. Nat Rev Neurosci 5:502–510
14. Gierer A, Meinhardt H (1972) A theory of biological pattern formation. Kybernetik 12:30–39
15. Meinhardt H (1982) Models of biological pattern formation. Academic, London (available at http://www.eb.tuebingen.mpg.de/meinhardt/82-book)
16. Meinhardt H (2008) Models of biological pattern formation: from elementary steps to the organization of embryonic axes. Curr Top Dev Biol 81:1–63
17. Turing A (1952) The chemical basis of morphogenesis. Philos Trans R Soc Lond B Biol Sci 237:37–72
18. Shimizu H, Fujisawa T (2003) Peduncle of Hydra and the heart of higher organisms share a common ancestral origin. Genesis 36:182–186
19. Meinhardt H (2006) Primary body axes of vertebrates: generation of a near-Cartesian coordinate system and the role of Spemann-type organizer. Dev Dyn 235:2907–2919
20. Duboc V, Rottinger E, Besnardeau L, Lepage T (2004) Nodal and bmp2/4 signaling organizes the oral-aboral axis of the sea urchin embryo. Dev Cell 6:397–410
21. Schier AF (2009) Nodal morphogens. Cold Spring Harb Perspect Biol. doi:10.1101/cshperspect.a003459
22. Huang X, Dong Y, Zhao J (2004) *HetR* homodimer is a DNA-binding protein required for heterocyst differentiation, and the DNA-binding activity is inhibited by *PatS*. Proc Natl Acad Sci U S A 101:4848–4853
23. Meinhardt H (1995) Growth and patterning – dynamics of stripe formation. Nature 376:722–723
24. Hörstadius S (1939) The mechanics of sea-urchin development studied by operative methods. Biol Rev 14:132–179

25. Müller WA, Plickert G (1982) Quantitative analysis of an inhibitory gradient field in the hydrozoan stolon. Wilhelm Roux Arch 191:56–63
26. Meinhardt H (1993) A model for pattern-formation of hypostome, tentacles, and foot in hydra: how to form structures close to each other, how to form them at a distance. Dev Biol 157:321–333
27. Bode PM, Awad TA, Koizumi O, Nakashima Y, Grimmelikhuijzen CJP, Bode HR (1988) Development of the two-part pattern during regeneration of the head in hydra. Development 102:223–235
28. Technau U, Holstein TW (1995) Head formation in hydra is different at apical and basal levels. Development 121:1273–1282
29. Meinhardt H (2009) The algorithmic beauty of sea shells, 4th enlarged edn (with programs on CD). Springer, Heidelberg/New York
30. Wilby OK, Webster G (1970) Experimental studies on axial polarity in hydra. J Embryol Exp Morphol 24:595–613
31. Meinhardt H, Gierer A (1980) Generation and regeneration of sequences of structures during morphogenesis. J Theor Biol 85:429–450
32. Holley SA, Jackson PD, Sasai Y, Lu B, De Robertis EM, Hoffmann FM, Ferguson EL (1995) The Algorithmic Beauty of Sea Shells. A conserved system for dorsal-ventral patterning in insects and vertebrates involving sog and *Chordin*. Nature 376:249–253
33. Schubert M, Holland LZ, Stokes MD, Holland ND (2001) Three amphioxus Wnt genes (amphiwnt3, amphiwnt5, and amphiwnt6) associated with the tail bud: the evolution of somitogenesis in chordates. Dev Biol 240:262–273
34. Reinhardt B, Broun M, Blitz IL, Bode HR (2004) *Hybmp5-8b*, a *BMP5-8* orthologue, acts during axial patterning and tentacle formation in Hydra. Dev Biol 267:43–59
35. Wacker SA, Jansen HJ, McNulty CL, Houtzager E, Durston AJ (2004) Timed interactions between the Hox expressing non-organiser mesoderm and the Spemann organiser generate positional information during vertebrate gastrulation. Dev Biol 268:207–219
36. Ben-Zvi D, Shilo BZ, Fainsod A, Barkai N (2008) Scaling of the *BMP* activation gradient in *Xenopus* embryos. Nature 453:1205–1211
37. Palmeirim I, Henrique D, Ish-Horowicz D, Pourquie O (1997) Avian *hairy* gene-expression identifies a molecular clock linked to vertebrate segmentation and somitogenesis. Cell 91:639–648
38. Kiecker C, Niehrs C (2001) A morphogen gradient of *Wnt/β-catenin* signalling regulates anteroposterior neural patterning in *Xenopus*. Development 128:4189–4201
39. Nordström U, Jessell TM, Edlund T (2002) Progressive induction of caudal neural character by graded Wnt signaling. Nat Neurosci 5:525–532
40. Ober EA, Schulte-Merker S (1999) Signals from the yolk cell induce mesoderm, neuroectoderm, the trunk organizer, and the notochord in zebrafish. Dev Biol 215:167–181
41. Meinhardt H (2001) Organizer and axes formation as a self-organizing process. Int J Dev Biol 45:177–188
42. Arendt D, Nübler-Jung K (1994) Inversion of dorsoventral axis. Nature 371:26
43. Levin M, Palmer AR (2007) Left-right patterning from the inside out: widespread evidence for intracellular control. Bioessays 29:271–287
44. Raya A, Izpisúa Belmonte JC (2006) Left-right asymmetry in the vertebrate embryo: from early information to higher-level integration. Nat Rev Genet 7:283–293
45. Vandenberg LN, Levin M (2009) Perspectives and open problems in the early phases of left-right patterning. Semin Cell Dev Biol 20:456–463
46. Nonaka S, Tanaka Y, Okada Y, Takeda S, Harada A, Kanai Y, Kido M, Hirokawa N (1998) Randomization of left-right asymmetry due to loss of *nodal* cilia generating leftward flow of extraembryonic fluid in mice lacking KIF3B motorprotein. Cell 95:829–837
47. Nakamura T, Mine N, Nakaguchi E, Mochizuki A, Yamamoto M, Yashiro K, Meno C, Hamada M (2006) Generation of robust left-right asymmetry in the mouse embryo requires a self-enhancement and lateral-inhibition system. Dev Cell 11:495–504

48. Meinhardt H (1978) Space-dependent cell determination under the control of a morphogen gradient. J Theor Biol 74:307–321
49. Nieuwkoop PD (1952) Activation and organization of the central nervous system in amphibians. III Synthesis of a new working hypothesis. J Exp Zool 120:83–108
50. Duboule D (2007) The rise and fall of Hox gene clusters. Development 134:2549–2560
51. Keller R (2005) Cell migration during gastrulation. Curr Opin Cell Biol 17:533–541
52. Rentzsch F, Guder C, Vocke D, Hobmayer B, Holstein TW (2007) An ancient *Chordin*-like gene in organizer formation of Hydra. Proc Natl Acad Sci U S A 104:3249–3254
53. Keynes RJ, Stern CD (1988) Mechanisms of vertebrate segmentation. Development 103:413–429
54. Stern CD, Keynes RJ (1987) Interaction between somite cells: the formation and maintenance of segment boundaries in the chick embryo. Development 99:261–272
55. Keynes RJ, Stern CD (1984) Segmentation in the vertebrate nervous system. Nature 310:786–789
56. Cooke J (1981) The problem of periodic patterns in embryos. Philos Trans R Soc Lond B Biol Sci 295:509–524
57. Dubrulle J, McGrew MJ, Pourquie O (2001) *FGF* signaling controls somite boundary position and regulates segmentation clock control of spatiotemporal Hox gene activation. Cell 106:219–232
58. Cooke J, Zeeman EC (1976) A clock and wavefront model for control of the number of repeated structures during animal morphogenesis. J Theor Biol 58:455–476

Cell Division and Hyperbolic Geometry

Misha Gromov

Preface This article was written more than 20 years ago, when I was not even aware of the existence of Lindenmayer systems, although they in fact preceded hyperbolic systems. The question discussed here, and which remains unsolved, is a mathematician's baby version of a biology problem: can *simple/universal* local cell division/interaction rules account for the observed features of organisms' development, e.g., for stabilization of the shape/architecture of animal organisms during their post-embryonic phase of growth. References of two mathematics papers related to the topic, which appeared during last decade, have been added ([4],[5]).

The idea of cell division comes, as everybody knows, from biology. Everything alive, from bacteria on, is built of cells; and as the cells divide, the living creatures grow. One might think that an individual cell divides according to a relatively simple program, and that this program is the same for all cells of a given type. Of course the execution of the program may depend on what happens around a given cell. In particular, it may depend on what goes on with the neighbor cells located in the immediate vicinity of our cell.

So far nobody has been able to find simple principles governing cell division which can explain the result of the divisions, that is, the variety of shapes observed in the living matter we see around us.

Now, I want to show how a pure mathematician would treat this problem. The essence of the mathematical approach is to extract what one could call *the idea* of cell division. This idea must simultaneously be formal, simple and viable. That is, the idea must have the potential to grow and develop into an aesthetically satisfying mathematical theory. (True mathematics always brings unexpected turns at every

M. Gromov (✉)
Institut des Hautes Etudes Scientifiques, Bures-sur-Yvette, France
e-mail: gromov@ihes.fr

V. Capasso et al. (eds.), *Pattern Formation in Morphogenesis*, Springer Proceedings in Mathematics 15, DOI 10.1007/978-3-642-20164-6_18,

step as one follows the development of an idea. One cannot tell in advance which direction will bring you to a viable theory and which will terminate in a dead end).

Ideas do not come cheap in mathematics. The price one should be ready to pay is the biological significance of the issuing theory. It may of course happen, by accident, that the mathematical development will be related to real life, but this is not a concern for a pure mathematician. A good way to start the search for an idea is to look at the simplest possible case. This is, of course, where our living thing consists of a single cell that divides into two cells identical to itself, and where the two baby cells have no link between them. It takes some time for the babies to grow to adult size, and then each of them divides again. The mathematics here is reduced to a dull sequence of numbers: 1, 2, 4, 8, 16 . . . In fact, the picture becomes not so dull when we consider longer stretches of time. For example, let the division occur every hour. Then in 2 h we have 2 cells, in 3 h there are 4, and so on. Does it take long to have more cells than there are elementary particles in the known universe? (Say in a ball with a radius of 20 billion light years). Is it a day, a month, a century, or a billion years? Of course, the answer may only impress a mathematically naive mind (about a month is the best estimate), but the very idea of unrestricted growth of time has proved very fruitful in many fields of pure (and not so pure) mathematics.

Now let us imagine that every daughter cell remembers its mother. It is amazing how much this "memory" enriches the mathematical structure. To visualize what happens, we draw line segments between each child and its parent, and then we see the following ancestral trees.

Moment 0 : there is a single cell, indicated by a point •;
Moment 1 : The original cell divides into two, which are joined to the mother by two edges (Fig. 1).

Then, each moment subsequently, new branches appear (Fig. 2).

Here there are at least two different ways of seeing what happens as the time eventually goes to infinity. In one setting, we imagine new branches getting smaller and smaller so that the limit tree, with infinitely many branches, is still contained in a bounded region of space. It looks something like this (Fig. 3).

Every infinite branch here has an end, and all these form what mathematicians call a *Cantor set.*

Another possibility for growth is where all edges remain the same length. The resulting tree is infinite in size, and its branches go to infinity, as shown in the following picture (Fig. 4).

Our next objective is to understand what happens if the sister cells do not drift apart but remain close together, so that after a stretch of time the descendants of a single cell form a true organism rather than a collection of disjoint cells. The simplest case of this is where the sister cells always have one point in common. Then the division process looks schematically as follows (Fig. 5):
which can be schematized further as Fig. 6.

Fig. 1 See text for details

Fig. 2 See text for details

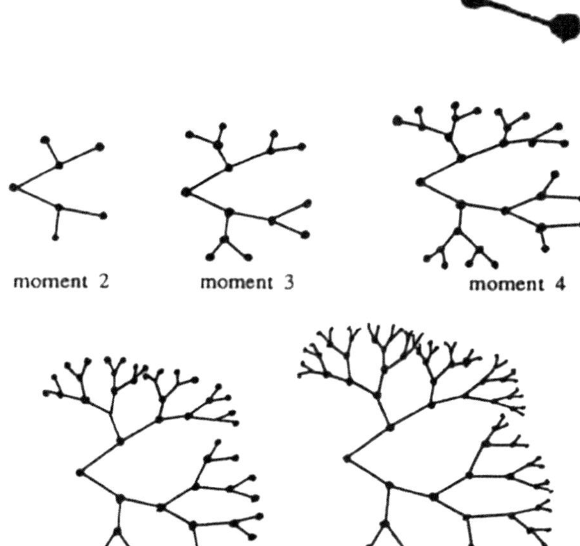

The limit object here (corresponding to the Cantor set in the free division picture) is an ordinary segment of real numbers, say [0,1], where the division process is equivalent to the binary representation of the numbers.

Next, we look at the picture where the mother is represented by the unit interval, and the babies are also unit intervals, which overlap over two subintervals, say [0,] and [,1] in [0,1]. The schematic division picture is as follows (Fig. 7).

The limit object here is a rather complicated one-dimensional space, a kind of infinite necklace.

Let us summarize the essential features of our examples. First of all, our cells have no internal structure, and so they are symbolically represented by points. Two or more cells in an organism may be neighbors and the *neighborhood* is the only relation we introduce between the cells. Graphically, one joins every two neighbors by an edge, the triple neighborhood is designated by a triangle, etc. Here is an example (Fig. 8).

Next we want to introduce the notion of a *cell division law*, such that all cells divide according to this law. At this point it is convenient to attach an extra structure to our cells in order to allow several possibilities for division. For example, we may have two kinds of cells, say black and white, such that every black cell divides into two white ones, whereas every white cell divides into one black and one white cell. Thus, mathematically speaking, our cell organisms are represented by *colored*

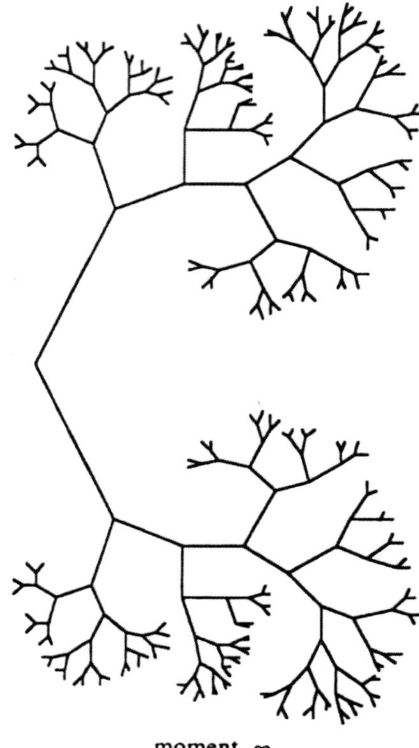

Fig. 3 See text for details

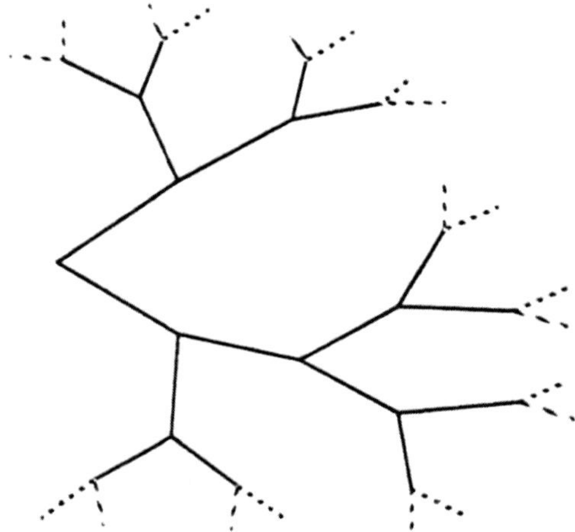

Fig. 4 See text for details

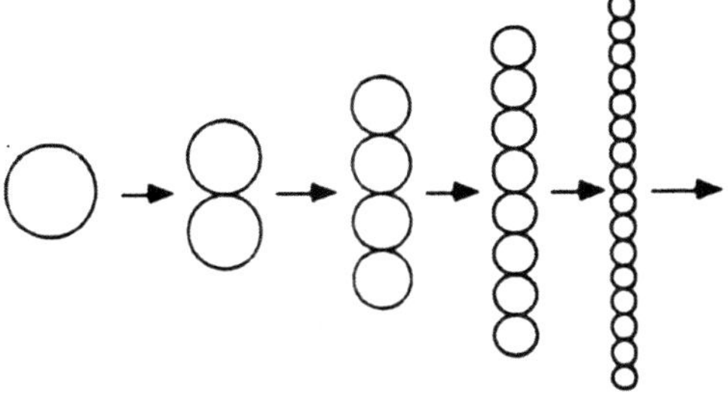

Fig. 5 See text for details

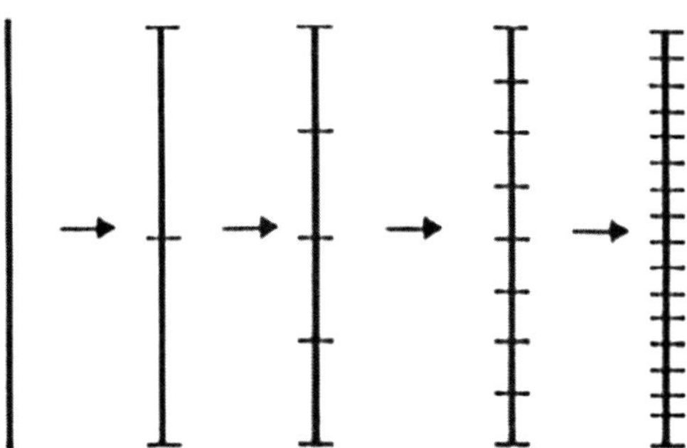

Fig. 6 See text for details

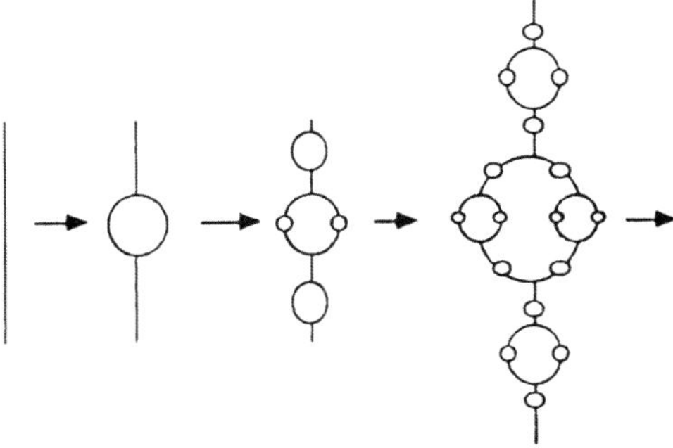

Fig. 7 See text for details

simplicial complexes ("simplex" is the generic name for segment, triangle, tetrahedron, etc.). Then the division law assigns to each color a colored simplicial complex, and the resulting division process of a given organism (colored complex) consists in replacing each vertex (the vertices of a complex are just the points representing the cells) by the complex corresponding to the color of the vertex. Notice that, here, a cell may divide into more than two baby-cells. Also notice that, if we want the result of a division to again be a simplicial complex, we must specify the neighborhood relation between the new (daughter) cells. This requires a slightly more precise and elaborate definition of a division law, which we do not explain here.

Now, we think of a division law as a transformation acting on the set of colored complexes (or on some subset of these), and our problem is to understand what

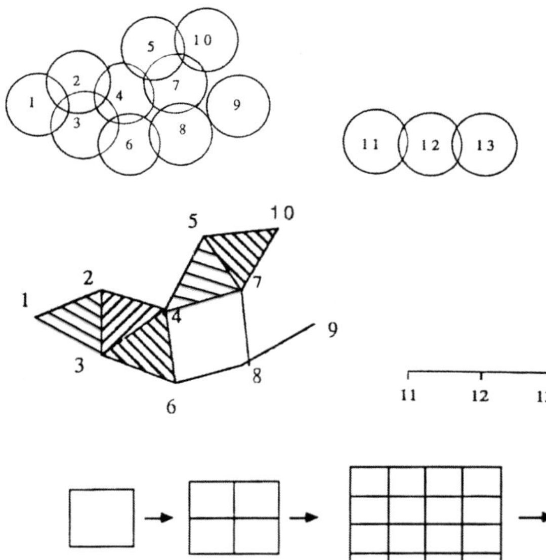

Fig. 8 See text for details

Fig. 9 See text for details

happens under infinite iteration of this transformation. In particular, one wants to classify the limit spaces resulting from infinitely many divisions of an original complex (organism). This general problem looks too hopelessly complicated for meaningful mathematical study. Yet there is an interesting class of division laws where one expects eventual answers to the questions. This class is constituted of what we call *hyperbolic* (or expanding) laws, where we, roughly speaking, require that the division makes every organism grow "in all directions". All our earlier examples were hyperbolic. Here are some more (Fig. 9):

The divided square above grows uniformly in both directions, but the hyperbolicity allows some directions to grow faster than others, as seen in Fig. 10.

On the other hand, the following division is non-hyperbolic (Fig. 11).

Whenever we have a hyperbolic law, we can define the limit space X, similar to the Cantor set in our first example, but now of an arbitrary dimension. This dimension is the basic invariant of the underlying division law; and there are many more topological and geometric characteristics of the limit space X reflecting the (asymptotic) properties of a particular division law. In fact, there is an adequate description of the cell division creating X in terms of the space X itself. Namely, the division law gives rise to a sequence of finite partitions of X into smaller subspaces, where the elements of the first partition correspond to the colors. Then the division process is "generated" by finitely many (partially defined) homeomorphisms of X which have a certain expanding property. The most interesting examples here arise from *hyperbolic dynamical systems* and their *Markov partitions* and from *hyperbolic groups* acting on their *ideal boundaries*. Unfortunately, any meaningful discussion becomes technical at this point and we refer the reader to the papers [1],[2] and [3].

Fig. 10 See text for details

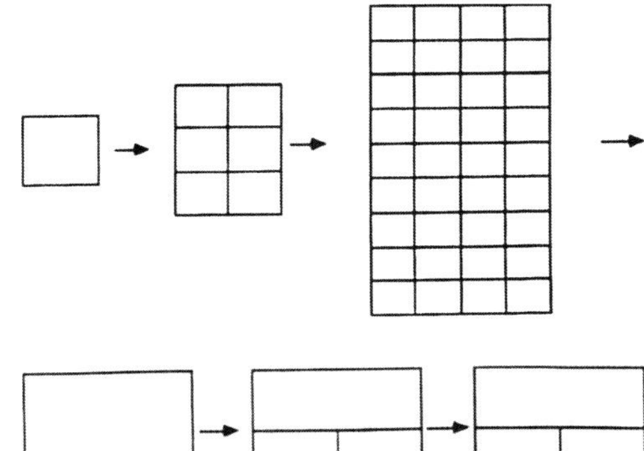

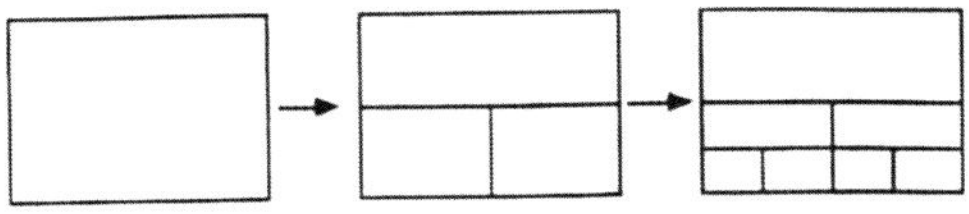

Fig. 11 See text for details

We conclude by indicating other models of cell division which also lead to interesting mathematics. One thing one can do is to allow the individual division to be a random rather than a deterministic event. One may think of this randomness as a kind of a secret color attached to a cell (now the color may take infinitely many values corresponding to the points of a probability space), and the division depends on this color.

Another possibility is to consider a continuous model, say a Riemannian manifold (e.g., the space we live in), and replace the division law by a differential evolution equation which would force a uniform expansion of the metric (which seems to happen to our universe). It is, however, unclear whether a model of this kind can lead to a persistent creation of local topology, as occurs in the discrete case (e.g., in the necklace example).

References

1. Bowen R (1970) Markov partitions for axiom A diffeomorphisms. Am J Math 92:725–747
2. Gromov M (1981) Hyperbolic manifolds groups and actions. Ann Math Stud 97:183–215, Princeton
3. Gromov M (1987) Hyperbolic groups. p.25-265, Math. Sci. Res. Inst. Publ., 8, Springer, New York.
4. Previte JP (1988) Graph Substitutions. Ergodic Theory and Dynamical Systems. Vol 18, Part 3. 661–686
5. Cannon JW, Floyd W, Parry W (2000) Crystal growth, biological cell growth and geometry. In: Pattern Formation in Biology, Vision and Dynamics, World Scientific, pp. 65–82

Formalistic Representation of the Cellular Architecture in the Course of Plant Tissue Development

Ivan V. Rudskiy

Abstract Transformation of biological inherent information into the spatial organization of an organism is a major challenge of modern developmental biology. The aim of this paper is to find a complete and sufficient description of the regular spatiotemporal structure of the multicellular organisms using plants as a basic model. The locally reiterative development of the real organism's structures was regarded as a combinatorial map from a globally regular space with a proper group action. Such a space was considered as a "developmental program" of the organism. The following questions are resolved here: (1) formalistic description of spatiotemporal structure of the real organisms at the cellular and tissue levels of organization in terms of graphs; (2) definition of regularities and flexibility in the structure development in terms of groups acting on graphs; (3) the basic species-specific properties and principles of reiterative work of "developmental program" spaces. We believe that the considerable and appropriate formalization of biological spatiotemporal data opens the possibilities for revealing the correspondence between structural mechanisms at incomparably different levels of organization, such as molecular-genetic, cellular and morphological levels.

1 Introduction

Plants and animals have regular structure not only in space but in time owing to the repetitive way in which their development is expressed in the life cycle and their metameric organization. Anatomical and morphological species-specific complexity of every multicellular organism is based on the species-specific regularity of cellular architecture. However, even in the present highly technological time it is not an easy task to perceive and to trace all the developmental changes in cellular arrangement

I.V. Rudskiy (✉)
Institut des Hautes Études Scientifiques, Bures-sur-Yvette, France
e-mail: rudskiy@ihes.fr

V. Capasso et al. (eds.), *Pattern Formation in Morphogenesis*, Springer Proceedings in Mathematics 15, DOI 10.1007/978-3-642-20164-6_19,

directly from microscopic observations. A precise and detailed description of tissue fragments is necessary for reconstructing the developmental history of the whole sample. This spatio-temporal data enables one to identify and compare the equivalent developmental blocks of the species-specific developmental program at the tissue level of the organism organization. Such a description should be implemented in a formalistic way with the explicit and transparent representation of biological structures and their transformations by virtue of the mathematical objects. The present work is dedicated to the elaboration of basic principles for the formalistic description of the cellular architecture in relation to its reiterative and deterministic development.

Higher plants are unique organisms in most cases enable us to trace back all cell divisions of their tissues according to the shape of cell walls. Most examples considered here are based on our knowledge of plant cellular architecture development; nevertheless this work can be applied to animals and in some cases the differences in plant and animal cellular architecture are particularly accented.

In next two sections we provide general information about geometry of plant tissues and show the basic principles of formalistic description of real organisms with respect to development of their species-specific cellular architecture. In the last and main section we introduce the notion of a regular space—complete genealogical tree $\Gamma(H_0)$ with species-specific information encoded in its labeling and its quotient space—developmental program P. We have constructed the subgroup topology for a free group H_0 acting on the space $\Gamma(H_0)$, and present a method of revelation and analysis of proper group action with respect to concrete questions of developmental biology. The results achieved should be useful in analysis of stem cell activity, programmed cell death, integrity and local-to-large-scale spatial patterns of tissue development. The following two sections have a final subsection "Omitted peculiarities", where several important, but omitted, properties are mentioned. These sections were included to provide additional relevant information that expands the basic constructions, but may require more detailed work.

2 Basic Assumptions

Cellular architecture. The cellular architecture of a tissue sample is regarded here as a cellular complex C, which consists of n-cells in the mathematical sense, where n means the dimensionality of the object. Each cell of the organism tissue corresponds to a 3-cell, i.e. to a three dimensional object. The whole body of an organism is regarded here as a compact three-dimensional manifold with n-cellular structure, such that each two-dimensional sphere S^2, i.e. the boundary of an organism's cell, confines a three-dimensional ball D^3 corresponding to the internal parts of the biological cell.

Every plant cell is adjacent to another cell or to the external environment along its boundary—the set of 2-cells (two dimensional facets) and only along them (Fig. 1, left). Plant cells are incapable of being adjacent along separate edges (1-cells) or corners (0-cells) owing to the mechanism of their division. A new

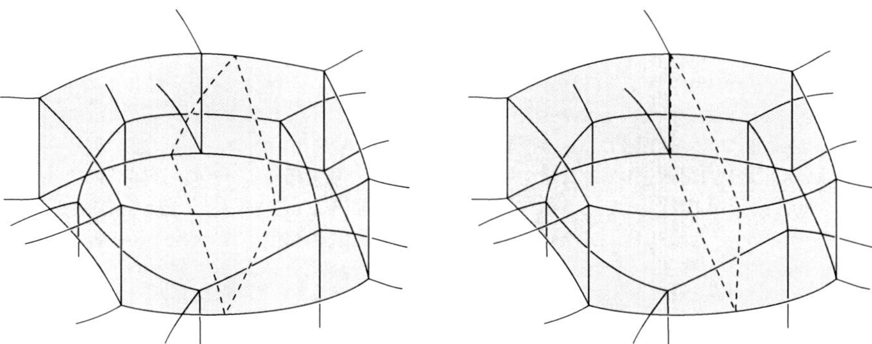

Fig. 1 Spatial arrangement of cells in the layered plant tissue. One of the possible (*left*) and impossible (*right*) positions of a new boundary (*dotted line*) between two daughter cells derived from the common maternal cell (*gray*)

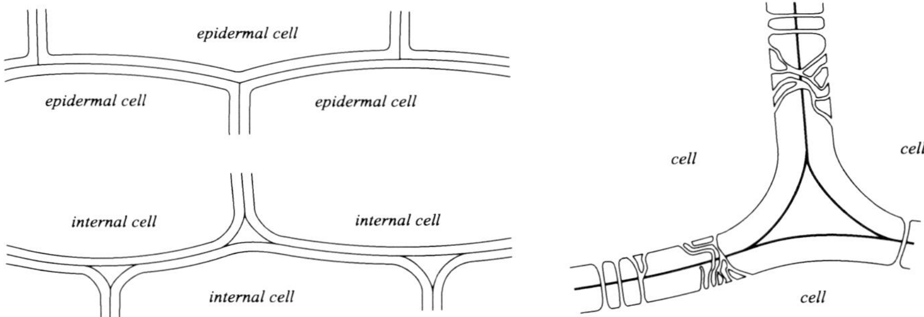

Fig. 2 Structure of the edges of adjacent cells in plant tissue. Intercellular spaces appear between the cells of internal tissues (*left, lower part*), and are absent in the epiderm and embryonic tissues (*left, upper part*). Plasmodesmata—intercellular cytoplasmic contacts between internal cells of plant tissue (*right*)

boundary between dividing cells is laid down inside the maternal cell and it attaches itself inwardly forming the edges. Those edges can intersect but never coincide with the edges formed by cells adjacent to the maternal cell (Fig. 1, right), since there are no such observations or known mechanisms of boundary coincidence. More precisely, the biological cell is represented here as a filled polyhedron with the edges and facets shared by adjacent cells or contacting with external environment. Most plant epidermal and embryonic tissues possesses exactly such a structure of the common cell walls between the adjacent cells. However, the majority of matured plant tissues have intercellular spaces (Fig. 2, left) caused by the local disintegration of the cell wall along the edges of cellular contacts [1]. This process diminishes the possibility of occasional edge or corner coincidence of a new cell wall that could give point-like or line-like sites of adjacency in plant tissues.

Developmental changes. Cellular spatial arrangement of the organism tissues is clearly a consequence of developmental processes active at a given moment in time.

Among the most important developmental processes are cell divisions, death and movements. With respect to these processes only the iterative way of morphogenesis can support the species-specific regularity of cellular spatial architecture. Cell divisions, as mentioned above, are regarded here as an attachment of the two-dimensional facet inside the existing 3-cell which splits it into two new cells. The inverse operation is a successive puncturing and deletion of the facets with respect to the order of their appearance in developmental history. Orientation and localization of a new boundary inside a dividing cell are deterministic processes related to the intrinsic properties of maternal and daughter cells, which are mostly unclear at present moment. For more than a hundred years there have been several, generally true, observations known as "laws of cell division". The most famous are those that state that the living cell divides into two equal masses or volumes (J. Sachs and O. Hertwig), and that the plane of cell division is always perpendicular to the longest axis of the cell (L. Errera), or in other words, the new boundary is a minimal surface dividing the maternal cell into two equal daughter cells (reviewed in [2, 3]).

Cell death is a common and necessary event in multicellular organism development. This process corresponds to the termination of further cell divisions and to the deletion of a 3-cell in case of full decomposition of dead cell remnants. Cell movements can be regarded here as a change in the cell's spatial neighbourhood, however this feature is not so essential for plants in comparison with animals.

Structural units. Identification of an organism involves establishing equivalence between a description of the sample organism and the data known about a species. In general, biological information about an object is represented as a hierarchically organized list of characters corresponding to subdivision of the organism's body into structural units of different order. Decisions about equivalence of two organisms or their parts are usually taken according to the presence of a peculiar and regular set of units, such as cells or tissues, and their spatial arrangement.

A rather more natural but complex way of establishing of organ and tissue identity was proposed by Sattler and Rutishauser [4]. Their approach, known as "dynamical" or "process morphology", is based on the search for equivalence of developmental processes rather than of structures, since latter is a result of the first's activity. Following their principles the associations of cellular spatial arrangement with developmental transformations are regarded here as basic and natural structural units of multicellular organism. Such units however can hardly be subordinated in a hierarchical manner.

Omitted peculiarities. The plant cell wall is an extracellular matrix generally composed of the cellulose microfibrils. The majority of plant cell walls between (living) adjacent cells preserves the numerous intercellular channels—plasmodesmata (Fig. 2, right). The plasmodesmata penetrate the cell wall, are covered with the cytoplasmic membrane and create a continuous and organized cytoplasmic network between cells in tissues [5]. Indeed, the two-dimensional boundary of the plant cell is rather more complex than S^2. The boundary of cells united into the plasmodesmal network represents the closed surface of genus $g \gg 0$.

Cells of animal tissues have a rather more rich range of intercellular contacts based on different means of cytoplasmic unity and cellular attachment to the extracellular matrix. Such a variability allows a cellular adjacency that is uncommon for plants, such as along edges (1-cells) and points (0-cells). This circumstance permits only the limited application of some of our assumptions to animal tissues.

Some plant tissues, such as endosperm, for example, as well as animal tissues, have the capability of simultaneous cellularization that occurs considerably later than division of nuclei. In this case the cell boundaries are preferably determined by regular subdivision of a space rather than by successive growth and divisions of cells in cell lineages.

3 Developmental History of the Individual Plant

Cellular genealogy. The developmental changes of individual plant are regarded here as a sequence of finite number z of developmental stages beginning from the state of the initial cell. The number of stages is naturally limited by the production of new initial cells at some stage close to the final one. Each stage is here represented as a genealogical tree, whose vertices correspond to the cells, and directed edges to their genealogical relations (mother to daughter cells). Pendant vertices of the tree correspond to a set of cell observed at the current developmental stage, while the remaining vertices represent the ancestral cells, i.e. the developmental history of the stage. Each tree is rooted in the vertex corresponding to the initial cell. The genealogical tree of a stage is obtained from the tree of a previous stage by the addition of two vertices to one of the pendant vertices, inasmuch as the minimal developmental transformation is (here conventionally) one cell division. Successive developmental stages beginning from the initial cell are clearly defined by the structure of their genealogical trees. Let, for example, the life cycle of the sample organism consists of $z = 5$ stages and the minimal developmental transformation is one cell division (Fig. 3a). Let two cells of the final stage transform into the two indistinguishable initial cells of plants of a new generation, while the remaining cells undergo no further divisions and die.

Definition 1 *The infinite genealogical tree constructed for all the ancestors and progenies of cells of an individual plant is a total genealogical tree $T_{\#}$, where the subscript "#" means the individual name of an organism of a given species. The tree $T_{\#}$ is built up on the countable infinite vertex set $V(T_{\#})$ and the set of directed edges $E(T_{\#})$, which connect any maternal cell to its two and only two daughter cells.*

Any subtree rooted in the vertex $v_i \in V(T_{\#})$ represents a cell lineage of the corresponding cell $L(v_i) \subset T_{\#}$. It is possible to say, that the vertex v_i is genealogically or temporally connected with all other vertices of $L(v_i)$. Hence the tree $T_{\#}$ (Fig. 3b) is a natural identification scheme of cells in a tissue sample with respect to their relations to a given cell v_i measured by a number of cell generations.

The total genealogical tree $T_{\#}$ contains two types of branches: branches with finite length, which correspond to cell lineages with limited developmental capacities, and infinite branches, which represent the potentially immortal cell lineages. Consider the cell lineages $L(v_i^{\alpha})$ and $L\left(v_i^{\beta}\right)$ induced respectively by v_i^{α} and v_i^{β}—daughter cells of the cell $v_i \in V(T_{\#})$. There are three general types of cells

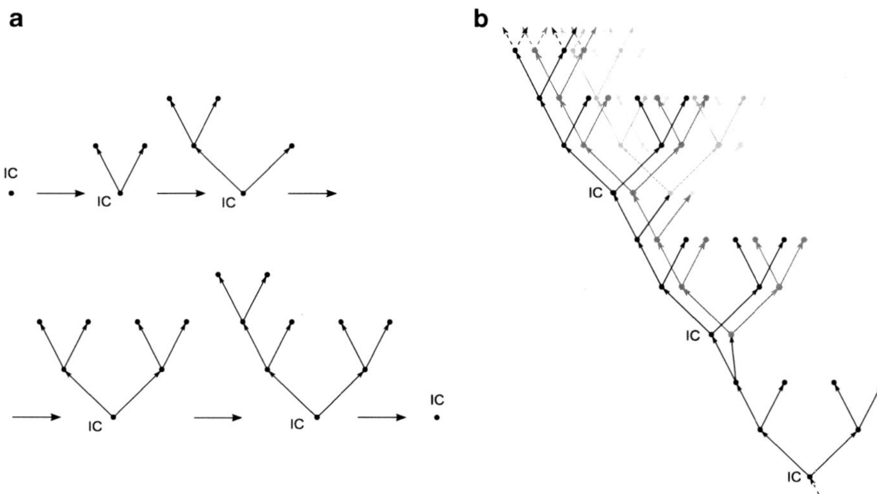

Fig. 3 Genealogy of cells of the sample organism. (**a**) Sequence of developmental stages represented as a sequence of genealogical trees. (**b**) Structure of the total genealogical tree $T_{\#}$ (fragment). *IC* initial cell

v_i in any organism independently of the $T_{\#}$ complexity: (1) the cells in which division produces two potentially immortal cell lineages are called α, β-infinite, (2) the cells in which division only supports the single potentially immortal cell lineage, are called α- or β-infinite and (3) the cells with the finite developmental capacities of its daughter lineages—finite cells. Individuality of the tree $T_{\#}$ is determined by internal and external factors those cause termination or induction of cell divisions in the lineages.

Cellular spatial adjacency. One of the most important features of any cell is its own microenvironment or spatial localization in the organism body. The spatial localization of a cell in the plant body at any moment of time, i.e. at any stage, can be unequivocally and species-specifically described by its adjacency to the other cells and to the outer surface as a graph of spatial adjacency $G_n, n \in z$.

Definition 2 *The graph of spatial adjacency* $G_n, n \in z$ *has the vertex set* $V(G_n) \subset V(T_{\#})$ *with respect to the set of cells peculiar to the stage n, and the set of edges* $E(G_n)$. *Any two vertices from* $V(G_n)$ *are connected by the edge from* $E(G_n)$, *if two corresponding cells are spatially adjacent.*

Every graph G_n is finite and connected, and corresponds to the integrity of the organism. It is possible to say that any vertex $v_i \in G_n$ is spatially path connected to any other vertex of the same graph. The graph of spatial adjacency G_n is a natural scheme of pair-wise identity of two-dimensional facets composing the boundary of cells (3-cells). In other words, G_n (Fig. 4a) is a natural scheme of the affine homeomorphisms between two-dimensional facets of 3-cells.

In plant tissue samples it is generally necessary to indicate the contact of cells with external environment. For example, the relation to external space is a general

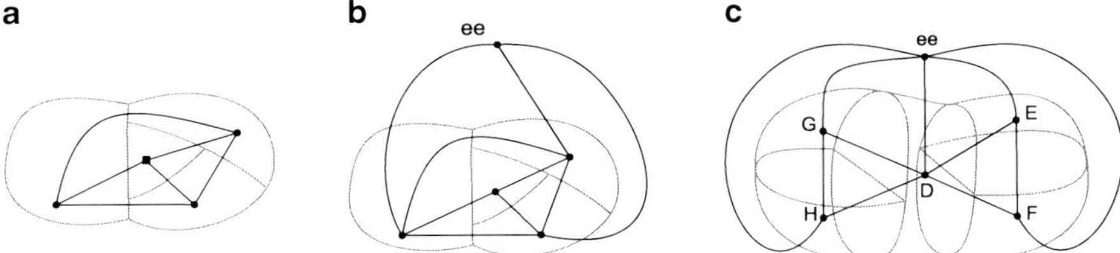

Fig. 4 Spatial relations of cells and structure of spatial adjacency graph G_n. (**a**, **b**) Two ways of indication of the cellular superficiality. (**c**) Cellular boundaries with respect to the graph G_n. *ee* external environment

character in defining of the superficial epidermal and the inner plant tissues. This relation is denoted here in two equivalent ways: by introduction of a new type of vertex (square vertex in Fig. 4a) or by addition of the vertex corresponding to the external environment adjacent to all superficial cells (Fig. 4b). The first variant is more convenient for graphical representations and morphological analysis of G_n, the second for matrices of adjacency or incidence.

Spatio-temporal organization. Developmental processes are regarded here as transformations or projections $g(G_n) : G_n \rightarrow G_{n+1}$ between two successive states of cellular spatial arrangements G_n and G_{n+1} according to the map of their temporal identity provided by the structure of the total genealogical tree $T_{\#}$. Transition $g(G_n)$ can be associated with an arbitrary number of cell deaths and divisions, which transform the vertex set $V(G_n) \in G_n$ to the vertex set $V(G_{n+1}) \in G_{n+1}$. Hence developmental processes can be regarded either as transformations of the spatial neighbourhood of cell lineages $\{L(v_i)\}$ induced by the set of vertices $\{v_i\} \in G_n$ from the initial state G_n, or as transformations of the "temporal" neighbourhood of several cells $\{v_i\} \in G_n, G_{n+1}$ identified in both initial and transformed states of cellular arrangements.

Since the developmental history and future of an individual plant are potentially infinite, one can construct the complex graph $G_{\#}$ of spatial adjacency and genealogy by means of addition of the non-directed edges between the pairs of vertices in the total genealogical tree $T_{\#}$ with respect to the graphs of spatial adjacency G_n for each stage.

Definition 3 *The complex graph of spatial adjacency and genealogy $G_{\#}$ is union of the total genealogical tree with the set of graphs of spatial adjacency: $G_{\#} = T_{\#} \cup \{G_n\}_{\infty}$ for each $n \in z$ stage, such that $V(G_{\#}) = V(T_{\#})$ and the set of edges $E(G_{\#})$ is a union of edges sets from $T_{\#}$ and $\{G_n\}$ $n \in z$.*

The spatial adjacency of every cell is described in a complex graph for each stage (Fig. 5a) in the course of organism development. Extending this principle to the all ancestor and progeny cells gives the infinite structure of $G_{\#}$ (Fig. 5b). Thus all cells of lineages from $T_{\#}$ are highly individual and become naturally labeled. The labeling of cells in a lineage according to their adjacency shows the much more unique composition, the bigger the lineage is studied, which can be easily verified after the comparison of *A* and *B* cell lineages in Fig. 5a. The graph G_n is always a

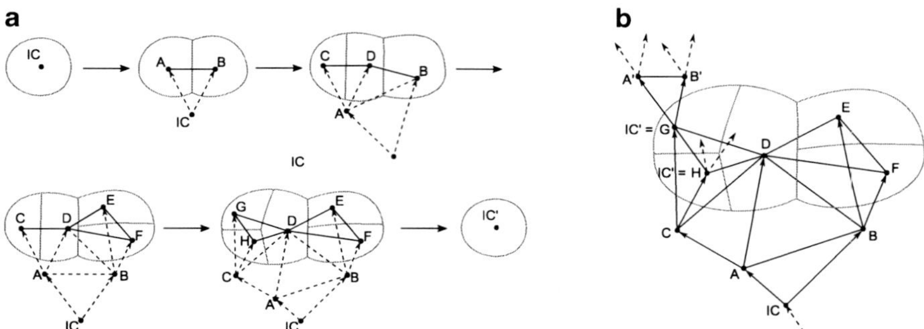

Fig. 5 Construction of the complex graph of cellular genealogy and adjacency $G_\#$ for the sample organism. (**a**) Representation of developmental stages as parts of the complex graph. The edges of spatial adjacency graph G_n are denoted by *solid lines*. (**b**) Structure of $G_\#$ (fragment), only cells of the final stage are shown

separating subgraph (separator) of $G_\#$ induced by the vertex set $V(G_\#)$. The genealogical tree $T_\#$ is always a maximal spanning tree of $G_\#$.

Omitted peculiarities. The graph of spatial adjacency G_n does not precisely reproduce the structure of the cellular complex $C_n, n \in z$ considered in the previous section. The edges $E(G_n)$ correspond to the facets—2-cells and each vertex to the 3-cell. The boundary of any cell of the multicellular organism consists of a set of adjacent facets, however, the facet structure can be different. In some cases it could correspond not to a disk, but to a tube, such as that between the cell D and external environment as shown in Fig. 4c. Moreover, among some types of animal cells, and in some plausible special cases plant cells, can have multiple separated facets of adjacency between two cells.

The fertilization process is an integral part of sexual reproduction, which is common for most organisms. It involves the fusion of gametes (sexual cells) and their genomes. Obviously, fertilization leads to the non-simple fusion of a set $\{T_\#\}$, that belongs to organisms capable of cross fertilization. Hence, the resulting graph $G_\#$ has a rather more complicated structure.

Cellular motility and particularly the developmental movements are basic features of animal tissues. Indirect cellular movements are also peculiar for plant tissues in form of intrusive growth of cells, in the case of postgenital (superficial) tissue fusion and cell separation processes [1, 6]. With respect to this, every edge set $E(G_n)$ should be regarded not as a single constant set, but as one from the set of dynamic states $\{E(G_n)\}$.

4 Regularity and Programming of Developmental Changes

Developmental program. The developmental program of cellular architecture of the organism represents a branching tree-like algorithm with a subset of cycles. Each cycle is a morphogenetic process such as cell growth, cell division and death, executed in an iterative manner. The most external cycle corresponds to the complete sequence of developmental stages peculiar to the organism and includes the unicellular stage or stage of the initial cell, which is universal for all multicellular organism. Any part of this cycle could be involved in a number of subordinated cycles that are usually composed of reiterative development of structurally and morphogenetically equivalent cell lineages.

The organism's previous, as well as its potential, developmental history is inscribed in the structure of the complex graph of spatial adjacency and genealogy $G_{\#}$. For the majority of living organisms this graph has no clear "crystallographic" periodic structure. Indeed, regular properties appear only locally at different scales of $G_{\#}$. Hence, the identification of regular formative processes is an analysis and comparison of the distinct iterative parts in $G_{\#}$. The graph $G_{\#}$ has a natural structure of a covering space graph $\widetilde{X}$ over some finite graph X, since in a most trivial case it can always be represented as a map of a singe cell division, i.e. triangle, translated infinitely many times with respect to the quantity of non-pendant vertices of the total tree $T_{\#}$.

Definition 4 *The program of the organism's cellular architecture development P is a compact space with a species-specific structure such that for any individual complex graph of spatial adjacency and genealogy $G_{\#}$ and any connected part $P' \subset P$ there is a unique map $p_{\#} : G_{\#} \to P$ with an unique lift $p_{\#}^{-1}(P')$ in $G_{\#}$.*

Flexibility of the developmental program execution via the map $p_{\#}^{-1}$ provides a uniqueness of the graph $G_{\#}$ for the individual organism. On the other hand, this individual flexibility is bounded by species-specific properties of the compact space P. It is convenient to consider the developmental program as a configuration space of all possible developmental events and $p_{\#}^{-1}$ as a projection of them into developmental history, i.e. to the graph $G_{\#}$. According to this principle the space P for the sample organism mentioned above can be represented as a cyclic graph of genealogy (Fig. 6a), where spatial adjacency is hidden, or as a three-dimensional cellular complex constructed after substitution of the third spatial axis by the temporal axis in the form of a "2D + 1" representation (Fig. 6b). External and internal cellular boundaries of the "time sections" generate longitudinal subdivision of 3-cells by the directed time-related "vertical" facets (2-cells) shown in Fig. 6b. The set of the maps $\{p_{\#}^{-1}\}$ for this species is generated by vertical translation and by choice of the cell G or H as an ancestor of a new organism. In this view the developmental program as a configuration space possesses a clear instructive function. Any object of the same species, such as the sample at some developmental stage, is a "temporal section" of this space. Any developmental changes $g(G_n) : G_n \to G_{n+1}$ are considered here as a result of directed motion from the lower section to the upper section in this configuration space. Unfortunately, this representation does not cover cases when the graphs of spatial adjacency are non-planar. Moreover, the space P of living organisms possesses a strikingly much higher complexity. At the present moment we can only try to determine some general principles of the species-specific composition of P. The physical and biological nature of the developmental program responsible for the spatial organization of multicellular organisms is still

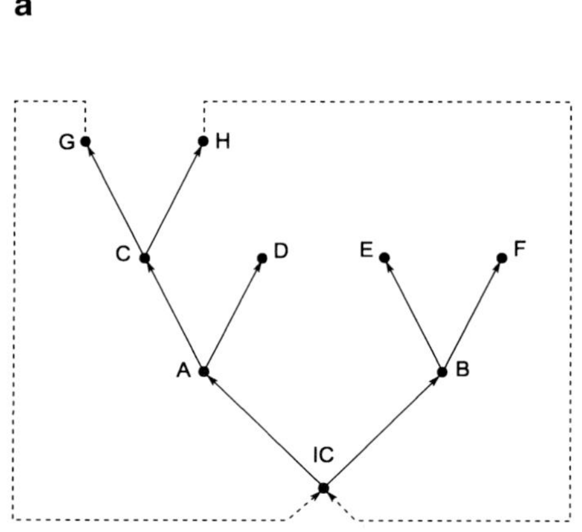

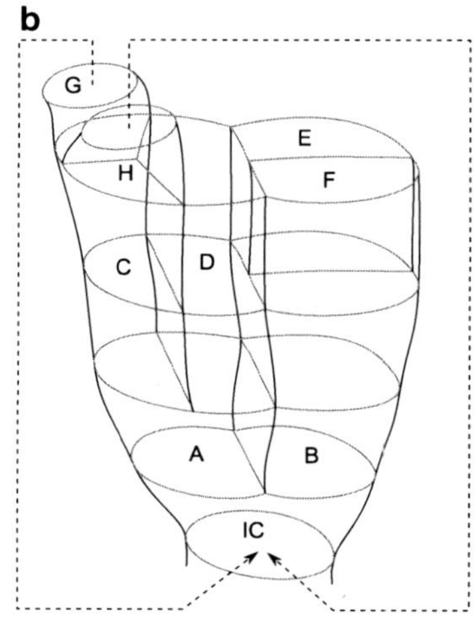

Fig. 6 Representation of the developmental program P for the sample organism. (**a**) As a cyclic graph of genealogy. (**b**) As a four-dimensional manifold with decreased dimensionality

to be discovered. In the current work we regard the basic properties of the developmental program P represented as a graph, and study a group action on it.

Regular cell divisions. The developmental processes we are looking for are sets of cell divisions and deaths associated with setting up, distribution and loss of spatial adjacency among descendant cells. In other words, the developmental process is a class of equivalence uniting some connected subgraphs $\{G'_{\#}\} \subset G_{\#}$ with determined similar structure. Since the initiation and termination of equal processes can be considerably distanced or overlapped in space and in time, the complex graph of spatial adjacency and genealogy $G_{\#}$ is a main space to be investigated for the presence of "spatio-temporal developmental units" and for the revelation of their relations.

Analysis of periodic structure or automorphisms of the total genealogical tree $T_{\#} \in G_{\#}$ is capable of providing us with some general information about the iterative developmental processes. For this reason, in this section we consider the maps $\{p_{\#}^{-1}\}$ restricted to the tree $T_{\#} \in G_{\#}$ only. Cell division is a way of cell reproduction by splitting into two new cells which corresponds to the two outcoming edges for every vertex in the tree $T_{\#}$. Being natural objects, the two daughter cells are always different, even though they are indistinguishable for the investigator. Let us denote these two cells as a and b. The process of cell generation is associated with the cell label by indicating the incoming directed edges of the genealogical tree with the same labels (Fig. 7a). No matter how complex this organism may be, every cell can be encoded with a (reduced) word from the free group $H_0 = \langle a, b \rangle$, where $S_0 = \{a, b\}$ is a set of group generators. According to this, every cell of an organism has an unique sequence (word) composed of a and b as characters, such as $ab, aba, a^2ba^3b^4$, as a label. Consider the embedding of $T_{\#}$ into the Cayley graph of this group $\Gamma(H_0)$ (Fig. 7b). Every power of a character is

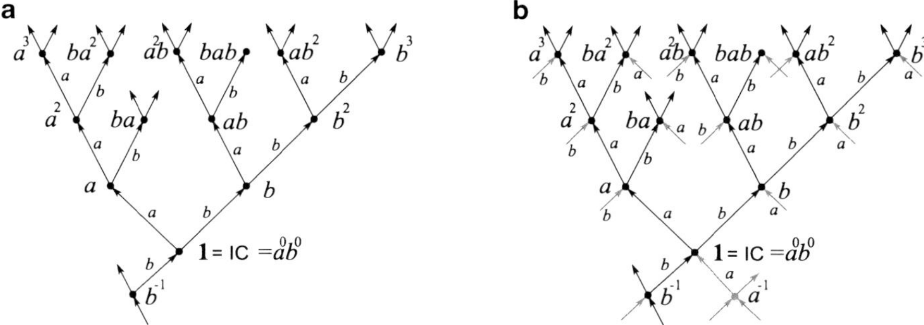

Fig. 7 Basic iterative structure of any genealogical tree expressed in the labeling principle. (**a**) Fragment of the total genealogical tree $T_\#$ with vertices labeled as elements of the group $H_0 = \langle a,\ b\rangle$. (**b**) Embedding of the tree $T_\#$ (*black*) into the complete genealogical tree $\Gamma(H_0)$ or the Cayley graph of the free group H_0 (*gray*)

positive with respect to the "identity" vertex—the initial cell, which can be chosen arbitrarily. Moreover, two cell types originating from a common maternal cell endow the organism body with a most basal metameric structure of cellular architecture since any two cells are sister cells with respect to the existence of their common maternal cell distanced in time. On the other hand, an arbitrary total genealogical tree $T_\#$ is constructed only from copies of the single subtree with vertex labels $[1, a, b]$. However, the same composition of characters in a cell label can be achieved after continuously many preceding combinations of a and b, that is why every vertex of the total genealogical tree $T_\#$ should have two incoming edges labeled as a and b. This does not correspond to the natural structure of $T_\#$, because, in common, a cell division occurs without preceding cell fusion. It occurs only in the case of cell fusion and fertilization. Thus, the Cayley graph $\Gamma(H_0)$ of the group H_0 corresponds to another more general entity in the biological sense.

Definition 5 *A complete genealogical tree* $\Gamma(H_0)$ *is a Cayley graph* $\Gamma(H_0)$ *of the group* H_0 *of cell labeling. It is a directed infinite tree composed of vertices of degree 4 each: with 2 incoming and 2 outcoming edges, with an arbitrary "identity" vertex.*

By definition, the complete genealogical tree $\Gamma(H_0)$ is a universal cover for the developmental program compact space P. Thus, the total genealogical tree $T_\#$ of arbitrary complexity belonging to any individual organism of a given species is always embeddable into $\Gamma(H_0)$. In a real organism some finite set of cells with labels from $S_1 \in H_0$ according to their distinct genealogy serves as a source for new initial cells of the new generation organism with potentially the same developmental history. For example, this set[1] is $S_1 = \{s_1 = a^3,\ s_2 = b^2ab,\ s_3 = b^4\}$ (Fig. 7b). Limitation of the organism's body growth and transformation of cells with the

[1] A group element or subgroup generator is written as a left product of group generators. For example, $aba^2 = (a(b(aa)))$.

codes from S into the initial cells leads to the "nullifying" of their division counter, for example $a^3 = b^2ab = ab^3 = 1 = a^0b^0$. Obviously, the absence of such a code in remaining cell lineages must cause them to enter into programmed cell death. On the other hand, it is possible to suppose the existence of a "programmed cell death code" set S'_1, such that it restricts further cell growth and divisions. Any mismatch between these two sets S_1 and S'_1 can cause uncontrolled cell divisions or developmental malfunctions. Similar mechanisms of activation and execution of developmental processes were intensively studied using the cell automata and L-systems approaches [7] based on the theory of sets, but not groups, since after application of limits such as S'_1 instead of relations such as S_1 to the number of elements and group operations, the group structure is destroyed. It could be reasonable and efficacious for some tasks, but limited in the sense of extrapolation to a larger scope of the cellular architecture organization. Therefore we have to leave the group H_0 free without any limitations and relations. In order to take into account the developmental process of a higher level of organization, it is necessary to introduce another group H_1—the subgroup of the group H_0.

In order to describe the regular patterns of cell divisions we need to assume below that all cell divisions are determined exclusively by genealogy of dividing cell. Thus, the individuality of tree $T_\#$ is a result of the influence of external factors leading to non-regular termination of some branches. The principle of cell labeling is extended to the arbitrary finite set of genealogically related cell lineages, instead of the two sister cells a and b, considered above, as follows. Let the plant cell lineages reproduce itself via regular sequences of cell divisions, which lead to the repeating appearance of regular cell types, i.e. precursors of the initial cells denoted by specific labels from H_0. In fact, a living organism can reproduce itself in full or in part via proliferation of specialized cells through iterative elaboration of its own cellular architecture according to the determined set $S_1 \in H_0$ of genealogical pathways starting from the initial cell stage. Thus, the reproductive capabilities of an organism mentioned above are represented as elements of a (free) subgroup $H_1 = \langle S_1 \rangle$ with the set $S_1 \in H_0$ as a set of generators. Introduction of a subgroup in the group H_0 is a manifestation of identity among cells uniquely labeled by elements from H_0. In our example, a subgroup H_1 has three generators $S_1 = \{a^3, b^2ab, b^4\}$, so the species-specific iterative unit or the metamer for this organization level reproduces itself infinitely only via these three possible ways. Other branches of tree $T_\#$ are assumed to be finite. This means that the total genealogical tree $T_\#$ peculiar to the organism considered above consists precisely of copies of some finite subtree attached to each other at the vertices labeled by 1 and by powers and products of elements from the set S_1 (Fig. 8a, thick lines). It is important that this organism is free to "choose" any way of reproduction, from any cell labeled as in the set S_1. Thus, the tree $T_\#$ always has a unique embedding into the graph $\Gamma(H_0)$ (Fig. 8b).

Consider a word metric of these two groups H_0 and H_1, which measures the distance, i.e. number of cell divisions, between any two cell vertices along their common genealogical tree. These two groups have the same (word) metric $d_{S_0}(x, y)$, $x, y \in H_1 \subseteq H_0$, since every generator from $S_1 \in H_1$ can be measured by the

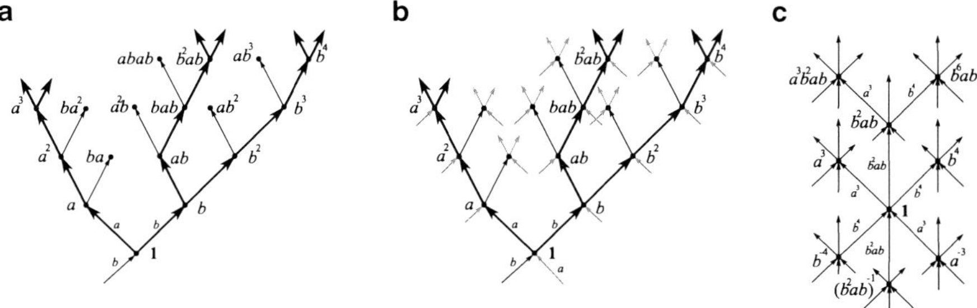

Fig. 8 Introduction of the iterative structure at a higher level of organization with respect to the labeling of genealogical tree. (**a**) Fragment of the total genealogical tree $T_{\#}$. Edges corresponding to generative elements of the subgroup $H_1 = \langle a^3,\ b^2ab,\ b^4 \rangle$ are *thickened*. (**b**) Embedding of the tree $T_{\#}$ into $\Gamma(H_0)$. Orbifolds of H_1 generators are *labeled* and *thickened*. (**c**) Cayley graph $\Gamma(H_1)$ in a metric of the subgroup H_1

metric of H_0 (Fig. 8b). The metric of the subgroup H_1 allows one to "forget" about finite cell lineages (Fig. 8c) and allows to analyze reiterative units at a higher level of organization. Hence the metric of a subgroup specified bears species-specific information about regular limited as well as possibly infinite proliferative growth of structures at a given level of organization. Below is a generalization of this principle.

Definition 6 (Based on Gromov [8]) *Every two subgroups* $[H_{i-1},\ H_i]$ *from a finite series* $H_i \subseteq H_{i-1} \subseteq \ldots \subseteq H_1 \subseteq H_0,\ i \in N$, *which describe the iterative development of cellular architecture at the corresponding (i−1)-th and i-th levels of organization, are* λ_i*-quasi-isometric, if there are a set of natural numbers* $\{\lambda_i > 0\}$ *and a map* f_i: $H_{i-1} \to H_i$, *such that an unequality* $\mathrm{dist}_{S_{i-1}}(x, y) \leqslant \lambda_i\, \mathrm{dist}_{S_i}(f_i(x), f_i(y))$ *holds for any* $x,\ y \in H_{i-1}$.

The set of quasi-isometric subgroups $\{H_i, S_i\}$ provides a scale-subordinated regular "tiling" of the basic structure of the complete genealogical tree H_0. Namely, the quasi-isometric groups $\{H_i\}$ with their generators $\{\langle S_i \in H_{i-1} \rangle\}$ induce a finite sequence of quasi-isomorphisms:

$$H_0 \to H_1 \to H_2 \to \ldots \to H_{i-2} \to H_{i-1} \to H_i \tag{1}$$

This series is a sequence of projections of groups into subgroups. Indeed it reflexes the successive inclusion of every cell vertex $h_0 \in T_{\#}$ of some regular type of genealogy labeled by the group element into the cell lineage of another cell from a preceding cell generation with respect to increasing level of organization: $h_0 \subset L(h_1) \subset L(h_2) \subset \ldots \subset L(h_{i-1}) \subset L(h_i) \in T_{\#}$ or, in other words, the cell with label h_0 is a part of the cell group of regular type originating from the cell h_1, the whole this cell group is part of a tissue of regular type originating from the common cell h_2 and so on. Thus, the cell lineage $L(h_{i+1})$ is a potentially immortal one with respect to the unlimited proliferation of its descendant cell with a label h_i. Moreover, the cell with a label $h_i \in H_i$ is α, β-infinite with respect to the i-th level of organization, if h_i is a power of any element from the generative set S_i. Hence α- or β-infinite cells correspond to the orbifold of generative elements with respect to some subgroup H_i. Namely these cells are labeled as subwords of H_i generators (Fig. 8b).

The action of every H_i group element is a translation along the complete genealogical tree $\Gamma(H_0)$. Each generator $s_k \in S_i$ of the group H_i maps the initial cell vertex to the vertex distanced in $\lambda_k = |s_k|_{S_0}$ cell divisions, where the distance is expressed as a word length of s_k with respect to the generating set S_0. Since the minimal distance of translation $\lambda_k \in \{H_i\}$ is non-zero and no graph edges are inverted, the action of any subgroup $\{H_i\}$ is one of hyperbolic automorphisms $\mathrm{Aut_H}(\Gamma(H_0))$ of $\Gamma(H_0)$ in the sense of a definition provided by Gromov [9] and Weidmann ([10], unpublished). Hence the λ-quasi-isometry of groups H_i and H_0 is defined by a maximal meaning of the minimal translations provided by the group generators $\lambda_i = \max\{\lambda_k\}$. We consider the finite series of quasi-isometric groups with their subgroup structure $H_i \subseteq H_{i-1} \subseteq ... \subseteq H_1 \subseteq H_0$ as one of general species-specific properties of the cellular architecture development of multicellular organisms.

This series of subgroups $\{H_i\} \subseteq H_0$ generally consists of non-normal subgroups of infinite index, which are incapable of describing precisely a whole species-specific set of regular cell types, particularly terminally differentiated finite cells. Since the set of cell types is finite for any species and the developmental program space P is compact, there is a normal subgroup H of finite index equal to a fundamental group $\pi_1(P, x_0)$ of the compact space P with some fixed point $x_0 \in P$, such that the subgroup H_i is contained in it. In order to find the normal subgroup H with species-specific properties, it necessary to find and describe regular patterns of appearance of cells of regular types with limited or locally limited proliferative capabilities provided by the existence of programmed cell death.

Determined cell death. Programmed cell death is a natural and determined limitation of proliferation in the majority of cell lineages committed to differentiation. This process is clearly reflected in the finite length of some branches of $T_{\#}$ and is responsible for the disintegration of the organism body in the case of different types of reproduction. According to the spatial structure of the sample organism shown above in the third section, the union of graphs $\bigcup^z G_n$ is disconnected between the vertex of the initial cell $IC' \in G_1$ and vertices of her daughter cells A' and $B' \in G_2$ (Fig. 5b). This property is natural owing to the separated state of the new organism's initial cells after their appearance inside the body of the parent organism and the finite life time of the matured tissues. In other words, only a part of the cell lineages induced by cells of the previous stage is included to the organism body at the next stage because of the programmed death processes. However, among higher plants, the tissues of a new generation organism are usually separated much later, approaching the thousand cells level, being integrated into the mother plant body in a different degree during embryogenesis inside the seed, or in case of vivipary, being attached to the shoot tissues.

Programmed cell death is a precise, tissue- and cell-specific process of cell elimination necessary for normal development of any multicellular organism. This process is provided by a specific sequence of events at the subcellular level

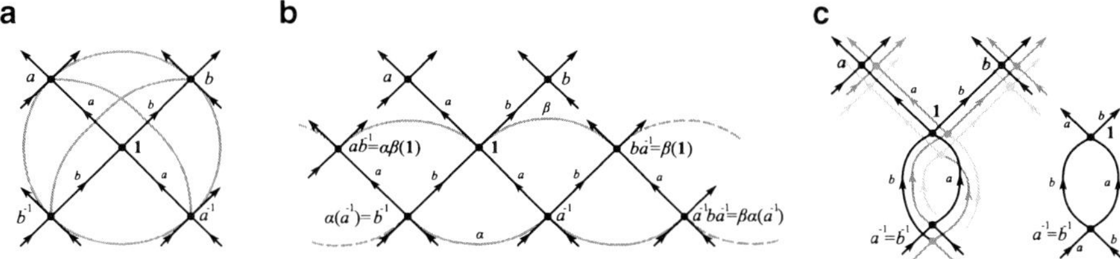

Fig. 9 Constructing of the quotient graph $\Gamma(H_0)/F$ under the action of elliptic automorphisms. (**a**) Neighbourhood of the arbitrary vertex in the graph $\Gamma(H_0)$. Vertex identification by possible automorphisms with a fixed point 1 are shown as *gray edges*. (**b**) Action of two elliptic automorphisms $\alpha\beta = F$ without a common fixed point in $\Gamma(H_0)$, $\alpha = (a^{-1}, b^{-1})$ and $\beta = (1, a^{-1}b)$ with fixed points 1 and a^{-1}, respectively. (**c**) Collapsing of the identification edges and folding of branches (denoted by *fading gray*) of the tree $\Gamma(H_0)$ according to two (be-)infinite chains of identities $[... = a^{-1}ba^{-1} = a^{-1} = b^{-1} = b^{-1}ab^{-1} = ...]$ and $[... = a^{-1}b = 1 = b^{-1}a = ...]$

of organization, which is very conservative for animals and highly variable in plants [11]. For example, programmed cell death is determinatively initiated in the course of perforated leaf development in plants such as *Monstera* [12] or in the case of xylogenesis—xylem tissue development [13]. In addition, programmed cell death is a mechanism of tissue defense against intracellular invasions [14] and is the final stage of tissue senescence [15]. Thus, termination of a branch in the total genealogical tree $T_\#$ is a result of developmental program implementation induced by the internal or external inductors. The location of terminated cell division sequences in the tree $T_\#$ along the potentially immortal lineages provides the individual "signature" of an organism. The question to be resolved here is a decomposition of the tree $T_\#$ into regular cell divisions and cell divisions peculiar to the individual organism according to the patterns of cell lineage termination.

Species-specific identification of all the cells of regular type in terms of $\Gamma(H_0)$ vertex labels is achieved by means of the method of graph folding provided by Stallings [16]. First of all, we will show what information can be taken out from analysis of $T_\#$ embedding into $\Gamma(H_0)$. Secondly, we will show how to reveal and construct the normal subgroup $H \subseteq H_0$, associated to the series $\{H_i\}$ preserving all species-specific properties of proper cell appearance in genealogy with respect to every level of organism organization.

Consider the non-hyperbolic, namely elliptic automorphisms $\mathrm{Aut}_{\mathrm{E}}(\Gamma(H_0))$ by Gromov [9] and Weidmann ([10], unpublished), which are characterized by zero distance of translation or by presence of a fixed point in $\Gamma(H_0)$. These groups of $\Gamma(H_0)$ automorphisms are rotational and reflexing symmetries with some vertex $x \in V(\Gamma(H_0))$ as a fixed point. The closest neighbourhood of the vertex $x = 1 \in \Gamma(H_0)$ distanced in one cell division consists of four cells $\{a, b, a^{-1}, b^{-1}\}$ and naturally provides the symmetry group of the tetrahedron (Fig. 9a). Increasing the distance for the neighbourhood being considered enables us to construct more complex groups of polyhedral symmetries.

The species-specificity of the automorphisms $\mathrm{Aut_E}(\Gamma(H_0))$ in every point of $\Gamma(H_0)$ are determined by the subgroup series $\{H_i\}$ and the embedding of $T_\#$ into the graph $\Gamma(H_0)$ as follows. Consider the map $r_\#^{-1} : T_\# \to \Gamma(H_0)$. Each individual map from $\{r_\#^{-1}\}$ is unique and injective. This map is injective as a result of decrease of vertex valence from 4 to 3 via the proper identity of terminal vertices of two incoming edges, since any non-zygote cell has a single ancestral cell and only two daughters, and because of unique termination of some branches. This map is unique and proper, because the same cell type can appear regularly after different sequences of cell divisions. Let each vertex of the $\{H_i\}$ orbifolds (α- or β-infinite cells) be considered as a fixed point of some elliptic automorphism. Identification of vertices in $\Gamma(H_0)$ to some regular cell type is provided by action of two elliptic automorphisms without a common fixed point (vertex). For example, let these automorphisms be two permutations $\alpha, \beta \in \mathrm{Aut_E}(\Gamma(H_0))$ with fixed vertices x, y, respectively. Both automorphisms cyclically permute two sets of vertices from the neighbourhood of the fixed vertex, such that $x = 1$, $\alpha = (a^{-1}, b^{-1})$ and $y = a^{-1}$, $\beta = (1, a^{-1}b)$ (Fig. 9b). As one can see in Fig. 9b–c, the product of two elliptic automorphisms $\alpha\beta$ is hyperbolic since $\alpha\beta$ naturally extends to the be-infinite directions and leads to the collapsing of two tree branches rooted in vertices a^{-1} and b^{-1} owing to the folding of the subtrees according to two (be-)infinite chains of identities $[... = a^{-1}ba^{-1} = a^{-1} = b^{-1} = b^{-1}ab^{-1} = ...]$ and $[... = a^{-1}b = 1 = b^{-1}a = ...]$. Indeed, these chains describe the translation $a^{-1}b$ in the tree $\Gamma(H_0)$. Two adjacent edges can be selected as a generator of a free group $F = \alpha\beta = <a^{-1}b>$ $\in \mathrm{Aut_H}(\Gamma(H_0))$ induced by the two elliptic automorphisms $\alpha, \beta \in \mathrm{Aut_E}(\Gamma(H_0))$. As a result, the quotient space $\Gamma(H_0)/F$ is constructed (Fig. 9c, to the right). The triple valence of the vertices $a^{-1} = b^{-1}$ and 1 is achieved by choosing the single edge a or b to connect them. The precise meaning of the group F action in the complete genealogical tree $\Gamma(H_0)$ is that the cell 1 has derived no matter from a^{-1} or b^{-1}.

It is possible to generalize the example considered above for an arbitrary pair of elliptic automorphisms providing the identity of other vertices from a neighbourhood of fixed points distanced to a different degree (Fig. 10). Obviously, any new pair $\alpha\beta$ of elliptic automorphisms of the complete genealogical tree $\Gamma(H_0)$ considerably expands the set of hyperbolic automorphisms $\mathrm{Aut_H}(\Gamma(H_0))$ for this tree. On the other hand, the set of hyperbolic automorphisms is bounded by the finiteness of the set of regular cell types. Hence $\mathrm{Aut_H}(\Gamma(H_0))$ should be a finite index subgroup of H_0 and should contain the subgroup H_i as a free factor, i.e. $\mathrm{Aut_H}(\Gamma(H_0)) = H_i * F_i$, where a second free factor F_i is a completion of H_i in H_0 up to the subgroup of the finite index. Thus, the total genealogical tree $T_\#$ has a unique embedding $f_\#$ into the quotient space $\Gamma(H_0)/F, F = \cup F_i$ which is an infinite but locally more compact space than $\Gamma(H_0)$. It is easy to see that the identification of vertices in $\Gamma(H_0)$ by any subgroup provides the cycles in the quotient space. Such cycles can have an arbitrary length, such as three, for example (Fig. 10b) that is specific for the current map.

Now we construct the space P. Consider a map $r : \Gamma(H_0) \to \Gamma(H_0)/\mathrm{Aut_H}(\Gamma(H_0))$. to be species-specific and determined by structural properties of

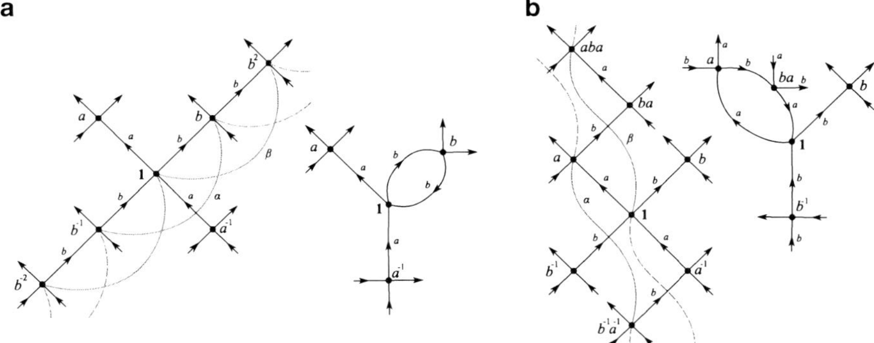

Fig. 10 Action of two elliptic automorphisms $\alpha\beta = F$ in $\Gamma(H_0)$ and construction of the quotient graph $\Gamma(H_0)/F$. Vertex identification by permutations is shown by *gray lines*. (**a**) $\alpha = (b^{-1}, b)$, $\beta = (1, b^2)$. (**b**) $\alpha = (b^{-1}a^{-1}, a)$, $\beta = (1, aba)$

the space P. In order to reveal the structure of P, it necessary to find automorphisms $\{F_i\}$ and the map r. Define the set $\mathrm{Aut_H}(\Gamma(H_0))$ as $\{H'_i = H_i * F_i\}$ for all $i > 0$. The map r is decomposable along two free factors of $\mathrm{Aut_H}(\Gamma(H_0))$ into maps $r'_i : \Gamma(H_0) \to \Gamma(H_0)/H_i$ and $r''_i : \Gamma(H_0) \to \Gamma(H_0)/F_i$ (see diagram below). This decomposition is unique for each F_i. Indeed, $F_i = \cup F_i$ for maximal meaning of i, because it includes all H_i subgroup completions in H_0 for every i. Existence of the finite index subgroup H'_i containing H_i for each i is guarantied by the theorem of M. Hall (cited in Kapovich, Myasnikov, [17]), as for any finitely generated subgroup of a free group. Moreover, there is only a finite number of such subgroups for any finite number k_i of conjugacy classes [18]. Each such subgroup H'_i contains a finite index normal subgroup of H_0, a normal core $\mathrm{Core}_{H'_{i-1}}(H'_i) = \cap\, h^{-1} H'_i h$ constructed via intersection of all H'_i cosets with representatives $\{h\} \in H'_{i-1},\ i > 0$. In the general case, $H'_i \not\subseteq \mathrm{Core}_{H'_{i-2}}(H'_{i-1})$, though $H_i \subseteq H'_{i-1}$, $H'_i \subseteq H'_{i-1}$ and $\mathrm{Core}_{H'_{i-1}}(H'_i) \subseteq \mathrm{Core}_{H'_{i-2}}(H'_{i-1})$ are always true for all $i > 0$. The construction of a normal core for each subgroup H'_i enables one to correct and value a "distortion" of embedding of H'_i into H_0. Thus, the species-specific series of subgroups $\{H_i\}$ becomes extended up to the series of finite index subgroups $\{H'_i\}$ and finite index normal subgroups $\{\mathrm{Core}_{H'_{i-1}}(H'_i)\}$ of H_0. Finally, for each $\{H_i\}$ it is possible to construct a map $p_i : \Gamma(H_0)/\mathrm{Aut_H}(\Gamma(H_0)) \to P_i$, where $P_i = \Gamma(H_0)/\mathrm{Core}_{H'_{i-1}}(H'_i)$. Then, the developmental program space P consists of the union of covering subspaces P_i quoted by normal subgroups such that P_i is a universal cover, P_r always covers P_{r-1} $(0 < r < i)$, and P_1 is a universal base for all coverings of P representations. According to A. Hatcher [19] these relations of covering spaces are called deck transformations. This term implies a symmetry of covering space provided by the action of such normal subgroups.

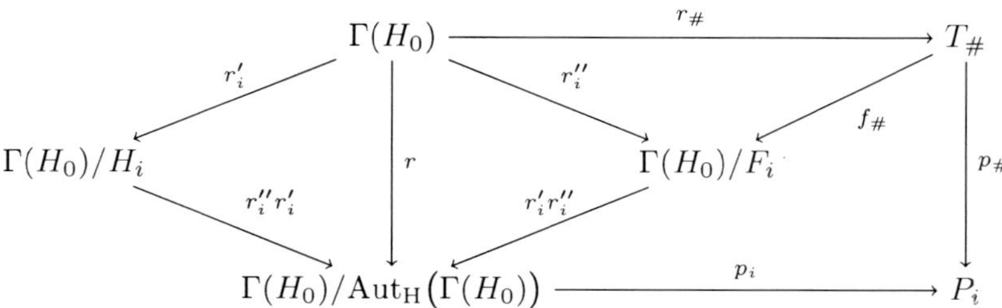

All maps discussed above are presented in the commutative diagram to above. It is interesting to trace on how the proper and species-specific maps r and $\{p_i\}$ should be revealed from realistic data obtained regarding embedding $r_{\#}^{-1}$ and the surjective projection $p_{\#}$. These allow us to establish deeper generalizations.

Consider the action of hyperbolic automorphisms $\mathrm{Aut_H}(\Gamma(H_0))$ for the sample organism considered previously with its structural details shown in Fig. 8. Assume that all labeled vertices in Fig. 8b correspond to the regular cell types, that means they are identically labeled in the quotient graph $\Gamma(H_0)/\mathrm{Aut_H}(\Gamma(H_0))$ and all of them are lifted to the tree $T_{\#}$ providing its unique embedding into $\Gamma(H_0)$. In other words, it is the same as saying that for this organism no other cell types exist and any possible cell vertex from the tree $V(T_{\#})$ can be identified with a single vertex from $V(\Gamma(H_0)/\mathrm{Aut_H}(\Gamma(H_0)))$, that is a finite set. This set is shown in Fig. 8a as beingembedded in the iterative fragment of the tree $T_{\#}$. The complete genealogical tree $\Gamma(H_0)$ has a specific labeling for this organism provided by the group $H_i = H_1 = \langle a^3,\ b^2ab,\ b^4\rangle$, $H_1 \subseteq H_0$, which was shown in Fig. 8b in the word metric of H_0 and in Fig. 8c in the metric of H_1. This labeling must be preserved under the action of $\mathrm{Aut_H}(\Gamma(H_0))$, since the tree $\Gamma(H_0)$ consists of the copies of $\Gamma(H_0)/\mathrm{Aut_H}(\Gamma(H_0))$ attached one to another along the vertices labeled by the combination of powers of the generative elements of H_1 written in the metric of H_0. The action of an element $h \in H_i$ is a translation of the vertices of the tree $\Gamma(H_0)$ with a label $h_0 \in H_0$ to the vertex with a label hh_0 along the orbifold h^N, $N \in N$ and h_0 as identity. We then construct the folded graph of the subgroup H_1 via identification of $\Gamma(H_0)$ vertices according to the subgroup generators. This graph is infinite, but possesses a bounded subset of vertex labels provided by identity and labels of orbifold generators (Fig. 11a). The folding explains the following identities of some graph labels: $b^2 = ab$, $b^3 = bab$, $ab^2 = a^2b$ and $ab^3 = abab$. Thus, we have found six labels of regular cell types. This gives a regular labeling only for a limited subset of vertices from the branches of subtree $\Gamma(H_0)$ induced by elements of H_1.

In order to complete $\Gamma(H_0)$ labeling and construct $\Gamma(H_0)/\mathrm{Aut_H}(\Gamma(H_0))$, one should find the subgroup F_1 such that $H'_1 = H_1 * F_1$ is a finite index subgroup of H_0. We have applied here an approach to constructing the subgroup graph based on the work of I. Kapovich and A. Myasnikov [17]. First, it is necessary to find H_0 group elements not from H_1, which correspond to other regular cell types. This question is clearly biological, namely, the identification of the remaining four terminally differentiated cells (ba, ba^2, $ab^2 = a^2b$, $ab^3 = abab$) from the regular

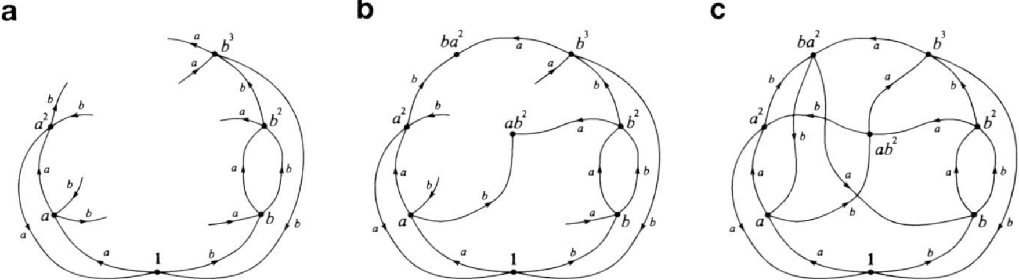

Fig. 11 Construction of the graph $\Gamma(H_0)/H'_1$. (**a**) Quotient space $\Gamma(H_0)/H_1$—a graph of the subgroup $H_1 = \langle a^3,\ b^2ab,\ b^4 \rangle$ with orbifolds of elements as regular cell types. (**b**) Addition of H_0 elements $ab^2, ba^2 \notin H_1$ as another two regular cell types. Identification of corresponding edges. (**c**) Addition of necessary edges in order to construct $\Gamma(H_0)/H'_1$

subtree of $T_\#$ to some regular types by arbitrary cytological character. Let us suppose that there are only two regular types $ab^2 = ba$ and $ba^2 = ab^3$. Secondly, the pathways of these elements are identified in the graph of H_1 (Fig. 11b). Finally, since there are no more regular cell types, several edges should be added in order to obtain two incoming and two outgoing edges with labels a and b, generators of H_0 for each vertex. In the current example we have found eight regular cell types with labels $\{1,\ a,\ a^2,\ b,\ b^2,\ b^3,\ ba^2,\ ab^2\}$ (Fig. 11c) and have constructed a graph of finite index subgroup $H'_1 \subseteq H_0$ with nine generating elements including three from H_1 and six from F_1. The basis of this subgroup can be read by tracing of loops in the graph $\Gamma(H_0)/H'_1$ along some spanning tree with a fixed point at the identity vertex (Fig. 11c). Thus, the structure of $\Gamma(H_0)/\mathrm{Aut_H}(\Gamma(H_0))$ is found.

The labeling of the complete genealogical tree $\Gamma(H_0)$ is an "unrolling" of the quotient space $\Gamma(H_0)/H'_1$ and was carried out as follows. Fix the initial cell vertex 1 and recreate a part of the tree $T_\#$ that corresponds to the regular cell types by moving from the initial vertex only in a positive direction along edges a and b fixing all H_1 orbifolds. The remaining cell vertices are the identities provided by subgroup F_1. Expansion of all possible cell vertex identities covers every vertex in the complete genealogical tree $\Gamma(H_0)$ (Fig. 12). Termination of the cell divisions should be expected only in the cell lineages not covered by the orbifolds of H_1 generators.

Construction of the developmental program spaces $P = \{P_i = \Gamma(H_0)/\mathrm{Core}_{H'_{i-1}}(H'_i) = \Gamma(H_0/\mathrm{Core}_{H'_{i-1}}(H'_i))\}$ is not a trivial task even for $i = 1$. However, there is a nice computational solution of this question. According to Kapovich and Myasnikov [17], the finite index normal subgroup of $H'_1 : \mathrm{Core}_{H_0}(H'_1) = \cap\ h^{-1} H'_1 h,\ h \in H_0, \notin H'_i$ can be computed as a graph product of all conjugacy classes $h^{-1} H'_1 h$ graphs. H'_1 has eight conjugacy classes with respect to eight vertices of its graph, each of which serves as an identity for the conjugate. A resulting graph of H'_1 normal subgroup can have up to 8^8 vertices, but unlike the first, it is vertex transitive.

In the current interpretation the regularities of cell division patterns are indeed uniform and very strict. The single "internal" flexibility of genealogy is concluded in the individual map $f_\# : T_\# \to \Gamma(H_0)/\mathrm{Aut_H}(\Gamma(H_0))$ implying the unique embedding of a genealogical tree into the quotient space of H_0 (see commutative diagram

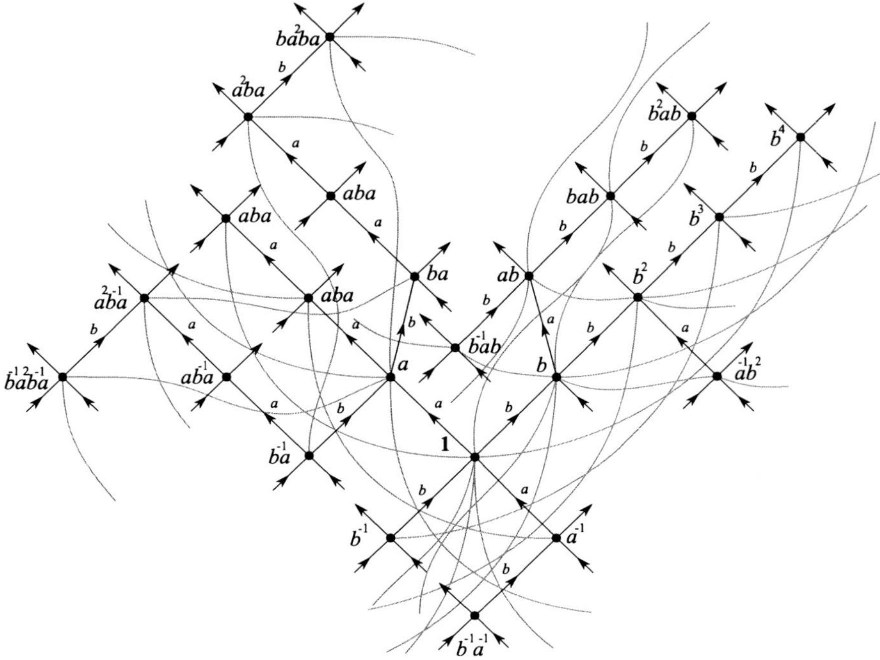

Fig. 12 Recreation of covering space structure for the tree $\Gamma(H_0)$ with respect to the map r^{-1} by lifting of H'_1 orbifolds. Only a part of vertex identities is shown by gray be-infinite lines

above). There is a continuum of such embeddings provided by fixing a unique infinitely distanced vertex in $\Gamma(H_0)$ as a root for $T_\#$. The morphogenetic role of developmental program space here, is just a uniform "counting" of subordinated developmental events. In general, this assumption is true for animals as a result of very strong committance of cells in the course of organism development. Plant cells are more flexible until they become terminally differentiated. For this case there is a more general construction based on the points considered above.

Now admit that at every stage of development the cells "feel" an organism integrity or, in other words, a connectedness of the spatial adjacency graph G_n. Let $T_n \subset T_\#$ be a subtree induced by graph $G_n \subset G_\#$ such that all pendant vertices are $V(G_n)$ and a root vertex is a mother cell vertex of all G_n cell vertices less distanced from them in $T_\#$. Denote this root vertex as an identity in the graph $\Gamma(H_0)$. It is obvious, that this vertex is stable in the course of development of the organism from the initial cell until senescence and properly shifts along $\Gamma(H_0)$ after separation of the initial cell of a new generation. Assume that some cells from G_n are dead before senescence (due to some incident, for instance) such that T_{n+1} has a not properly shifted root vertex. Let this organism still be viable since G_{n+1} has some α- or β-infinite cell vertices. Any $\Gamma(H_0)$ vertex translation is an element $h \in H_0$. This translation is not species-specifically stage proper or just species-specifically proper if $h \notin \{H_i\}$, for some or for all $i > 0$, respectively. In the case of absence of integrity control, the remaining cells (vertices of G_{n+1}) will continue their proper species-specific development (group action) as the organism is intact. In the case of such control, further cell fate will be properly changed for all in G_{n+1}. This changing is predictable and corresponds to the action of the quotient subgroup of H_0, namely some one from the set of developmental program spaces $\{P_i\}$. All these actions are elliptic with vertices of $\Gamma(H_0)$ corresponding to G_{n+1} as fixed points, such that the new root vertex with label h becomes a new (proper) identity. Geometrically, these actions are reflections or rotations of $\Gamma(H_0)$ subtrees around fixed points provided by permutations of tree edges. In the case of terminal differentiation of some cells from G_{n+1}, establishment of a new cell fate is impossible, which inevitably leads to a new root vertex shift for T_{n+2}. Redifferentiation of cells become naturally constant only in the case of precise coincidence of new cell fates with their viability. Otherwise, it will lead to the total disintegration of the organism body up to viable (if they exist) unicellular cell states. The last scenario is quite common in nature for some aquatic macrophytes (Lower plants), for example. Thus, the induction of programmed cell death in cell lineages can have an internal combinatorial nature and, on the other hand, a regular "reaction" to the different external factors.

Real plants, however, provide us with a more delicate structure of their reiterative patterns of cell divisions. Most Higher plants basically possess two systems of organs with potentially infinite growth capabilities—shoots and roots. More generally, these properties bear all primary and secondary plant meristems. The apical meristems of shoot and root have their peculiar stem cell lineages which follow their specific iterative patterns. Reflexing to our constructions, the species-specific sequence of subgroups established above (1) should be a tree-like diagram with "families" of $\{H'_i\}$ subgroups at each node. Only the fist one H'_1 will preserve its uniqueness since it will give its repertoire of identities to the whole set of subgroups. Changing the subgroups involved in organ and whole body construction can lead to drastic changes known as somatic metamorphosis, which are peculiar for many animals such as cnidarians, trematodes and insects as well as for plants. Complexity and uniqueness of subgroup topology defined by specific families of subgroups is a well known tool in group theory. It was intensively studied by Hall [20], who called these subgroup families as topological lattices of a group. In this paper we have attempted to apply some ideas from his works. By analogy, such lattice topology is a more precise and complete description of regular cell types instead of canonical hierarchical biological classifications. The revelation of deep and delicate reiterative mechanisms on the large-scale level of cellular architecture will undoubtedly advance our efforts to resolve many questions of modern developmental biology as well as our understanding of the developmental malfunctions associated with many important medical issues such as cancer induction and stem cell activity.

Acknowledgments The author would like to thank M. L. Gromov and N. Morozova for their valuable remarks and discussion of many questions touched in this work. The existence of similar species-specific "hyperbolic laws" of cell divisions was suggested earlier by M. Gromov [21], whose work has considerably influenced to the creation of this paper.

References

1. Roberts JA, Whitelaw CA, Gonzales-Carranza ZH, McManus MT (2000) Cell separation processes in plants—models, mechanisms and manipulation. Ann Bot 86:223–235
2. Priestley JH (1929) Cell growth and cell division in the shoot of the flowering plant. New Phytol 28(1):54–84
3. Souèges R (1936) Exposés d'embryologie et de morphologie végétales. IV. La segmentation. Premier fascicule: I.—Les fondements. II.—Les phénomènes internes. Hermann et Cie, Paris. 90 p
4. Sattler R, Rutishauser R (1997) The fundamental relevance of morphology to plant research. Ann Bot 80:571–582
5. Kragler F, Lucas WJ, Monzer J (1998) Plasmodesmata: dynamics, domains and patterning. Ann Bot 81:1–10
6. Jarvis MC, Briggs SPH, Knox JP (2003) Intercellular adhesion and cell separation in plants. Plant Cell Environ 26:977–989
7. Speller TH Jr, Whitney D, Crawley E (2007) Using shape grammar to derive cellular automata rule patterns. Complex Syst 17:79–102
8. Gromov ML (1993) Asymptotic invariants of infinite groups. In: Geometric group theory II. LMS lecture notes 182. Cambridge University Press, Cambridge, 194 p
9. Gromov ML (1981) Hyperbolic manifolds groups and actions. Ann Math Studies (Princeton) 97:183–215
10. Weidmann R (2011) Lectures geometric group theory. http://www.math.uni–kiel.de/algebra/weidmann/GGT2_2012/bassserre.pdf
11. Lam E (2004) Controlled cell death, plant survival and development. Nat Rev 5:305–315
12. Gunawardena AHAN, Sault K, Donnely P, Greenwood JS, Dengler N (2005) Programmed cell death and leaf morphogenesis in *Monstera obliqua* (Araceae). Planta 221:607–618
13. Fukuda H (2000) Programmed cell death of tracheary elements as a paradigm in plants. Plant Mol Biol 44:245–253
14. Heath MC (2000) Hypersensitive response–related death. Plant Mol Biol 44:321–334
15. Thomas H, Ougham HJ, Wagstaff C, Stead AD (2003) Defining senescence and death. J Exp Bot 54:1231–1238
16. Stallings JR (1983) Topology of finite graphs. Invent Math 71:551–565
17. Kapovich I, Myasnikov A (2002) Stallings foldings and subgroups of free groups. ArXiv:math/0202285v1[math.GR]
18. Hall M (1949) Coset representations in free groups. Transl Am Math Soc 67(2):421–432
19. Hatcher A (2002) Algebraic topology. Cambridge University Press, Cambridge, 556 p
20. Hall M (1950) A topology for free groups and related groups. Ann Math (Second Series) 50 (1):127–139
21. Gromov ML (1990) Cell division and hyperbolic geometry. Preprint IHES. 9 p

The Geometry of Morphogenesis and the Morphogenetic Field Concept

Nadya Morozova and Mikhail Shubin

This paper presents a set of ideas concerning the connection between biological information, encoded in the cells, and the realization of the geometrical form of a developing organism. Some suggestions are made for mathematical formalization of this connection. The active discussion of this and similar questions at the IHES Workshop (2010) indicated a strong interest in the subject. Therefore, we believe that, although this work is still far from finalized, it is worthwhile publishing this paper in order to stimulate further discussion. Hence, we strongly encourage the reader to independently consider the ideas, concepts and statements presented here in the framework of an integrated model, as it is possible that some of them may turn out to be unviable, while others may be of some interest and importance.

Abstract The process of morphogenesis, which can be defined as an evolution of the form of an organism, is one of the most intriguing mysteries in the life sciences. The discovery and description of the spatial–temporal distribution of the gene expression pattern during morphogenesis, together with its key regulators, is one of the main recent achievements in developmental biology. Nevertheless, gene expression patterns cannot explain the development of the precise geometry of an organism and its parts in space. Here, we suggest a set of postulates and possible approaches for discovering the correspondence between molecular biological information and its realization in a given geometry of an organism in space–time.

First, we suggest that the geometry of the organism and its parts is coded by a molecular code located on the cell surfaces in such a way that, with each cell, there can be associated a corresponding matrix, containing this code. As a particular model, we propose coding by several types of oligosaccharide residues of glycoconjugates.

N. Morozova (✉)
Laboratoire Epigenetique et Cancer, CNRS FRE 3377 and Université Paris-Sud, CEA Saclay, 91191, Gif-sur-Yvette, France
e-mail: morozova@vjf.cnrs.fr

M. Shubin
Department of Mathematics, Northeastern University, Boston, MA, USA

V. Capasso et al. (eds.), *Pattern Formation in Morphogenesis*, Springer Proceedings in Mathematics 15, DOI 10.1007/978-3-642-20164-6_20,

Second, we provide a notion of *cell event*, and suggest a description of development as a tree of cell events, where by *cell event* we understand the changing of *cell state*, e.g. the processes of cell division, cell growth/death, cell shifting or cell differentiation.

Next we suggest describing *cell motion laws* using the notion of a "morphogenetic field", meaning an object in an "event space" over a "cell space", which governs the transformation of the coded biological information into an instructive signal for a cell event for a given cell, depending on the position of the cell in the developing embryo. The matrix on a cell surface will be changed after each cell event according to the rule(s) dictated by the morphogenetic field of an organism.

Finally, we provide some ideas about the connections between the morphogenetic code on the cell surface, cell motion law(s), and the geometry of an embryo.

1 Biological Problem and Main Postulates for its Formalization

1.1 Biological Background

All studies on the mathematical/theoretical formalization of morphogenesis can be divided into two main categories: some based on a principle of self-organization and others based on a deterministic concept. We consider the concept of predetermination of a geometrical shape/form of living species to be the most appropriate, and to support the need to build the formalization on this basis, we cite here some important experimental observations.

1. Transplantation experiments in Amphibians (from a donor organism to an acceptor) and in Drosophila (from one part of an organism to another part of the same organism) show the following rule: the specification of the *positional information* is determined by the position of the cell in the organism, and the determination of the *specific cell type* is determined by the specific genomic information of the given cell. (*Positional information* is the information on the location of cell(s) in space related to given reference points inside a developing organism). For example, transplants from a presumptive stomach of one organism (a triton) transferred to the region of the presumptive mouth in another organism (a frog), will develop into a mouth, but the form (geometry) of this mouth will be that of the donor organism (triton) [27].

2. There are two types of regeneration: morphallaxis and epimorphosis. During morphallaxis the absent parts of the regenerating organism are developed from the existing cells by the changing of their specification and a redistribution in the regenerating organism. An example of this type of regeneration is a redistribution of the cells of the body column of a hydra, forming a new head with tentacles after losing the original head. During epimorphic regeneration, the first step for the cells in the boundary area is dedifferentiation—the formation of the regeneration

blastema by rapid cell proliferation; the cells in the blastema become totipotent (i.e., they are able to differentiate into many different cell types) and then start to form a proper organ (e.g. a limb or tail) according to a plan of the whole organism [15].

In both cases of regeneration, during this process cells "move" (change their internal state or location) in order to reach a predetermined morphological and functional structure of the given organism.

3. A cell from a 4–16 cell embryo will behave (grow, divide, differentiate, etc.) differently in a case of further development as a part of a whole embryo, versus a case when it is isolated at this stage and cultivated in cell culture. However, in both cases the final result of development will be the same—the predetermined species-specific form of the given embryo (or larva) [9]. In addition, at the early stages, changing, during the course of an experiment, the position of cells inside the embryo does not prevent normal embryo formation, although initially these cells have been committed to different developmental pathways [17].

Thus, the deterministic concept will be the main basis of our formalization of the geometry of development. Interestingly, R. Thom, working on the mathematical formalization of a fundamental law(s) underlying morphogenesis, based on catastrophe theory, ultimately came to the conclusion that it is impossible to create a formalization of morphogenesis that is not based on a "deterministic concept" [30, 31].

1.2 Morphogenetic Field Notion

Next, we need to elucidate the notion of "morphogenetic field". The term *morphogenetic field* was suggested by embryologists decades ago [34, 35], to describe the morphology of the developing embryo. According to this biological notion, a morphogenetic field can be defined as a group of cells, the location and the future fate of which have the specification within the same boundary [36]. For example, a zone in the early embryo, including all cells that can potentially participate in the formation of the limb, was named "the field of the limb". Such a field acquires the ability for internal regulation in the case of loss or addition of its parts. In experiments on *Amblystoma* embryogenesis it was shown that each half of the limb disc can regenerate the whole limb at the new position of its transplantation [16]. The same phenomena of full regeneration of the whole limb from each of several segments of limb disc, divided by impenetrable padding, was shown in experimental and in native conditions. In nature, the phenomenon of frogs and salamanders with multiple legs has been observed as the result of limb disc invasion by the eggs of a parasite trematode, thus cutting the limb field into several parts. This means that the limb field can be considered as an equipotential system, i.e. all cells in this system are committed to form any part of the structure corresponding to this field [14]. More recently, the term morphogenetic field has been used, with the same meaning, in a number of works on embryology [2, 6–8].

In addition, the notion of morphogenetic field for the mathematical formalization of developmental processes was suggested in some works by R. Thom [30, 31], but in a very general manner, with no connection to real biological information.

Despite the differences between biological and mathematical concepts of morphogenetic field, they both reflect two very important observations, namely, that the developmental behavior of a cell depends on the instructive signals from the surrounding space (area), and that different areas in a developing embryo contain precise instructions about the shape of corresponding organs. Because these features cannot be ignored in any model aiming to formalize the developmental process, we continue to exploit the morphogenetic field term in the framework of our model, as a possible convenient tool to describe the connection between biological information, encoded in the cells, and the realization of the geometrical form of a developing organism.

We will use the notion of morphogenetic field in a mathematical sense, meaning by "field" a structure containing a space–time dependent mechanism, which mediates the transformation of biological information, contained in cells, into the corresponding geometrical form of an organism in space–time; or, more precisely, into an instructive signal for a cell motion (cell event), depending on the position of this cell in the developing embryo. In a way, this is an elaboration of the existing (and very general) notion, to give a more concrete meaning.

When introducing mathematical formalization of biological laws, it is important to note that in biology the main law(s) are represented as law(s) based on *coding of biological information* (e.g. protein sequence is coded in DNA). Thus, if the mathematical description of morphogenesis can be made in the framework of field theory, then it should be modified so as to consider the behavior of the objects, whose nature and evolution depends on law(s) of coding of the information (i.e. geometrical information) rather than on energy minimization laws.

1.3 Main Conjectures for Mathematical Formalization

We will start formalization with the following **conjectures**:

1. During embryogenesis, each cell undergoes cell divisions, growth, shifts and expression of specific molecules according to a **determinate plan**, invariant for each living species.
2. This **determinate plan** for development of an organism can be considered as a tree of *cell events* from the initial state of the first cell (zygote) to a final predetermined state of an organism, where under *cell event* we understand "developmental events", such as cell divisions, cell growth (death), cell differentiation and cell shifts.
3. This **determinate plan** is coded by a set of specific biological markers, which, most likely, may exist and be transmitted as a set of cell membrane markers. Our main assumption is that such a code may be provided by a pattern of short

oligosaccharide residues of glycoproteins (glycoconjugates) on a cell surface, changing in time and space. It is possible that some other cell surface markers, e.g. specific proteins, may play this coding role; however, short oligosaccharide residues of glycoconjugates have several specific features which make them the most plausible substances for such coding.

4. The general laws for cell events (*cell motion laws)*, namely, the dependence of cell events on coded and positional biological information, have to be the same for all living species, leading to different forms and shapes resulting from different species-specific molecular parameters. Our main goal is to describe these *cell motion laws*.
5. Cell motion laws can be mathematically formulated using the notion of a morphogenetic field.

It is important to note that, in the framework of our model, the cascades of specific molecular events that correlate with pattern formation (e.g. differential gene expression, directed protein traffic, etc.) appear not be the reason for a cell event. Rather, they are, along with the cell event itself, associated with the "coding information" on a cell surface, or, using another terminology, realized due to an instruction for the cell from the morphogenetic field of an organism. The concrete signal transduction pathways connecting the "coding information" on a cell surface and the expression of the given sets of genes need to be elucidated.

2 Morphogenetic Model

2.1 Sets of Morphogenetic Markers on the Cell Surface can be Written in the Form of a Matrix

We assume that the information regarding the geometry of an organism is contained on the cell surface, in the form of a code composed of biological molecules of a special type. Our prevailing assumption is that, most likely, such a code can consist of oligosaccharide residues of glycoconjugates on the cell surface. There are 12 types of monosaccharide that exist in oligosaccharide residues of glycoconjugates (with six of them being of hexose type), and there have been numerous observations indicating that these oligosaccharide residues are connected with the determination of cellular morphogenetic pathways [1, 3, 5, 10, 11, 12, 13, 18, 19–21, 22–26, 28, 29, 32, 33, 37, 38]. We will use this "oligosaccharide code" assumption in our examples below; however, it is clear that this is not the only possible "cell surface code", and the real biological code (if it exists) should be elucidated by additional joint theoretical and experimental research.

Our next assumption is that this information can be written in the form of a matrix A_{nm}, in which every element corresponds to the level of a certain type of oligosaccharide (or monosaccharide) in a given region (section) of the cell surface.

For an easy illustration of this idea, we may consider two sections on the cell surface (*left* and *right*), and six types of monosaccharide to be used in a coding, and in this case we will obtain a 2×6 matrix associated with each cell (Fig. 1).

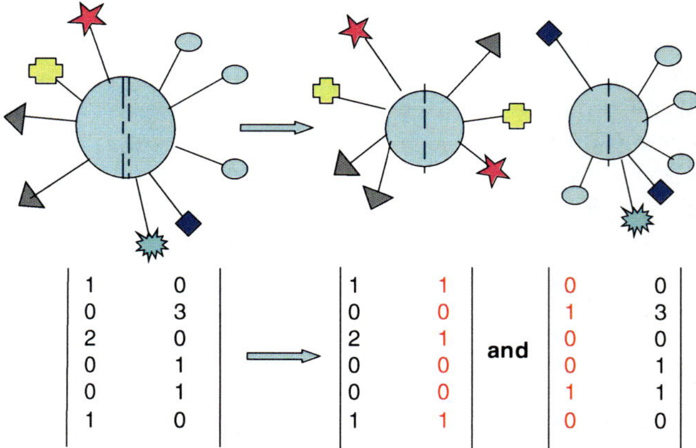

Fig. 1 A simplified two-section example of the matrices corresponding to a mother cell and two daughter cells following cell division. One column of each matrix corresponding to the daughter cells is equal to the mother cell (columns in *black*), and each new column in the daughter cell matrix is created according to a rule R_d (columns in *red*)

The vector corresponding to each section (i.e. an order, in which we will place the numbers corresponding to the quantity of each type of monosaccharide residue in a section) will have the following structure:

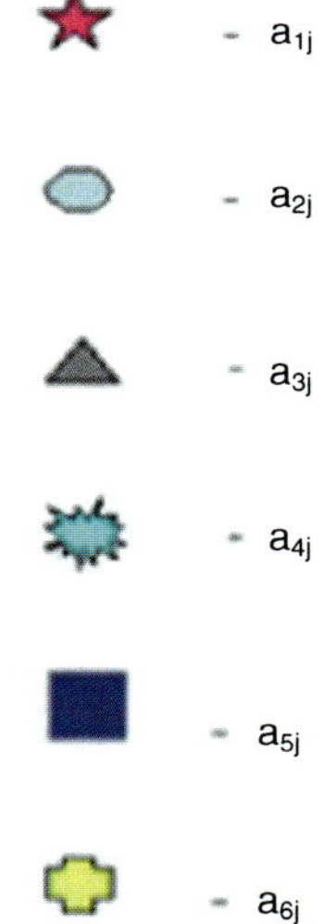

where each symbol defines one type of "coding" residue (mannose, glucose, galactose, rhamnose, fucose or xylose).

When a cell undergoes division, one part of each of the two newly generated matrices, corresponding to two new cells, will be identical to the parent matrix, while the other part will be generated de novo, i.e. filled with new numbers (Fig. 1).

We next suggest that filling in the daughter matrices with the new A_{nm} elements after/during each cell division occurs according to a rule R, the general principle of which may be the same for all living organisms:

$$R\left(A^{i}_{nm}\right) \rightarrow \{A^{j}_{nm}, A^{k}_{nm}\} \tag{1}$$

where the matrices A^{j}_{nm}, A^{k}_{nm} of two new daughter cells are created by application of the rule R to the matrix of the mother cell A^{i}_{nm}.

In biological terms this means that there exists a rule (R), according to which new glycoconjugate residues will appear on the cell surface in specific quantities and in specific locations.

It is clear that a *set of "conservative" elements* $\{A_{nm}\}$ in new daughter cells is determined by the position of the plane of cell division, dividing cell membranes (and thus all sets of its markers) into two parts. We can illustrate it with a more realistic example, considering 8 sections on the cell surface with respect to 3 spatial axes, X, Y and Z, and 6 types of monosaccharide to be used in a coding. Then for this case we will obtain a 6×8 matrix associated with each cell. (Here X, Y, Z axes represent the anterior–posterior (AP), dorsal–ventral (DV) and left–right (LR) axes, normally used in embryology).

If a cell divides symmetrically with respect to the membrane surface, and only in planes $X = 0$, $Y = 0$, or $Z = 0$ with respect to the coordinate system defined for the zygote, then during each cell division one set of 4 columns from the 8-column matrix A^{i}_{nm} of the mother cell ***i*** will be equal to the corresponding set of matrix A^{j}_{nm} of a daughter cell ***j***, and the other set of 4 columns of A^{i}_{nm} will be equal to matrix A^{k}_{nm} of a daughter cell ***k***. The remaining 4 columns (of a total of 8) in each new cell j and k will be created *de novo* after each cell division.

For example, if the plane of cell division is $X = 0$, columns 1, 2, 3 and 4 of the mother cell matrix will be equal to the same columns of the "left daughter cell" and columns 5, 6, 7 and 8 to the "right daughter cell"; if the plane of cell division is $Y = 0$, columns 1, 3, 5 and 7 of the mother cell matrix will be equal to the same columns of the left daughter cell and columns 2, 4, 6 and 8 to the right daughter cell; etc.

Thus the proposed model implies that the determination of the position of the cell division plane is realized by the rule R_d, by defining the new matrices of new daughter cells. The elements A^{j}_{nm}, which, according to R_d, will be equal to those in the mother cell A^{j-1}_{nm}, will thus determine the border between two new daughter cells, and thus, the position of the cell division plane.

Generally speaking, a cell division plane can divide a cell into two parts not symmetrically related to the membrane surface, and the cell division plane can be under any set of angles α,β,γ to embryo axes X, Y, Z. On the other hand, it is clear that, in the framework of the very simple matrix representation of "cell surface coding" used above, it must be postulated that only those locations of the cell division plane that cut the cell surface at a boundary between two "coding sections" are permitted. Thus, for considering the realistic potentiality of cell division plane orientation, the suggested meaningful quantity of "coding sections" on a cell surface should definitely be greater than 8 (16, 32, or more). If one wishes to avoid such a "boundary postulate", then, instead of a matrix, cell surface information should be written as a 3-dimentional array, with the additional dimension corresponding to the location of each marker inside each "coding section".

Changes in the quantity and composition of glycoconjugate residues on the cell surface during cell growth between its birth and vanishing (when it divides or dies) will occur according to another rule(s) R_l. These rules have to account for the processes of both appearance and disappearance of the residues on the cell surface.

We may suggest, therefore, that there is one rule R_l for each type of *cell event* S_l.

2.2 Cell Event Definition

We will assume that each cell i can be described by the state function of a cell $\psi^i(t)$,

$$\psi^i(t) = \left(A^i_{nm}(t), \mathbf{D}^i\right)$$

where $\mathbf{D}$ is a parameter showing the position of cell i in the embryo with respect to the zygote. This can be given in several ways, e.g. as a cell lineage and generation number, or, alternatively, as $\mathbf{D} = \sum \mathbf{d_k}$ where $\mathbf{d_k}$ is a vector perpendicular to the cell division plane and $\mathbf{k}$ is the number of cell divisions that have occurred from the zygote to the given cell (in the case of no other displacements, such as shifts).

In addition, we may propose an alternative possibility to mark the position of cell i in the embryo. It makes sense to assume that the cell surface markers on the zygote may have some additional specificity, discriminating them from all other markers of the same type, synthesized later in the course of development. In this case, the position of each cell in the embryo can be marked with respect to zygote membrane/surface, in relation to these specific markers.

By *cell event* we will understand the changing of the *cell state* $\psi^i(t)$, i.e. as a result of the cell event, cell matrix $A^i_{nm}(t)$, cell position $\mathbf{D}^i$, or both will be changed, generating a new cell state $\psi^i(t_1)$. This change can occur as a result of the processes of cell division, cell growth/death, cell shifting or cell differentiation.

We can formalize all possible cell events S_l as follows, subdividing them into three main types:

S_1 = cell division in plane W, with angles α,β,γ to embryo axes X,Y,Z (AP, DV, LR)
S_2 = cell growth, characterized by cell growth rate q.
S_3 = cell shift, characterized by cell shift vector $\mathbf{h}$.

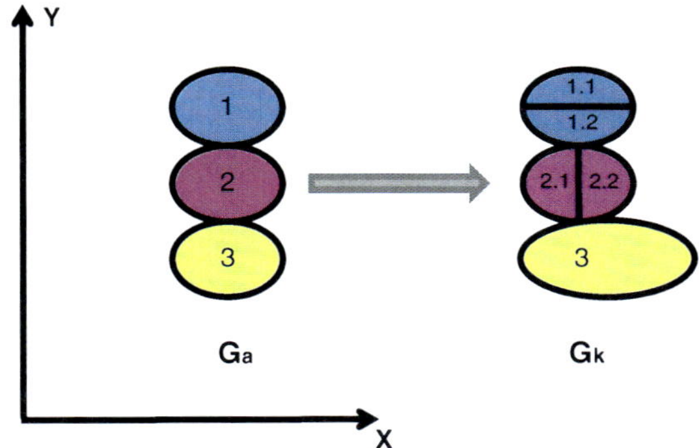

Fig. 2 Illustration of *cell event* concept in 3-cell embryo development. See text for details

Importantly, S_2 corresponds to 3 different processes:

S_{21}—cell grows (cell growth rate q > 0)
S_{22}—cell undergoes apoptosis (cell growth rate q < 0)
S_{23}—cell differentiates (cell growth rate q $= 0$)

depending on the sign of a cell growth rate q, q $= \frac{dV}{dt} = \oiint a dA$, where a is a local growth rate at point x on a cell surface C and dA is an area element on that surface. A cell growth vector $\mathbf{B(x)}$ on the cell surface is then defined as $\mathbf{B}(\mathbf{x}) = a(\mathbf{x})\mathbf{n}$, where $\mathbf{n}$ is the unit normal to cell surface at point x.

In a simplified case of development with an assumption that cells divide symmetrically in relation to their membrane surface and with only three possible division planes (X $= 0$, Y $= 0$, Z $= 0$), we will have only 3 types of cell division event:

S_{11}—cell division in plane X $= 0$
S_{12}—cell division in plane Y $= 0$
S_{13}—cell division in plane Z $= 0$

Figure 2 presents an illustration of this formalization to one step in 3-cell embryo development between two time points, corresponding to the stages $G_\mathbf{a}$ and $G_\mathbf{k}$ of an embryo, and for this step the cell event for each cell will be:

For cell number 1: S_{12} (cell division in the plane X $= 0$)
For cell number 2: S_{11} (cell division in the plane Y $= 0$)
For cell number 3: S_{21} cell growth with integral growth rate q > 0 ($q \approx 2$), where the growth vector has direction parallel to the y-axis.

2.3 Evolution of Embryo Domains' Surface Geometry Depends on Cell Events Within a Given Domain

The growth and development of an organism in space can be described as the dynamics of the boundaries of the corresponding **domains** of the developing embryo, morphologically distinct from its other parts (e.g. tissue, organ, limb bud or growth point). We will denote as $G_Y(t)$ the surface of the domain Y, which will be a smooth closed surface in $\boldsymbol{R}^3$, depending on time t as a parameter. For any x at $G_Y(t)$, we will denote as an external unit the normal vector to G_Y(t) at x by **n**, so that $\mathbf{n} = \mathbf{n}(x, t)$.

Given $G_Y(t)$, we will assume that, at the next moment, $t + dt$, the form of the surface $G_Y(t + dt)$ is such that it consists of all points $x + \sigma(x, t)n(x, t)dt$, where $\sigma(x, t)$ is a scalar function on $G_Y(t)$ at every given moment of time t.

To avoid a gauge transformations indeterminacy (existence of many equivalent presentations of the same family of surfaces which differ by choice of coordinates), we may present a growing embryo domain surface in the form

$$G_Y(t) = \{x : \mathrm{u}(x) = \mathrm{t}\}$$

where u(x) is a real-valued function on $\mathbf{R}^3$ with sufficiently regular level sets S(t). In this case the distance between infinitesimally close surfaces $G_Y(t + dt)$ and $G_Y(t)$ at point x is $\sigma(x, t)dt$ (by definition of σ) on the one hand, and using the geometric meaning of the gradient, it is $|\nabla \mathrm{u}(x)|^{-1}dt$ on the other hand.

This leads to the equation for the evolution of the embryo domains' surfaces:

$$\sigma(x, t)|\nabla \mathrm{u}(x)|^{-1} = 1 \text{ provided } \mathrm{u}(x) = \mathrm{t}. \tag{2}$$

which may be written as a system of equations.

We can then see the interconnection between the shape of the domain Y, the form of its surface $G_Y(t)$ and the set $\{\mathrm{S}_l\}_{G_Y}$ of all cell events that have occurred inside the domain Y enclosed by surface $\mathrm{G_Y}$ during time period dt:

$$\sigma(x, t) = \sigma(\{\mathrm{S}_l\}_{G_Y})$$

Thus, knowing all cell events in the domain Y during the time period dt, we would be able to calculate $\sigma(x, t)$, and next, given $G_Y(t)$ at moment t and $\sigma(x, t)$, we can find $G_Y(t + dt)$.

Correspondingly, the volume V_{G_Y} of the domain Y may be calculated as:

$$\frac{\mathrm{dV}_{G_Y}}{\mathrm{d}t} = \oiint_{G_Y} \sigma(x, t)dG$$

Cases of branching of a morphological domain into two new domains or the appearance of a separated domain inside an existing one are more difficult and require additional conditions for defining boundaries between domains to be included.

3 Formalization Using the Notion of Morphogenetic Field

Next we will discuss the connection between the morphogenetic code on a cell surface, cell events and the geometry of an embryo using the notion of a morphogenetic field (MF).

We will mean by the term "morphogenetic field" an object governing a law (or a set of laws) for transforming coded morphogenetic information into an instructive signal for a cell event S_l for a given cell, depending on the position of this cell in the developing embryo.

We may consider the existence of a hierarchical structure between the morphogenetic field of the whole organism and the local morphogenetic fields of the organ, the tissue, the morphologically distinct domain of the developing embryo, and of a single cell, assuming that the morphogenetic field of each tissue/domain in the developing embryo is a function of the morphogenetic fields of its cells, the MF of each organ is a function of the MFs of its tissues, and the MF of the whole organism at each stage is a function of the MFs of its organs (or morphologically distinct domains during early embryogenesis). The sense/scope of the instructive signals in each type of MF in this hierarchy (i.e. MF of organ, MF of tissue, etc.) should be different. Possibly, for MFs of higher levels in the hierarchy than the initial "morphological domain", the notions of *tissue event* and *organ event* should be introduced, indicating the signal for new tissue/organ formation. For example, the formation of each somite could be considered as a tissue event, and limb regeneration, as an organ event.

We will start our proposition with the description of the MF of a morphologically distinct domain of the developing embryo, as the initial step in this hierarchy.

3.1 *Morphogenetic Field can be Defined as a Function of a Set of Morphogenic Markers*

We conjecture that the morphogenetic field of any morphological domain of the developing embryo at a given moment in time is a function of the morphogenetic information of all cells located in that domain.

Thus, a mechanism influencing the evolution of the domain in space–time depends on (is a function of) the whole set of all matrices, corresponding to all cells belonging to the given domain, together with a factor of interactions between neighbor cells and a factor depending on the external media.

Let us denote a function corresponding to MF influence as M_F.Thus, we can write:

$$M_F(t) = M_F\left(\{A^i_{nm}(t)\}, M_{int}(t), M_m(t)\right) \tag{3}$$

where $\{A^i_{nm}(t)|i \in Y\}$ is a set of matrices for all cells i within embryo domain Y, $M_{int}(t)$ is a factor of interactions between neighbor cells, influencing M_F formation, and M_m is a factor depending on the external (media) conditions.

Anyhow, it is possible that a set of cell surface markers will be changed depending on the external conditions of a cell. In this case, information regarding the effects of $M_{int}(t)$ and $M_m(t)$ on M_F formation can be included in $\{A^i_{nm}(t)\}$, and we can write the simplified version of (1):

$$M_F(t) = M_F\left(\{A^i_{nm}(t)\}\right). \qquad (3')$$

Next we postulate an existence of non-linear operators R_l of MF, corresponding to each cell event S_l, such that filling in the new matrices with the new elements after/during each cell event S_l occurs according to operator R_l. For example, for event S_1 (cell division) it will be operator R_d

$$R_d\left(A^i_{nm}\right) \rightarrow \{A^j_{nm}, A^k_{nm}\}$$

where A^j_{nm}, A^k_{nm} corresponds to 2 new matrices of 2 new daughter cells after the division of the cell A^i_{nm}.

For event S_2 (cell growth) it will be operator R_g

$$R_g\left(A^i_{nm}(t_1)\right) \rightarrow A^i_{nm}(t_2)$$

where $A^i_{nm}(t_1)$ corresponds to the matrix of cell i just after cell division, and $A^i_{nm}(t_2)$ corresponds to the matrix of the same cell i after a period of growth. Thus, R_l is a rule for creating new matrices depending on the matrices of the previous state of the cell.

3.2 *Morphogenetic Field Defines Cell Event for Each Cell*

The next, very important, idea in our model is that, having a rule R_l for filling in the new matrices depending on the matrices of the previous state of the cell, we must consider the influence of MF and cell position **D** in the embryo for this process.

$$S_l = S_l(M_F, \mathbf{D})$$

This means that a cell with the given set of coding molecules on its surface will undergo different cell events and thus will have different rules R for creating the next matrices for the next cell state of this cell, depending on the position of the cell in the embryo. The mechanism regulating this is provided by function M_F of MF. The best example of such regulation is transplantation experiments in Amphibians (see Biological Background section), where cells of a triton committed for producing a stomach, after transplantation into a new (mouth) location in another organism (frog), develop a mouth. This can be explained as a command by MF of a tissue in a new cell position to implement a new chain of *cell events* for these cells, resulting in the proper structure for this organism. The fact that the form (geometry) of this mouth will be that of the donor organism (triton) means that the appropriate cell events depend both on MF field instructions, and on the content of coded biological information (matrices) of the cells A^i_{nm}.

Thus, the morphogenetic field, depending on the set of all information A^i_{nm} of its cells, defines a cell event for each cell depending on its position in the embryo (and, possibly, also on A^i_{nm} of the given cell) and R_l, corresponding to each type of cell event, works for filling in the new cell matrices A^i_{nm}. Two biological events: cell motion in MF (cell event) and the changing of the oligosaccharide code (matrix) on the cell surface in the frame of the model are supposed to take place during the same time interval, but in three different steps inside this interval: matrix $\rightarrow$ MF action $\rightarrow$ cell motion $\rightarrow$ new matrix.

3.2.1 Important Comment

There is another possible way to present the idea of the A^i_{nm}, R_l, S_l and MF connection based on a different understanding of the nature of MF and thus, resulting in a different way of describing it. The set A^i_{nm} of oligosaccharides on the cell surface (their quantity and composition) may unambiguously determine R_l, which also unambiguously determines the corresponding cell event S_l. In this case, MF can be understood as a field of cell matrices A^i_{nm} determined for each point (cell) in $\mathbf{R}^{3+1}$ with a non-linear operator $\boldsymbol{R}_l$ acting in this field. However, this representation cannot explain some of the biological data.

3.3 General Model

Our main assumption here will be that the MF of a morphological domain (tissue or organ), being a function of coded biological information (a set of matrices of all cells within this domain), also contains a possibility of "measuring" the discrepancy between the current geometrical form of the domain (tissue, organ) in $\mathbf{R}^{3+1}$ and the final geometrical form of the surface of this domain in $\mathbf{R}^{3+1}$. One possibility is that such a measurement may be connected with our previous assumption, that cell surface markers of the zygote may have some additional specificity, thus determining the position of each cell in the embryo with respect to the zygote membrane/surface.

Let us decide that this discrepancy for the domain Y can be measured as a *correspondence function* F_c between surface $G_Y(t)$ at moment t and surface $G_k = G_Y(t_{final})$:

$$F_c = F_c(G_k, G_Y).$$

Next we assume that the influence (effect) M_{F_Y} of the morphogenetic field of the domain Y on the cell c belonging to this domain depends on the *correspondence function* F_c

$$M_{F_Y} = M_{F_Y}(F_c(G_k, G_Y)), \tag{4}$$

and results in one of the possible cell events S_l, changing the cell state ψ_c:

$$M_{F_Y}\psi_c(t_1) = S_l : \psi_c(t_1) \rightarrow \psi_c(t_2). \tag{5}$$

Using these notations, we can formulate the following **statement**:

MF determines the morphogenetic effect M_F applied to the given cell, such that the resulting cell event will cause the minimization of the corresponding function F_c between the current state of the surface of the domain G(t), containing the given cell, and the final surface of this domain $G_f(t_{final})$ (Fig. 3).

This means that a developmental process can be understood as a set of cell motions in MF, where each cell will "move" (undergo a cell event) in the direction of minimizing the discrepancy between the encoded final and current geometrical forms of the morphological domain—by spatial motion, cell division(s) or by changing its "molecular state", meaning the processes of cell differentiation. This assumption is consistent with the experiments showing that a part of the corresponding domain can regenerate a whole structure corresponding to this domain (see Sect. 1.2 *Morphogenetic field notion*).

F_c is connected with M_F in the following way: F_c determines the pathway from the current to the final morphological state of the given domain as a tree of potential cell events in $\mathbf{R}^{3+1}$, and M_F determines the local motion law (S_l at the moment t) along this way (real cell event in each real time point).

The proposed idea means that, for example, in a situation when "cell 1" of the embryo presented in Fig. 2 is destroyed at the stage $G_{\mathbf{a}}$, the cell event for cell 2 will be different—it will be S_{12} (cell division in a plane $X = 0$), thus replacing the cell 1, and the growth rate q for cell 3 will be considerably decreased. Next, all 3 cells of a new 3-cell embryo will undergo 3 corresponding cell events, as described above for the normal 3-cell embryo, thus reaching the proper stage $G_{\mathbf{k}}$.

3.4 Biological Observations Supporting the General Model Hypothesis

Our general model, to some extent, can be confirmed in its basic form by an important biological observation. The model is based on the hypothesis that matrices corresponding to a cell surface concurrently play a role in the coding of the final

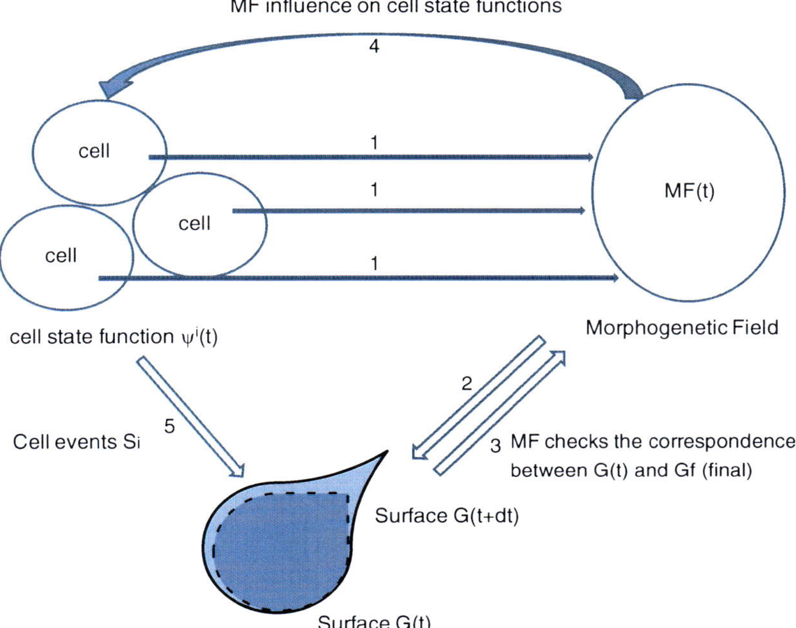

Fig. 3 Schematic representation of the general model 1. Set of cells determines morphogenetic field (MF) 2, 3. MF evaluates the geometry (position in space) of the surface G(t) (*dashed line*) and the correspondence between G(t) and G_f (final). 4. Influence of MF on the cells (M_F). 5. As the result of MF instruction, each cell from the set undergoes the corresponding cell event for creating a proper new object with a determined surface $G(t + dt)$ (*solid line*)

geometry of an organism and in the signaling for cell events at each moment in time. This proposed dual function of matrices A^i_{nm} can be confirmed by the following biological observation, which may be formulated as the following **Statement**: ***the precision of determination of the geometrical form of an organism by biological information is in inverse proportion to the level of totipotency of its cells.***

For example, the animal body is determined unambiguously (e.g. for mammalians: 1 head, 4 paws and 1 tail, and everything at the specified locations) and normally animal cells have a very low level of totipotency (this means that differentiated animal cells cannot undergo de-differentiation followed by re-differentiation into other organs and tissues in cell culture). The forms of plants are not unambiguous, and only the form of specific plant structures, such as flowers or leaves is determined uniquely, but not the final form of the whole plant (for example, in most cases, leaves and branches can appear in random quantities and in multiple possible places). This correlates with the fact that differentiated plant cells have a very high level of totipotency (i.e. one can easily obtain de-differentiation and re-differentiation of plant cells to form a new organism (or any plant organ) in cell culture).

In the framework of our suggested model, we can explain this phenomenon in the following way: in plants the cell surface code may have more redundancy (degeneracy); this influences, at the same time, the non-stringency for the final geometry of an organism, and the flexibility of conformity between a given cell matrix and its corresponding R, which allows, accordingly, a wide spectrum of possible meaningful cell events in the frame of the same initially coded information, thus increasing the cells' potency to differentiate into all the specialized tissues of the developing organism. For animals, the cell surface code may have much lower redundancy, and this causes a stricter determination of the final form of the organism, together with stringent individual cell fates for their cells, giving no possibility for alternative cell events.

Another important confirmation of our hypothesis is the behavior of protoplasts in cell culture. Plant cells have a cell wall, which covers the plant cell outside the cell membrane. Due to this fact, in plant cells all oligosaccharides of glycoconjugates of a cell surface are located on the surface of the cell wall, while in animals they are on the surface of the cell membrane. Protoplasts are plant cells without a cell wall, which can be removed from cells in cell culture using a specific enzyme. After cell wall removal, protoplasts produced from terminally differentiated cells of a plant (leaves, fruits, etc.) start to divide, forming at the same time a new cell wall. However, the new cell wall does not have any morphogenetic information, thus all cells continue to divide eternally, producing a callus (an unorganized mass of undifferentiated cells). This means that the morphogenetic information required for proper development was removed with the cell wall, supporting the hypothesis regarding the significance of molecules on the cell surface for coding morphogenetic information.

Cells in a callus can then be induced by different plant hormones to produce different organs (roots, shoots, etc.). Importantly, plant hormones, which exist in an organism in extremely small concentrations, are able to change the whole direction of development of cells in a callus: for example, depending solely on the ratio between two plant hormones, all callus cells will be governed to produce roots or shoots. This can be interpreted as a restoration of part of the morphogenetic information on a cell surface by specific activation by the hormones of some of the molecular signaling pathways responsible for turnover of oligosaccharides on the cell surface; the final pattern of this morphogenetic information (spectrum of oligosaccharides), which determines the subsequent organ development, depends on the interplay between hormones controlling signaling pathways responsible for each oligosaccharide. The importance of plant hormones for glycoconjugates production was already shown experimentally [32, 33].

Finally, in the framework of our model, the process of carcinogenesis can be understood as a loss by cancer cells of the ability to accept the correct signals of the MF to which they belong. The explanation for this may be that the mutations that lead to cancer progression influence the pathways responsible for the correct turnover of oligosaccharides on the cell surface, thus breaking from normal cell behavior dictated by MF. Substantial experimental data showing that some cell surface proteins, primarily glycoproteins, are altered in cancer cells, may be considered as additional support for our suggested hypothesis.

4 Some Ideas on a Possible Mathematical Description of MF

4.1 *Mathematical Description of MF as a Section of a Fiber Bundle*

Now we come to the question: what is the nature of the MF field and how can we describe it? We understand by MF **a species-specific object in the space of events (cell events) over the space of cells (cell space)**, having the capability of decoding biological information and converting it into the corresponding cell event depending on the location of the cell in the space–time of a developing embryo.

We suggest that it may be possible to formalize this idea using a description of a morphogenetic field as a section of a fiber bundle over a base, where the base contains all the cells existing in $\mathbf{R}^{3+1}$ space during the entire life of the embryo, while the possible cell events are considered as a fiber (Fig. 4a).Thus, we will introduce:

1. A space-time X which consists of points $\tilde{x} = (x, t)$ where x is a cell, t is the time; we consider as X all sets of cells existing in the embryo's entire developmental history;
2. A fiber bundle over X with the bundle space E and the projection map $p : E \to X$, where the fiber $E_{\tilde{x}}$ over a point $\tilde{x}$ in X consists of all possible cell events S_l for the cell $\tilde{x}$. This means that each fiber has the following structure (Fig. 4b): a union of 3 vector spaces $\mathbf{R}^3$, corresponding to 3 types of cell events, united by a singular point at $S_1 = S_2 = S_3 = 0$, corresponding to the process of differentiation (see Sect. 2.2 *Cell event definition*).

In this way, a section s of a fiber bundle

$$M_F(\tilde{x}) \to S_i$$

corresponds to a law for choosing/prescribing a cell event S_l to each cell $\tilde{x}$ in a base, taking into consideration that each cell $\tilde{x}$ is associated with the state function of a cell $\psi^i(t)$, defined by a matrix and a position of cell i in the embryo, while M_F depends on a set of matrices in the embryo domain, corresponding to the given MF. We may consider a space of the sections of the fiber bundle (MF space). Next we will consider an ordered set of cell events in a time interval, corresponding to section s, as a motion

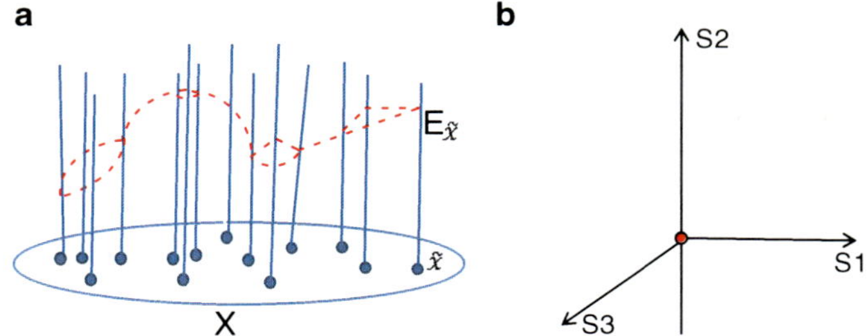

Fig. 4 Schematic representation of a morphogenetic field as a section of a fiber bundle over a base. (**a**) Base X ($\mathbf{R}^{3+1}$) represents all cells during the entire life of the embryo, all possible cell events are considered as a fiber $E_{\mathbf{X}}$. The morphogenetic field is represented as a section of the fiber bundle (in *red*), choosing the given cell event (*point*) on each fiber corresponding to each cell $\tilde{x}$. (**b**) A fiber structure. Axes S_l represent three vector spaces $\mathbf{R}^3$, corresponding to three types of cell events (cell division, cell growth and cell shift, respectively). The singular point in red corresponds to cell differentiation

in morphogenetic field, and try to elucidate the law(s) of cell motion. We will assume that function C, determining cell motion, is a function on the bundle space E of a point S_l on the fiber over the corresponding point $\tilde{x}$, and $\tilde{x}$ itself:

$$C = C(\tilde{x}, S_l). \tag{6}$$

We will choose those local sections s of a fiber bundle corresponding to the given MF that will satisfy the equations of motion in MF, which may be found, according to our general model, from the condition:

$$\delta F_c[s] = 0 \tag{7}$$

where the correspondence function $F_c[s]$ will be defined as a real-valued function of the sections of the bundle:

$$F_c[s] = \sum_{\tilde{x} \in \mathrm{X}} C(\mathrm{s}(\tilde{x})), \tag{8}$$

and the variation $\delta F_c[s]$ is taken with respect to all possible variations of the section s. Here, for discrete parameters, the variation is to be understood as the set of all differences $F_c[s] - F_c[s']$, where for every $\mathrm{x} \in \mathrm{X}$ the (variable) section s takes values that are the closest possible to the corresponding values of s', a section which corresponds to "coded" MF (MF').

If the most general form of C were known, then imposing conditions (7) on the (8) would in principle result in a set of differential equations, assuming the sum converges to a well-defined integral. These differential equations could be interpreted as the equations of (cell) motion.

Thus, in the framework of the suggested model, a cell event, corresponding to each cell, is dictated by the morphogenetic field by selecting a point on each fiber corresponding to each point on the base. The section of "coded" MF of the whole organism consists of numerous sections for all morphological domains and tissues, finally combined in organs. The development of the embryo in time corresponds to a t-dependent 3-dimensional object moving along a 4-dimensional MF section of the fiber bundle. The real organism attempts to make its MF identical to its corresponding coded MF (MF') by minimizing the discrepancy between them at each time point. Thus, at each time point **t** all cell events leading to the next step (**t** + 1) will be dictated by minimization of the discrepancy. Using this formalization, we can define a *correspondence function* F_c not as a discrepancy between surfaces $G_Y(t)$ and $G_k = G_Y(t_{final})$ of the embryo domain, as in Sect. 3.3, but as the discrepancy between corresponding real and "coded" sections of MF:

$$F_c = F_c(MF', MF).$$

The correlation between these two definitions will be discussed later.

It is important to note that additional conditions may be imposed on the MF structure and cell motion laws by analysis of characteristics of a natural graph structure on X with two types of edges described as follows:

1. Edges connecting neighbor (adjacent) cells on the same time level;
2. Edges connecting each cell with its daughter cells.

Cell motion in MF, which is based on the trees of corresponding cell events (shown on the Fig.5, upper graph), is tightly connected with the genealogical graph(s) of cells in the developing organism (Fig.5, lower graph, where each vertex corresponds to a cell in an organism); however these two graphs (or their parts) may be isomorphic only in the cases when a developmental program consists of rapid cell divisions occurring without any other cell events. On the other hand, there are cases when cell motion in MF is completely independent of genealogical graphs, e.g. during a process of regeneration in Hydra (morphallaxis, see section 1.1), where cells undergo a changing of their positions inside an organism accompanied by changing of their morphology, thus creating a new shape without additional cell divisions. In this case a cell motion in MF will be represented as a set of individual trajectories in MF for each cell, consisting of shift, growth and differentiation cell events, integrated in a section(s) due to the corresponding graph structure on X which includes edges between neighbor (adjacent) cells (for different time levels). Thus, each section of MF should be designated with relation to information borne by both types of edges (1 and 2) in the graph structure on X (Fig. 5).

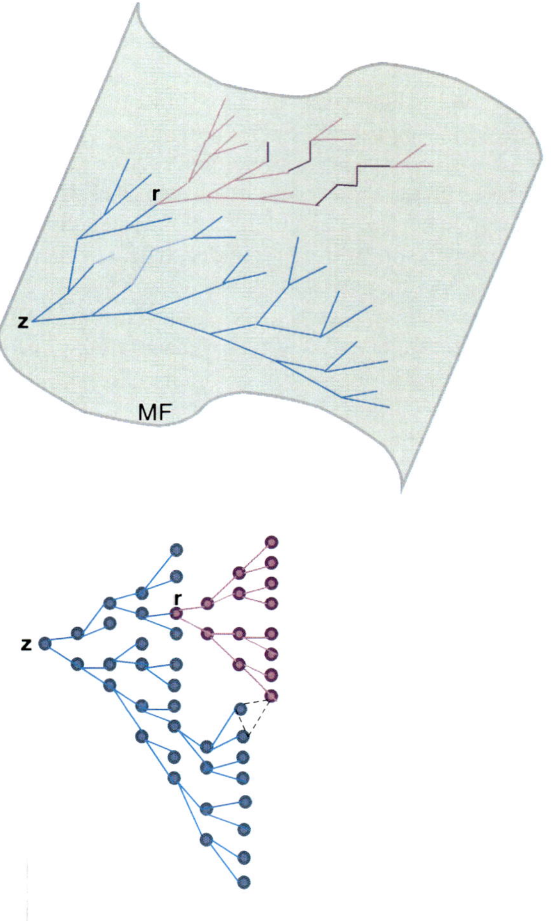

Fig. 5 Tree of Cell Events in morphogenetic field. Each vertex on the upper graph corresponds to a cell event. Cell motion (a path) for the cell *z* is the whole tree, while cell motion (a path) for the cell *r* is a tree marked in pink color. Outgoing edges from the vertices, which correspond to cell events other than cell division ones (e.g. cell growth, apoptosis, differentiation), are shown in different tints (light-blue or purple-pink). Each vertex on the lower graph corresponds to a cell in an organism; the graph represents a genealogical tree of cells. The triangle in black dotted lines shows an example of edges bearing an information about neighbor cells in an organism

4.2 Probabilistic Character of Cell Events

An additional important suggestion may be to consider next the possibility of a description taking into account a probabilistic character of each cell event S_l. The probabilistic character of cell behavior can be seen from the two points:

1. In normal development the final geometry of an organism is coded not unambiguously, but rather as a multitude of possible trajectories within a given defined "corridor", i.e. the form and size of organs may have variations but only within a certain range. This can be represented by a near-constant probability density of cell event paths (MF configurations) inside a corridor, such that the total probability of a path to be inside a corridor is close to unity, with near-0 density outside it.
2. In the case of extreme cell fate changes (via transplantation, placing cells into cell culture, injury followed by subsequent regeneration, etc.), cells, depending on the level of their totipotency, have the possibility of differentiating into all (or

at least several) specialized tissues of the developing organism. This means that at each step of development a cell has many possible alternatives of further behavior, from which only one is realized under given conditions.

Thus, we propose to consider these possibilities as the probability of realization of different cell fate programs. A sequence of cell events that transform an initial state of a tissue ψ_1 into a final state ψ_2, which we will refer to as a "path", is represented by a particular section of the fiber bundle, s, and there is a certain probability that this particular section, among the many possible ones, will be realized. Let us denote the probability of transition from ψ_1 to ψ_2 via a section s by

$$P_s(\psi_1 \to \psi_2).$$

As mentioned above, these probabilities are assumed to have a value that is roughly constant inside the corridor of possible paths, and rapidly falls off outside it. The total probability of reaching a given final state from a given initial one is then given by the sum of individual probabilities for all conceivable paths that start at ψ_1 and end at ψ_2. That is to say,

$$P = \sum_{s \in \Gamma(E)} P_s(\psi_1 \to \psi_2)$$

where $P = P(F_c[s])$, Γ(E) is the set of all possible sections s of a bundle space E. Clearly, the probability of a particular path depends on the probabilities of individual cell events comprising it, in the usual way:

$$P_s(\psi_1 \to \psi_2) = \prod_i P_i(S_l)$$

where $P_i(S_l)$ is the probability of an individual cell event occurring. However, since not every path constructed in such a way is possible, e.g. a cell that has divided at a previous moment in time cannot perform growth in the next (simply because it no longer exists), the cell events S_l in the above product must all belong to the same section connecting the initial and final states.

Therefore, it would be advantageous, if possible, to find a way to characterize the probabilities of entire paths, rather than having to build them up from individual events. For this, we propose an idea to account for the probabilistic nature of cell events using a formalism where the amplitude for a given path is a function of the action evaluated along that path. In this case, there exists a "classical path" of events (corresponding to the initially coded pathways of cell events), and all the paths in its vicinity contribute the most to the probability of transitioning from the initial state to the final state. Paths that lie further away from the "classical" one also contribute, but with quickly declining weights. Thus, by taking the sum of $P(F_c[s])$ on Γ(E), where Γ(E) is the set of all possible sections of E, one will get the situation where the main contribution to probability will be given by the sections that satisfy the equations following from (7) and (8), but other possible trajectories of cell motion will also potentially exist. Finding such a characterization is a work in progress.

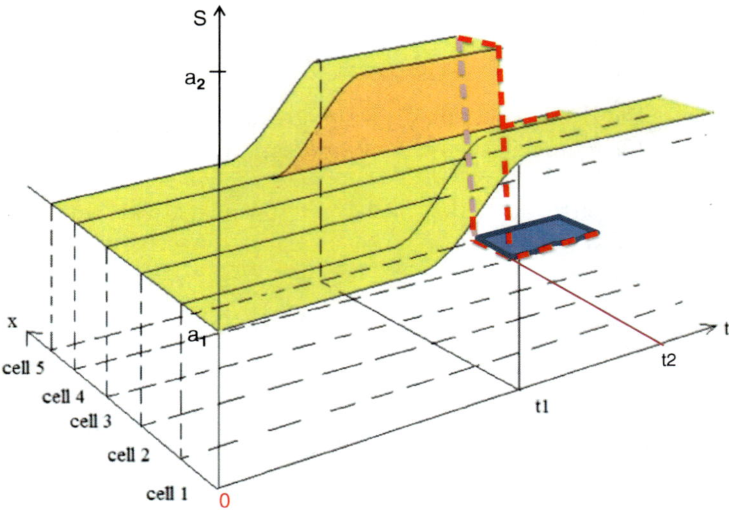

Fig. 6 A simplified model of MF in a fiber bundle representation. Five cells are located in 2-dimentional space (x,t), and correspond to the same section of MF above them (colored in *yellow*). Until the point in time t_1, all five cells are performing the same event (e.g. growth in the same direction with the same growth rate a_1). At time t_1, cell 1 and cell 5 start to grow with different rate, and the section over them obtains a new value on the fibers S (a_2). At time t_2, cell 1 still continues to grow, thus the section above it is still the same one (*yellow*), while cell 5 starts to perform an event, differentiation, which means the appearance of a new section (in *blue*). The borders between different sections are shown by a *red dotted line*

4.3 Evolution of MF as a Tree of Hypersurfaces

If the MF of a developing organism consists of many local sections in the total space of the bundle, corresponding to all morphological domains, then there exist species-specific surfaces in this space, determining a branching of the initial MF section to new ones, corresponding to the appearance of new MFs of new domains, tissues and organs. These surfaces correspond to the border between cell events, differentiation (and maybe also apoptosis), and cell events of other types (Fig. 6).

It is possible that, for each living species, not only the final geometrical form is determined, but also the set and sequence of appearance of local MFs, corresponding to each morphologically distinct domain of the embryo, that is the plan of boundary surfaces in MF space corresponding to species development is determined. It can be understood as a "tree of surfaces". When the MF of a tissue, at some point, reaches a determined surface in total space, the cell on its border undergoes one of the following processes: non-symmetric division (meaning that two daughter cells will be committed to two different MFs (two different tissues)), death, or final determination. New daughter cells after non-symmetric division "on" this surface will belong to two different new fields, or one to a new and one to an "old" field. This means that, if we present development as a tree of cell events, then there exist specific areas on this tree, belonging to different MFs, and there is a law determining the branching and/or separation of these areas.

5 Open Questions and Further Work

5.1 Determination of R_l, C and M_F

The next step in formalization is to discover equations/rules/laws determining F_c, C, R_l and M_F, associated with a given biological information of the object.

We assume that the general form of the functions C, M_F and R_l should be the same for all organisms, while its realization for each species should depend on the concrete biological information of an organism (on the corresponding matrices of the oligosaccharide code). We propose that determination of equations for the function C may be possible by using the understanding of a morphogenetic field as a section of a fiber bundle over the cell space. The general form of the function C may be inferred by comparison of experimental data from early steps of embryogenesis for different types of organisms (e.g. sea urchin or dicotyledonous plant (Fig. 7 a, b)), with the idea of finding common general features between them using cell events formalization.

To this end, the question about the "distance" between cells in the embryo is of the utmost importance. One possible proposition could be to use as a "biological measure" for such a distance the distance between "specific cell surface markers of the zygote", which, as was suggested in Sect. 2.2, may mark the position of cell i in the developing embryo. The existence of these specific zygotic cell surface markers is, in itself, an open biological question. In any case, in the framework of this model, there should be found a way to calculate a distance between cells as a distance between matrices/arrays corresponding to cells, which may be an interesting mathematical question in this concrete case.

The main open question for creating a real formalization of pattern formation in the framework of the suggested model is the one regarding interrelation between functions F_c, C and M_F. Our working hypothesis is that this interrelation could be found mathematically using the language of differential geometry (e.g. the analogs of the notion of connection, etc.), and the analogs of the notion of potential, adapted for the proposed model. However, this work is only in the preliminary stages and strongly depends on all other details of the model.

After getting the general equation for M_F, it might be possible to find the equation for the corresponding cell event S_l:

$$S_l = S_l(M_F)$$

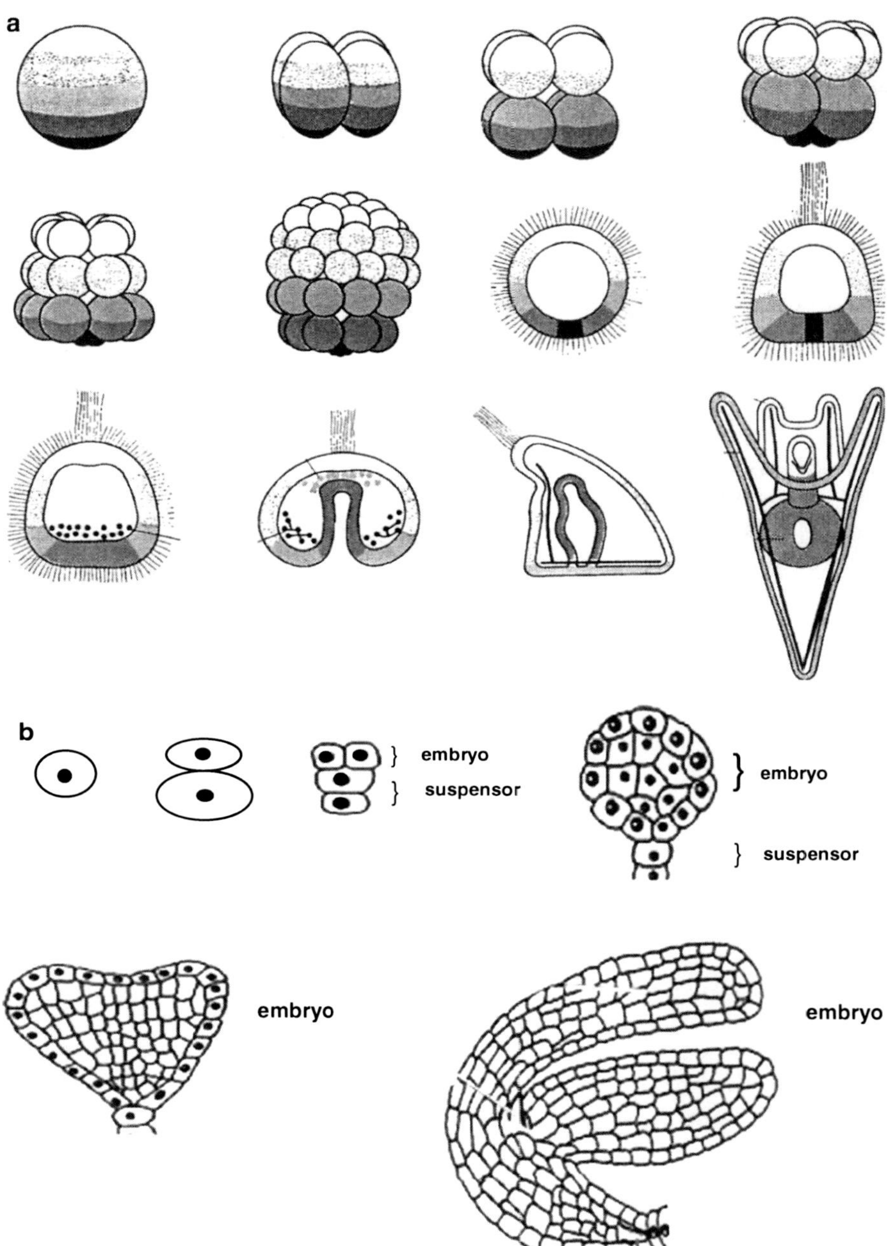

Fig. 7 (**a**) Schematic representation of the initial steps of sea urchin development, starting from one cell (zygote) (after Horstadius, 1939). (**b**) Schematic representation of the initial steps of dicotyledonous plant development, starting from one cell (zygote)

On the other hand, the correspondence between cell event S_l and the rule R_l for changing the oligosaccharide code (matrix A^i_{nm}) on the cell surface might be found experimentally. All these steps may give the final result—the law of creation of a corresponding MF (for example, its M_F) by a set of A^i_{nm} of corresponding cells.

An additional question, which may also be an important and non-trivial one, is about the interrelation between the shape of an embryo domain, described here as an equation for its surface *G(t)*, and the suggested MF formalism. The current model proposes that the influence of the MF laws on the evolution of an embryo's domain surface occurs via an intermediate "cell events" step. However, it is possible to suggest that the vector function $\sigma(x,t)\mathbf{n}(x,t)$ has a potential, in which case the family of surfaces G(t) would be described by the contours of this potential. On the other hand, it is possible to suggest that a function M_F, corresponding to MF influence, and a cell motion function *C* might be expressed in terms of the potential of MF. In this case we may have a question: is it possible to find a direct interconnection between these two, generally speaking, differently originated potentials? For example, the basis for such interconnection might be the following.

When viewed from a more intuitively tangible perspective, the effect of a morphogenetic field is to cause growth and development of cells at a given location. Thus, it corresponds to a change in form. The potential of such a field, if it exists, would therefore correspond to form itself (where by "form" one may understand both geometrical shape and the type of tissue comprising it). That is to say, that a potential, which is time-dependent, would represent the form that an organism tends to achieve at a particular time. This is in close parallel with the idea of the potential of the vector field of the rate of growth, whose contours would also correspond to the shape of the embryo, or domain surface at particular times. The mathematical verification of this parallel is another question for consideration.

Although rather loosely defined, such a potential satisfies the requirement of being independent of path: under most ordinary conditions, the final form of an organism is not dependent on the path that it took to arrive at it. This is because, in the above interpretation, the potential is associated with form itself, rather than with an analog of physical work. Indeed, the "work" (an analog of work, defined in terms of cell events or otherwise) done to achieve one and the same form through different paths may be significantly different. A rather obvious example would be the process of limb formation in the case of normal development and in the case of full regeneration of the whole limb from a small segment of limb disc, where different "work" will be done on the same set of cells in the segment in order to achieve the proper form of the limb. Meanwhile, the same final shape is achieved through two different paths.

However, it is clear that the appropriateness of this analog is an open question, important for further development of the proposed formalization.

5.2 Suggestions for Biological and Numerical Experiments

It is very important to note that, with or without considering the suggested MF theory, the work presented here contains several assumptions that would be interesting to test experimentally and numerically.

The first problem which we would like to suggest for experimental study is the assumption of the existence of an oligosaccharide code on the cell surface, and next, if this is shown to be the case, to determine the most meaningful way of presenting this information in matrix (or other formalized) form. It is possible that a biological code on a cell surface similar to that proposed can be found, but in another biological form (i.e. not as oligosaccharide residues). The experiments, monitoring the changing of the different types of cell surface cell markers in the course of different cell events, could be very promising.

The second experimental problem is to find the correspondence between a cell event S_l and the rule R for changing the biological code (matrix) on the cell surface. Both these tasks require experimental research and mathematical work (matrix analysis) to be carried out in tight collaboration.

The third problem is to give a mathematical description of the dynamics of the boundaries (surfaces) of the morphological domains of an organism in a space-time as a function of the set of all cell events that have occurred inside the volume enclosed by this surface during a time period dt. This investigation can be undertaken in a computational experiment based on the detailed descriptions of the early stages of embryogenesis for several types of organism, taken from the published data. To this end, the equation of evolution of a domain/embryo surface will be an important step. The equation for the smooth evolution of a surface, corresponding to $\sigma(x, t)$ approximated by (2), generally cannot be solved by regular methods. However, with some simplifications, regular methods might give good results [4]. Furthermore, as differential equations deal with continuous functions, and cell events are discrete, "difference equations" may be a possibility.

6 Conclusions

In this work we suggest several important assumptions/notations for discovering the interconnection between biological information and the geometrical form of an organism:

1. The geometry of an organism and its parts is coded by a biological code located on the cell surfaces. As a particular model, we suggest a code of six types of oligosaccharide residues of glycoconjugates.
2. For each cell this code can be presented as a corresponding matrix.
3. The notations of *cell state* and *cell event* are given.

4. The matrix, corresponding to a cell, is changed after each cell event (cell division, growth, etc.) according to a specific rule, R_l, for which we have suggestions for decoding.
5. The geometry of an organism during its growth and development can be described in terms of dynamics of the boundaries (surfaces) of the morphological domains of an organism.
6. The dynamics of the surfaces of the domain depends on the set of all cell events that have occurred inside the domain enclosed by this surface.
7. The notion of morphogenetic field is given as an object governing a law(s) for decoding the biological (cell surface markers) information and transforming it into corresponding instructions for each cell event depending on the positional information (location) of a cell in the developing organism.
8. A mathematical formalization of morphogenetic field is suggested.
9. A set of biological experiments for investigating the proposed theoretical hypothesis is suggested.

References

1. Albersheim P, Darvill A, Augur C, Cheong J, Eberhard S, Hahn MG, Marfa V, Mohnen D, O'Neill M (1992) Oligosaccharins: oligosaccharide regulatory molecules. Acc Chem Res 25 (2):77–83
2. Bolouri H, Davidson EH (2010) The gene regulatory network basis of the "community effect," and analysis of a sea urchin embryo example. Dev Biol 340(2):170–178
3. Bourrillon R, Aubery M (1989) Cell surface glycoproteins in embryonic development. Inter Rev Cytol 116:257–338
4. Courant R, Hilbert D (1962) Methods of mathematical physics, vol 2. Interscience, New York ea
5. Crossin KL, Edelman GM (1992) Specific binding of cytotactin to sulfated glycolipids. J Neurosci Res 33(4):631–638
6. Davidson E (1993) Later embryogenesis: regulatory circuitry in morphogenetic fields. Development 118(3):665–690
7. Davidson E (1998) Specification of cell fate in the sea urchin embryo: summary and some proposed mechanisms. Development 125:3269–3290
8. Davidson EH, Rast JP, Oliveri P, Ransick A, Calestani C, Yuh CH, Minokawa T, Amore G, Hinman V, Arenas-Mena C, Otim O, Brown CT, Livi CB, Lee PY, Revilla R, Schilstra MJ, Clarke PJ, Rust AG, Pan Z, Arnone MI, Rowen L, Cameron RA, McClay DR, Hood L, Bolouri H (2002) A provisional regulatory gene network for specification of endomesoderm in the sea urchin embryo. Dev Biol 246(1):162–190
9. Driesch H (1892) The potency of the first two cleavage cells in echinoderm development. Experimental production of partial and double formations. In: Willer BH, Openheimer JM (eds) Foundations of experimental embryology. Hafner, New York
10. Edelman GM (1988) Topobiology: an introduction to molecular embryology. Basic Books, New York
11. Friedlander DR, Hoffman S, Edelman GM (1988) Functional mapping of cytotactin: proteolytic fragments active in cell-substrate adhesion. J Cell Biol 107(6 Pt 1):2329–2340
12. Fry SC, Aldington S, Hetherington PR, Aitken J (1993) Oligosaccharides as signals and substrates in the plant cell wall. Plant Physiol 103(1):1–5
13. Fry SC (1994) Oligosaccharins as plant growth regulators. Biochem Soc Symp 60:5–14

14. Gilbert SF (1991) Developmental biology, 2nd edn. Sinauer Associates, Sunderland, Massachusetts
15. Gilbert SF (2000) Developmental biology, 6th edn. Sinauer Associates, Sunderland, Massachusetts
16. Harrison RG (1918) Experiments on the development of the forelimb of Amblystoma, a self-differentiating equipotential system. J Exp Zool 25:413–461
17. Horstadius S (1939) The mechanics of sea urchin development studied by operative methods. Biol Rev 14:132–179
18. Ito M, Takata K, Saito S, Aoyaki T, Hirano H (1985) Lectin binding pattern in normal human gastric mucosa: a light and electron microscopy study. Histochemistry 83:189–193
19. Johnson SW, Alhadeff JA (1991) Mammalian alpha-L-fucosidases. Comp Biochem Physiol B 99(3):479–488
20. McNeil M, Darvill AG, Fry SC, Albersheim P (1984) Structure and function of the primary cell walls of plants. Ann Rev Biochem 53:625–663
21. Mohnen D, Hahn MG (1993) Cell wall carbohydrates as signals in plants. Semin Cell Biol 4 (2):93–102
22. Morozova N, Bragina E, Vasiljeva V (2006) Dynamics and role of glycoconjugates of plant cell wall in embryoidogenesis. Phytomorphology 56(3&4):1–7
23. Nemanic MK, Whitehead IS, Elias PM (1983) Alterations in membrane sugars during epidermal differentiation: visualization with lectins and role of glucosidases. J Histochem Cytochem 86(N4):415–419
24. Ponder BA (1983) Lectinhistochemistry. In: Polak IM, van Noorden S (eds) Immunocytochemistry:practical applications in pathology and biology. Wright-PSG, Bristol, pp 129–142
25. Riou IF, Darribere T, Boucaut IC (1986) Cell surface glycoproteins change during gastrulation in Pieurodeles waltlii. J Cell Sci 82:23–40
26. Suprasert A, Pongchairerk U, Pongket P, Nishida T (1999) Lectin histochemical characterization of glycoconjugates present in abomasal epithelium of the goat. Kasetsart J Nat Sci 33:234–242
27. Spemann H (1938) Embryonic development and induction. Yale University Press, New Haven
28. Tan SS, Crossin KL, Hoffman S, Edelman GM (1987) Asymmetric expression in somites of cytotactin and its proteoglycan ligand is correlated with neural crest cell distribution. Proc Natl Acad Sci U S A 84(22):7977–7981
29. Taatjes DJ, Roth J (1991) Glycosylation in intestinal epithelium. Int Rev Cytol 126:135–193
30. Thom R (1983) Mathematical modeling of morphogenesis. Ellis Horwood, Chichester
31. Thom R (1989) Structural stability and morphogenesis: an outline of a general theory of models. Addison-Wesley Publishing Company, Advanced Book Program, Redwood City, CA
32. Tran Thanh Van K, Tonbart P, Cousson A, Darvill AG, Gollin DJ, Chelf P, Albersheim P (1985) Manipulation of the morphogenetic pathways of tobaco by oligosaccharins. Nature 314 (6012):615–617
33. de Vries SC, Booij H, Janssens RVR, Saris L, LoSchiavo F, Terzi M, van Kammen A (1988) Carrot somatic embryogenesis depends on phytohormone-controlled presence of correctly glycosylated extracellular proteins. Genes Dev 2:462–476
34. Waddington CH (1956) Principles of embryology. Macmillan, New York
35. Weiss P (1939) Principles of development. Holt, New York
36. Wolpert L (1977) The development of pattern and form in animals. Carolina Biological, Burlington, NC
37. Zablackis E, York WS, Pauly M, Hantus S, Reiter WD, Chapple CC, Albersheim P, Darvill A (1996) Substitution of L-fucose by L-galactose in cell walls of Arabidopsis mur1. Science 272 (5269):1808–1810
38. Zuzack IS, Tasca RI (1985) Lectin-induced blocage of developing processes in preimplantation mouse embryos in vitro. Gamete Res 12(3):275–290

Randomness and Geometric Structures in Biology

Vincenzo Capasso

Research in biomathematics has played an important role in identifying the biological principles that are responsible for patterns. What has been missing is the link between biological phenomena that occur at different scales.

Biological systems are naturally multiphysics-multiscale problems involving complex systems whose investigation demands a multidisciplinary approach.

At present, a multiple scale approach seems to be the preferred one for dealing with these challenging problems; modelling individual behaviour at the microscale may require the inclusion of stochastic fluctuations, while at the macroscale a deterministic approach, involving integro-differential equations, is usually preferred. Consequently, there has been a surge of interest in methods of up-scaling.

Geometric patterns such as vessel networks in vasculogenesis and patterns on bivalve shells, exhibit stochastic fluctuations which cannot be neglected when dealing with experimental data. Further, statistical methods for the estimation of geometric densities may offer significant tools to a) validate computational models of the studied process; b) monitor the efficacy of possible; c) diagnose; etc.

In previous articles about the future of Mathematical Biology [24], it was stated that "when physicists and mathematicians self-confidently moved on to biology, they equipped themselves with the arsenal of Applied Mathematics [Mathematical Physics] in an attempt to repeat the same success story. Mathematical Biology became almost equivalent to deterministic approaches in terms of differential equations on the level of biological systems or collectives (typically populations)."

In the early days, before the 1970s, biomathematics and biostatistics, though naïve with respect to modern mathematical and computational methods, were developed as bridges between biological data and mathematical modelling (see e.g. [2], and references therein). By contrast, the typical approach taken by mathematicians between the 1970s and the 1980s was discipline (biology)

V. Capasso (✉)
Dipartimento di Matematica, Università degli Studi di Milano, Milan, Italy
e-mail: vincenzo.capasso@unimi.it

V. Capasso et al. (eds.), *Pattern Formation in Morphogenesis*, Springer Proceedings in Mathematics 15, DOI 10.1007/978-3-642-20164-6_21,

independent or, worse, inappropriately trans-disciplinary [39, 40]. Even so, research in biomathematics has played an important role in identifying the biological principles that are responsible for patterns.

Indeed, the leading schools of Biomathematics were using mostly PDE models, or better reaction–diffusion equations, including the Turing-Murray school where pattern formation had been modelled via nonlinear reaction–diffusion systems [27], and the Keller-Segel [32] school had given rise to the explosion of chemotaxis models for a variety of biological applications.

What had been missing is the link between biological phenomena that occur at different scales. Nowadays it is current awareness of the limits of such a pure PDE based approach. Clearly, PDE modelling is macroscopic. In biological modelling, though a good source for insight, such models smooth out the discrete individual nature of cell division and, therefore, do not yield a natural framework in which to relate to the genetic data available from experiments. An obvious weakness of such models is that they are non-cellular and, thereby, ignore the signalling and switching that occurs between developing and dividing cells. Furthermore, in order to recover useful information from the available data, the coefficients in the model must have a meaningful connection to the biology under investigation. Turing [36] was aware of such limitations when he commented, in the Abstract, that "The theory ..... merely suggests that certain well-known physical laws are sufficient to account for many of the facts". This was the year before the publication of the Watson and Crick paper on the double helix [37]. On the basis of the information available today, Turing would write a completely different paper. Mathematically, the importance of his 1952 paper is more about his discoveries related to the behaviour of the solutions of coupled non-linear diffusion equations [1].

Given the large amount of genomic, transcriptomic, proteomic, metabolic information that is currently available, a major challenge is to establish how this genetic information is eventually converted into a spatial-temporal pattern of protein expression which defines the physical form of a cell, tissue and organism. A fundamental question in Biology, Biotechnology and Medicine is how the interplay between the genome and the physical environment drives pattern formation and morphogenesis.

In order to recover the required link between mathematical modelling and biology, the conversion of the genetic information into a spatial-temporal expression pattern needs to be studied by mathematical models together with relevant methods for the analysis of real data, such as geometric statistics and inverse problems [1].

A main goal would be to bring together developmental biologists and mathematicians in order to favour a multidisciplinary approach, from the mathematical modelling, analysis, numerical simulation and visualization, to the validation of models based on real experimental data.

According to Murray, the greatest growth of new areas of involvement will be in medical applications. ....Enormous progress has been made in the past two decades in a wide spectrum of areas, but the detailed mechanisms in practically all pattern-formation situations are still unknown, which of course is another reason the field is so fascinating [28].

Final scope of the research would be the development of significant mathematical models suitable for understanding the basic biological issues that may lead to diagnosis and prevention of tissue malfunction. Illustrative examples can be found in studies of carcinogenesis, angiogenesis, and embryogenesis [16]. The main features of the process of formation of a tumour-driven vessel network are [8, 12, 30, 34 and references therein] (1) vessel branching; (2) vessel extension (chemotaxis in response to a generic tumour angiogenic factor (TAF), released by tumour cells; haptotaxis in response to fibronectin gradient, generated in the extracellular matrix, as the endothelial cells migrate (a combination of degradation and production)); (3) anastomosis, when a capillary tip meets an existing vessel, thus forming loops; (4) blood circulation. A correct modelling of the capillary network naturally leads to a random network of endothelial cells, thus to a stochastic fibre system.

In many cases like this, the diagnosis of a pathology, or the description of a biological process is heavily dependent on the shapes that appear in images of cells, organs and biological systems. Consequently, mathematical models which relate the main features of these shapes to the correct diagnosis, or to the main kinetic parameters of a biological systems need to be studied. This will require interplay between mathematical morphology, statistical shape analysis and biomedical imaging [5, 6, 19, 25, 31, 33, 43 and references therein].

On the basis of this discussion, what can mathematics presently offer to solutions of problems in biology and medicine? Biological systems naturally are multiphysics-multiscale problems involving complex systems whose investigation demands a multidisciplinary approach with respect to the following aspects: (1) multiple scales (homogenisation, boundary layers, asymptotics, etc.); (2)variable stochastic geometries (interfaces, domain decomposition, networks, etc.); (3) nonlinear reaction-advection-diffusion systems; (4) stochastic differential systems; (5) mathematical and numerical methods for simulation and visualisation.

A new model of research is strongly required to handle the increasing complexity of biological systems. Biology is becoming itself increasingly multi-disciplinary, using information from different branches of life sciences; genomics, physiology, biochemistry, biophysics, proteomics, and many more. Further, this model of research places biology within a context that must necessarily engage other disciplines (such as Mathematics, Statistics, Statistical Physics, Chemistry, etc.) that are more strongly aware of the importance of collective phenomena than biology has been until now. Both mathematicians and biologists need to deepen specific areas of competence with the added value of communication in an interdisciplinary setting; i.e. the vertical expertise in a single discipline has to be revisited in an horizontal multidisciplinary application area, in order to function successfully [38, 41].

In a very simplified way we may claim that nowadays in Biology we are facing a situation which is paralleling the passage from the massive data base of available astronomical data of Tycho Brahe to the knowledge extraction due to Copernicus, Kepler and Galileo, which lead to the mathematical understanding of the universe by Newton's laws. The increasing availability of a massive amount of biological data requires a unifying theory that may eventually cause a revision of our understanding of the living organisms, and systems.

Indeed, the application of mathematical and computational tools to biology is producing equally exciting results, is providing insights into biological problems too complex for traditional analysis. On the other hand we cannot ignore the opportunity deriving from the availability of enormous computing power. Advances in computing architecture and scale are enabling simulations of complex biological processes at various organizational levels from atomic to cellular and beyond. High performance computing that takes full advantage of massive parallelism is a necessary means to obtain the performance needed to tackle the complexity associated with such problems [23, 44].

Even so, a major challenge in computational methods is tackling multiscale problems, and the development of modelling techniques capable of bridging the enormous scale gap , from the underlying molecular genetic information up to cells. organs, organisms. A major problem is to capture the main features of interactions at the scale of individual entities (cells, vessels, animals, insects, etc.) that are responsible for the emergence of more complex behaviour at a larger scale, including the formation of patterns (tumours, blood vessel networks, swarms, etc.) [21]; in such cases it is clear that the behavior of a system as a whole can be "more that the sum" of its parts [41].

For example, to understand and to control the spread of malaria, the whole network of the system including the genome of the mosquito Anopheles transmitting the infectious protozoon Plasmodium falciparum, the dynamics in hosts and parasites as well as the climate, and the environmental and socio-economic conditions influencing the spread of the disease must be taken into account [23].

Even in a simplified problem of collective behaviour of biological species, "A modelling problem is to bridge the finer scale description based on the (stochastic) behaviour of individuals (microscale) to the larger scale description based on the (continuum) behavior of population densities (macroscale)" [21].

At present, a multiple scale approach seems to be the preferred one for dealing with these challenging problems; modelling individual behaviour at the microscale, may require the inclusion of stochastic fluctuations, while at the macroscale a deterministic approach, involving integro-differential equations, is usually preferred. Consequently, there has been a surge of interest in methods of upscaling, this interest revealing the need for innovative and sophisticated homogeneization techniques, both deterministic and stochastic [9]. In this latter case, at the basis of such methods are Laws of large Numbers and Central Limit Theorems [13, 18, 20, 22, 29, 42 and references therein]. Both in direct interaction, as in swarm behaviour, or in indirect interaction, as in chemotactic models, bridging the two scales, micro and macro, may be obtained via averaging at a meso-scale, thus leading from fully stochastic individual based models to hybrid models, better known as mean field models, in which the kinetic parameters of the stochastic models for individuals are coupled to mean values of the driving fields (either population densities as in aggregation models, or biochemical reactants as in chemotaxis, or even both), keeping anyhow simple stochasticity at the microscale, which eventually leads to random patterning. The advantage of using averaged quantities at the larger scale is convenient, both from a theoretical point of view, and for computational affordability which otherwise would be far from reality when dealing with very large numbers of individuals in the relevant population [14, and references therein]. We wish to stress that anyhow substituting mean densities of populations or mean geometric densities of capillary networks to the corresponding stochastic quantities leads to an acceptable coefficient of variation (percentage error) only when a law of large numbers can be applied, i.e. whenever the relevant numbers per unit volume are sufficiently large; otherwise randomness cannot be avoided, and in addition to mean values, the mathematical analysis and/or simulations should provide confidence bands for all quantities of interest [7, 14].

Geometric patterns such as vessel networks in vasculogenesis, patterns on bivalve shells, and alike exhibit stochastic fluctuations which cannot be avoided when dealing with experimental data; indeed, for natural reasons which are typical of biological systems, we may be faced with a double stochasticity, one level due to the intrinsic randomness of phenomena at the level of the cell, organ or organism, and another due the randomness of parameters from organism to organism. In drug design this is a well know fact, so that sophisticated mathematical methods are required to face the identification of a typical dose (mean value and confidence band) [11].

To complete the scene, statistical methods for the estimation of geometric densities may offer significant tools to (a) validate computational models of the studied process; (b) monitor the efficacy of possible treatments (by quantifying the dose/response ratio); (c) diagnose (by describing in a quantitative and objective way the morphological differences of different fibre systems); etc. (see [4, 10], and references therein).

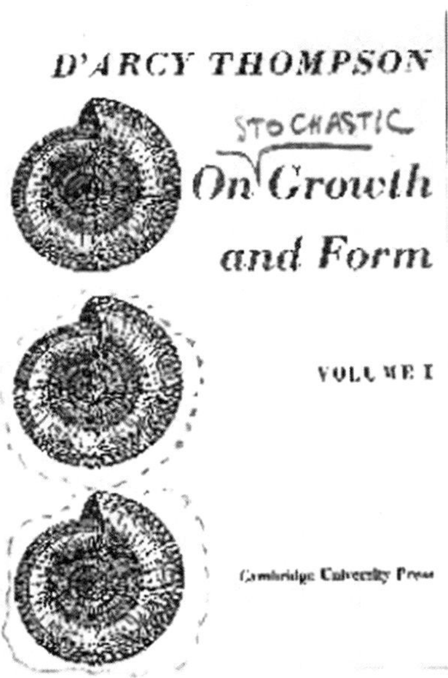

Bibliography

1. Anderssen B, Capasso V (2008) SSQBM – pattern formation and functional morphology (PFFM). Personal notes of the debate on the turing methodology is the only way to model pattern formation in biology. RICAM, Linz, Jan 2008
2. Bailey NTJ (1975) The mathematical theory of infectious diseases. Griffin, London
3. Bellomo N, De Angelis E, Preziosi L (2004) Multiscale modelling and mathematical problems related to tumour evolution and medical therapy. J Theor Med 5:111–136
4. Benes V, Rataj J (2004) Stochastic geometry. Kluwer, Dordrecht
5. Bookstein FL (1978) The measurement of biological shape and shape change, vol 24, Lecture notes in biomathematics. Springer, Heidelberg
6. Bookstein FL (1991) Morphometric tools for landmark data. Cambridge University Press, Cambridge
7. Burger M, Capasso V, Pizzocchero L (2006) Mesoscale averaging of nucleation and growth models. SIAM J Multiscale Model Simul 5:564–592
8. Byrne H, Chaplain M (1995) Mathematical models for tumour angiogenesis: numerical simulations and nonlinear wave solutions. Bull Math Biol 57:461–486
9. Canic S, Mikelic A (2003) Effective equations modeling the flow of a viscous incompressible fluid through a long tube arising in the study of blood flow through small arteries. SIAM J Appl Dyn Syst 2:431–463
10. Capasso V, Dejana E, Micheletti A (2007) Methods of stochastic geometry, and related statistical problems in the analysis and therapy of tumour growth and tumour-driven angiogenesis. In: Bellomo N, Chaplain M, De Angelis E (eds) Mathematical methods in cancer diagnosis and therapy. Birkhauser, Boston
11. Capasso V, Engl HW, Kindermann S (2008) Parameter identification in a random environment exemplified by a multiscale model for crystal growth. SIAM J Multiscale Model Simul 7:814–841
12. Capasso V, Micheletti A, Morale D (2008) Stochastic geometric models, and related statistical issues in tumour-induced angiogenesis. Math Biosci 214:20–31
13. Capasso VM (2008) Rescaling stochastic processes: asymptotics. In: Capasso V, Lachowicz M (eds) Multiscale problems in the life sciences from microscopic to macroscopic, vol 1940, Lecture notes in mathematics/C.I.M.E.. Springer, Heidelberg
14. Capasso V, Morale D (2011) A multiscale approach leading to hybrid mathematical models for angiogenesis: the role of randomness. In: Friedman A et al (eds) Mathematical methods and models in biomedicine. Springer, New York, In Press
15. Carbone A, Gromov M, Prusinkiewicz P (2000) Preface. In: Carbone A, Gromov M, Prusinkiewicz P (eds) Pattern formation in biology, vision and dynamics. World Scientific, Singapore
16. Carmeliet P (2005) Angiogenesis in life, disease and medicine. Nature 438:932–936
17. Carroll SB (2005) Endless forms most beautiful. The new science of evo-devo. Baror International Inc, Armonk, New York, USA
18. De Masi A, Luckhaus S, Presutti E (2007) Two scales hydrodynamic limit for a model of malignant tumor cells. Annales H Poincare, Probabilite et Statistiques 43:257–297
19. Dryden IL, Mardia KV (1997) The statistical analysis of shape. Wiley, Chichester
20. Durrett R (1999) Stochastic spatial models. In: Capasso V, Diekmann O (eds) Mathematics inspired by biology, vol 1714, Lecture notes in mathematics/C.I.M.E. Springer, Heidelberg
21. Durrett R, Levin SA (1994) The importance of being discrete (and spatial). Theor Pop Biol 46:363–394
22. Fournier N, Meleard S (2004) A microscopic probabilistic description of a locally regulated population and macroscopic approximations. Ann Appl Probab 14:1880–1919
23. Jaeger W (2001) Private discussion about MBIOS, a project of the Ruprecht-Karls-Universitaet. Heidelberg
24. Jagers P (2010) A plea for stochastic population dynamics. J Math Biol 60:761–764

25. Kendall DG, Barden D, Carne TK, Le H (1999) Shape and shape theory. Wiley, Chichester, N.Y
26. Miles RE, Serra J (eds) (1978) Geometrical probability and biological structures: buffon's 200th anniversary, vol 23, Lecture notes in biomathematics. Springer, Berlin-New York
27. Murray JD (1989) Mathematical biology. Springer, Heidelberg
28. Murray JD et al (2003) Foreword. In: Sekimura T et al (eds) Morphogenesis and pattern formation in biological systems. Springer, Tokyo
29. Oelschlaeger K (1989) On the derivation of reaction–diffusion equations as lilit dynamics of systems of moderately interacting stochastic processes. Prob Th Rel Fields 82:565–586
30. Plank MJ, Sleeman BD (2003) A reinforced random walk model of tumour angiogenesis and anti-angiogenic strategies. IMA J Math Med Biol 20:135–181
31. Schiaparelli GV (1898) Studio comparativo tra le forme organiche naturali e le forme geometriche pure. Hoepli, Milano
32. Segel LA (1984) Modelling dynamic phenomena in molecular and cellular biology. Cambridge University Press, Cambridge, New York
33. Serra J (1984) Image analysis and mathematical morphology. Academic Press Inc, London
34. Sun S, Wheeler MF, Obeyesekere M, Patrick CW Jr (2005) A deterministic model of growth factor-induced angiogenesis. Bull Math Biol 67:313–337
35. Thompson DW (1917) On growth and form. Cambridge University Press, Cambridge
36. Turing AM (1952) The chemical basis of morphogenesis. Philos Trans Roy Soc London B 237:37–72
37. Watson JD, Crick FHC (1953) Molecular structure of nucleic acids: a structure for deoxyribose nucleic acid. Nature 171(4356):737–738
38. Levin SA (ed) (1992) Mathematics and biology: the interface, challenges and opportunities. Lawrence Berkeley Laboratory, University of California, Berkeley (CA)
39. Capasso V et al (eds) (1985) Mathematics in biology and medicine, vol 57, Lecture notes in biomathematics. Springer, Heidelberg
40. Jaeger W, Murray JD (eds) (1984) Modelling of patterns in space and in time, vol 55, Lecture notes in biomathematics. Springer, Heidelberg
41. Capasso V, Periaux J (eds) (2005) Multidisciplinary methods for analysis, optimization and control of complex systems, vol 6, Mathematics in industry. Springer, Heidelberg
42. Capasso V, Lachowicz M (eds) (2008) Multiscale problems in the life sciences from microscopic to macroscopic, vol 1940, Lecture notes in mathematics/C.I.M.E. Springer, Heidelberg
43. Chaplain et al (eds) (1999) On growth and form. Spatio-temporal pattern formation in biology. Wiley, New York
44. PITAC (2005) Report, computational sciences ensuring America's competitiveness. Executive Office of the President of the United States of America

Printed by Publishers' Graphics LLC
BT20121011.19.22.13